CANNABIS Physiopathology, Epidemiology, Detection

CANNABIS Physiopathology, Epidemiology, Detection

From the proceedings of the Second International Symposium
organized by the National Academy of Medicine,
with the assistance of The City of Paris,
April 8–9, 1992

Edited by
Gabriel G. Nahas
Colette Latour
New York University
and Hôpital F. Widal, Paris, France

 **CRC Press**
Taylor & Francis Group
Boca Raton London New York

CRC Press is an imprint of the
Taylor & Francis Group, an **informa** business

First published 1993 by CRC Press
Taylor & Francis Group
6000 Broken Sound Parkway NW, Suite
300 Boca Raton, FL 33487-2742

Reissued 2018 by CRC Press

© 1993 by Taylor & Francis
CRC Press is an imprint of Taylor & Francis Group, an Informa business

No claim to original U.S. Government works

A Library of Congress record exists under LC control number: 92049005

Publisher's Note
The publisher has gone to great lengths to ensure the quality of this reprint but points out that some imperfections in the original copies may be apparent.

Disclaimer
The publisher has made every effort to trace copyright holders and welcomes correspondence from those they have been unable to contact.

ISBN 13: 978-1-138-10497-6 (hbk)
ISBN 13: 978-1-138-55780-2 (pbk)
ISBN 13: 978-1-315-15034-5 (ebk)

Visit the Taylor & Francis Web site at http://www.taylorandfrancis.com and the CRC Press Web site at http://www.crcpress.com

LIST OF CONTRIBUTORS

Peter Allebeck : Karolinska Institute, Stockholm, Sweden.
Michel Aussedat : Metz-Thionville Medical Center, France.
Henri Baruk : Académie Nationale de Médecine, Paris.
Henri Baylon : Académie Nationale de Médecine, Paris.
Nils Bejerot : Karolinska Institute, Stockholm, Sweden.
Peter Bensinger : Drug Enforcement Administration, Washington, D.C.
Jonathan Buckley : University of Southern California, Los Angeles.
Guy Cabral : Medical College of Virginia, Richmond.
Eric Cairns : Otago Medical School, New Zealand.
Taieb Chkili : Rabat Medical Center, Morocco.
William Clark : University of Maryland, Baltimore.
Adams Cowley : University of Maryland, Baltimore.
Bernard Defer : Hôpital Paul Guiraud, Villejuif, France.
Paul J. Donald : Univeristy of California, Davis.
Joanna Fowler : Brookhaven National Laboratory, Upton, N.Y.
Anna Fugelstad : St Göran, Stockholm, Sweden.
Lionel Gaillaud : Impact Médecin, Paris.
Karel F. Gunning : Rotterdam, The Netherlands.
Richard Howerd : Otago Medical School, New Zealand.
Juhana Idanpaan-Heikkila : W.H.O., Geneva, Switzerland.
Raul Jeri : San Marcos, Medical School, Lima, Peru.
John Jonsson : Linköping, Sweden.
Pierre Juillet : Académie Nationale de Médecine, Paris.
Denise Kandel : Columbia University, New York.
J.E. Ktiouet : Rabat Medical Center, Morocco.
Paul Lafargue : Gendarmerie Nationale, Rosny-sous-Bois, France.
Georges Lagier : Hôpital Fernand Widal, Paris.
Colette Latour : INSERM, Hôpital F. Widal, Paris.
Paul Lechat : Académie Nationale de Médecine, Paris.
Von O Leirer : Decision Systems, Stanford, California.
Pierre Levillain : Hôpital Fernand Widal, Paris.
David B. Menkes : Otago Medical School, New Zealand.
Jean Michaud : Cour de Cassation, Paris.
Daniel Morrow : Decision Systems, Stanford, California.
Paul Mulloy : U.S. Naval Academy, McLean, Virginia.
Gabriel Nahas : New York University, New York.

Juan C. Negrete : Mc Gill University, Montreal.
S. Niziolek-Reinhardt : Metz-Thionville Medical Center, France.
William D. Paton : Oxford University, Oxford.
Hélène Peters : Macalaster College, Saint-Paul, Minn.
Jovan Rajs : Karolinska Institute, Stockholm, Sweden.
Mitchell Rosenthal : University of New York, New York.
Belavadi Shankar : University of Maryland, Baltimore.
Richard H. Schwartz : Georgetown University, Washington.
Carl A. Soderstrom : University of Maryland, Baltimore.
George Spears : Otago Medical School, New Zealand.
Anna Trifillis : University of Maryland, Baltimore.
Renaud Trouvé : Hôpital Fernand Widal, Paris.
Henri Tuchmann-Duplessis : Académie Nationale de Médecine, Paris.
R. Vasquez : Medical College of Virginia, Richmond.
Nora Volkow : Broohaven National Laboratory, Upton, N.Y.
Peter Waser : University of Zurich, Switzerland.
Jerome Yesavage : Stanford University, California.

FOREWORD

This volume assembles selected proceedings from the Second International Colloquium on the "Physiopathology of Cannabis and the Detection of Illicit Drugs", which was held at the National Academy of Medicine with the assistance of the City of Paris.

The physiopathology of cannabis is a subject of major importance for the future of our society. It so happens that the acute impairment of cannabis on mental function was described in 1845 by a French physician Joseph Moreau, the father of modern psycho-pharmacology. While Moreau described the acute impairing effects of hashish on affective and cognitive functions of the brain, he had no way of assessing the physiopathological effects of this drug.

To day, thanks to the new techniques of medical imaging, it is possible to visualize, as illustrated by the studies of Dr. Nora Volkow, the impairment of brain functioning in areas which are connected with coordination, balance, memory and coherent behavior. Other studies describe the noxious effects of cannabis on reproductive function, and immune surveillance resulting in cancer and leukemia.

The second theme of this colloquium, detection of illicit drugs in body fluids is also of great importance. We have to protect our communities from the use of substances which impair the integrity of attention, memory, coordination which are essential for the proper performance of professionnal tasks. Testing for drugs are already practiced in competitive sports for detection of anabolic substances, and alcohol in road accidents. Unfortunately the present epidemic of drug abuse has resulted in the consumption of other drugs. Like cannabis, they impair persistently information processing by the brain and prevent it from performing correctly in driving a car, monitoring equipment or using a computer. It is therefore important to review methods and results already obtained in the detection of illicit drugs in different populations.

At a time when strong voices are advocating the relegalization of illicit drugs, and public health is threatened by the progression of illicit drug consumption, the City of Paris is proud to support outstanding scientific studies which should help to promote prevention programs for our youth. To day, I am eager to listen to scientists who are attempting to evaluate all of the consequences of modern technology on man and society. The control of the emission of toxic substances which pollute the environment has been initiated thanks to the efforts of concerned scientists. Some of them are now seeking methods to curtail the use and trivialisation of substances which pollute the internal milieu of man's brain and other vital organs. Our concern today is therefore to preserve mental and physical health and specially that of future generations. Such is our hope and such is our goal.

Jacques Chirac

Mayor of Paris
Former Prime Minister

PREFACE

It is quite fitting that this first international colloquium on the Physiopathology of Cannabis should have taken place under the aegis of the National Academy of Medicine. Indeed it is in 1845, that Dr Jacques Joseph Moreau, called Moreau de Tours, the father of psychopharmacology, presented to this Academy his classical treatise "Hashish and Mental Illness".

This text which is summarized in this volume may be considered as the "princeps" experience ever recorded on the clinical pharmacology of marihuana. Moreau, using himself as an experimental subject, was the first physician to study the role of "psychotoxic" substances on the development of acute mental pathology, first ingesting hashish and later inhaling nitrous oxyde. He summarizes his experience with hashish as follows : "As the effects of hashish take hold of the brain, one experiences profound modification in all aspects of thought processes. A true "dream-state", but a dream without sleep does occur." And he added : "It appears that two modes of moral existence, two lives have been imparted to man. Our first life results from our relationship with the external milieu, which we call the great Universe. Our second life is but a stored compilation of the first. Sleep is like a barrier separating these two lives.

Moreau demonstrated that delirium was a sort of dream. We have made similar observations in our investigations of catatonia which may often be considered as a state of dreaming while being awake. We have also reported a state of mental toxicity and delirium which resulted from bacterial infections especially with enterobacterium coli. These studies have been reported in the Proceedings of the Society Moreau de Tours, founded twenty years ago.

And today, all the clinical symptomatology of cannabis intoxication described so accurately by Moreau and which has been confirmed by many other psychopharmacologists, is being reinvestigated and correlated with biochemical and neurophysiological markers of the brain. Such studies will be discussed in this colloquium, which should be another landmark in our understanding of the human brain.

Henri Baruk
Professeur Honoraire de la Faculté de Médecine,
Membre de l'Académie Nationale de Médecine,
Président Fondateur de la Société Moreau de Tours.

INTRODUCTION

It is arduous, by definition, either to convince skeptics, or to change the opinion of those who refuse a priori to admit the truth. To recognize that the usage of cannabis is damaging contradicts these two attitudes. That is why we must be appreciative of those who have scientifically demonstrated that this drug is neither soft nor harmless, but that its toxicity is manifested in multiple ways among its consumers. The International Colloquium on Illicit Drugs, which was the basis for the present publication, took place in Paris in April of 1992. A number of leading authorities specialized in different aspects of drug dependence, participated in this colloquium and their reports are assembled in this volume. One panel of scientists reported that cannabis impairs psychomotor function, decreases cognitive processes, induces psychiatric disturbances, impairs immunity and reproduction and facilitates the appearance of certain cancers. The second panel reported that the accurate detection of the metabolites of the elusive cannabinoids is possible, thanks to the very sensitive techniques now available; this detection helps law enforcement authorities and magistrates in their task. The third panel discussed recent epidemiological data reporting the usage of cannabis in America and Europe.

Society must not capitulate to the destructive forces which are threatening its fabric, and prominent among these must be included drugs of dependence. The greatest help that scientists may offer to save drug addicts from self-destruction, is to define and spread as early as possible, up-to-date information concerning the depredation which threatens their lives. Every type of research must be relentlessly carried out in this field. Indeed, it is only on the basis of unquestionable data, that policy makers will be able to make informed decisions in order to define the most adequate measures to fight the bondage of drug dependence.

Paul Lechat
Membre de l'Académie Nationale de Médecine,
Former President, IUPHAR (International
Union of Pharmacological Sciences)

MESSAGE FROM THE ACADEMY

Professor Henri Baylon
Président, Académie Nationale de Médecine

For the second time in two years, the National Academy of Medicine had the privilege to host a colloquium on the grave problem of public health created by the illicit use of psychotropic substances.

The first colloquium, sponsored like this one, by the City of Paris, was devoted to the Physiopathology of Cannabis, Cocaine and Opiates. The present colloquium emphasized problems related more particularly to cannabis.

There are diverging opinions concerning the medical effects of cannabis. Many intellectuals consider it a "soft drug" and advocate its relegalization. On the opposite side, others don't distinguish this addiction from that of "hard drugs" and claim that it is responsible for delinquent behavior. American studies tend to indicate that cannabis consumption is the first step towards the use of heroin or cocaine. Furthermore it is apparent that the use of cannabis among adolescents occurs more and more precociously, starting at age twelve. This early experience with cannabis creates the risk of an escalation in the following years to the consumption of hard drugs.

The responsible leaders of the large cities have realized the grave threat which cannabis smoking represents for their future. In this regards, the City of Paris, which we wish to thank for its generous contribution towards the organization of this colloquium, has expressed an interest in devoting a large part of its scientific reports to cannabis. Internationally recognized scientists have been invited to present the results of their experiments and let us benefit from their special expertise.

The problem raised by cannabis are indeed international, and it is important that different experiences be compared and discussed.

In the first part of this colloquium, the general toxicity and the effects of the consumption of cannabis on behavior, on the genesis of mental illness were reported. Imaging techniques (positron emission tomography "PET") performed on cannabis users, illustrated the alterations occurring in brain metabolism, suggesting impairment of neurotransmission.

The second part of the colloquium reported broad surveys and discussions concerning different attitudes towards the use of cannabis in countries of Europe, North and South America. The view point of WHO concluded this discussion.

The third part of the colloquium was devoted to problems of detection of illicit drugs and their consequences in different situations, thus generalizing the theme of the meeting to all of the addiction. Sampling, methods of measurement, interpretation of results were discussed. Consequences of detection in victims of traffic accidents in cars or trucks, or railway conductors were reported. Detection in school population, industry and the armed forces were discussed. Finally problems of legislation of illicit drug use were analyzed.

This volume assembles the bulk of the papers presented to the colloquium. In addition, other articles written by specialists of the main subjects under consideration have been included. Older communications of lasting general interest have been reprinted in an appendix.

CONTENTS

III. DETECTION, IDENTIFICATION AND TESTING

General Conclusions - Henri Baylon

Appendix

Index

Section I

PHYSIOPATHOLOGY

1. Toxicology

GENERAL TOXICITY OF CANNABIS

Gabriel Nahas, M.D., Ph.D.

New York University, Medical Center, New York and
Laboratoire de Toxicologie Cellulaire, Hôpital F. Widal, Paris.

ABSTRACT

The physiopathological effects of marihuana smoke and of its constituent cannabinoids were reported first from in vitro and in vivo experimental studies. Marihuana smoke is mutagenic in the Ames test and in tissue culture and cannabinoids inhibit biosynthesis of macromolecules. Exposure of animals to THC or marihuana produces symptoms of neurobehavioral toxicity, disruptive effects on all phases of gonadal or reproductive function and is fetotoxic. Smoke inhalation produces symptoms of airway obstruction and squamous metaplasia. Ten years after completion of these experimental studies, clinical manifestation of marihuana physiopathology are now reported. These include : long term impairment of memory storage in adolescents; prolonged impairment of psychomotor performance, resulting in lethal train and car accidents; residual impairment of plane piloting; a six fold increase in incidence of schizophrenia; cancer of mouth, jaw, tongue and lung in 19 to 30 year old; non-lymphoblastic leukemia in children from marihuana smoking mother and fetotoxicity.

KEY WORDS

Marihuana, mutagenicity, fetotoxicity, neurobehavioral toxicity, impairment memory, performance, schizophrenia, cancer, leukemia.

INTRODUCTION

The recreational smoking of products derived from *Cannabis sativa*, mainly its resin (hashish) or the chopped flowering tops of the plant (marihuana) has become widespreaded in western industrialized countries since 1960. Because of its lack of acute life threatening effects, cannabis has been called a "soft drug.", no less damaging than coffee or tobacco (1). This designation might have to be revised, in view of its prolonged impairing effects on memory and learning and its residual neuropsychopharmacological properties and its effects on lung, on immunity system and reproductive function, which have only been recently reported (2) and which confirm earlier experimental observations (3).

GENERAL PROPERTIES OF CANNABIS

Products extracted from the *Cannabis sativa* plant for purposes of smoking originate from the drug-type of the plant, which grows in warm climes of Africa, Americas, South East, Asia, Australia and New Zealand. The flowering tops of cannabis contain an intoxicating material, delta-9-tetrahydrocannabinol (THC), which may vary from 1 to 6% of the total weight: 1 to 3% in grass or marihuana, up to 8% in hashish and exceeding 50% in hash oil. In contrast, the fiber-type of cannabis, which grows in cool climates and is used for the manufacture of rope and twine, contains very little THC. In addition to THC, over 60 other cannabinoids have been identified in cannabis, mainly cannabiniol (CBD) and cannabinol (CBN) which are not psychoactive but are biologically active. Cannabis also contains 8 different classes of compounds (4), numbering 421 and which are for the most part xenobiotics, such as (Table I) : Alkaloid derivatives of spermidine, sterols, terpenes and flavanoid glucosides.

Under the influence of heat, cannabinoids rapidly decarboxylate. At the temperature of pyrolysis (200-400°C), aromatization of the cannabinoids occurs. Some 150 polycyclic aromatic hydrocarbons have been identified in marihuana smoke and the proportions of the higher molecular weight compounds, particularly the carcinogen benzo[a]pyrene, are greater in marihuana than in tobacco smoke, suggesting higher toxicity. The most likely sources of these hydrocarbons are the cannabinoids. Other constituents of marihuana smoke include phenols, phytosterols, acids and terpenes. In other respects the smoke of tobacco and marihuana are similar, as can be seen from Table 1. Toxic substances such as carbon monoxide, hydrogen cyanide, and nitrosamines are present in equivalent concentrations in both smokes, and the "tar" yield is also similar.

THC and other cannabinoids are very fat-soluble and have a half-life of 8 days in fat. It therefore takes one month to completely eliminate a single dose of THC, which is stored in liver, lung, spleen and mostly in neutral fat. Consumption of cannabis at less than one-week intervals will result in storage of THC in the body. THC is a polar compound and is slowly metabolized into more water-soluble, non-psychoactive metabolites, 80 of which have been identified to date. The bioavailability of inhaled THC is 20 %, and when ingested, it is 6%. Less than 1% of the bioavailable THC reaches the brain, a fact which illustrate the psychoactive potency of this drug (6). Excretion of metabolites is via the liver and intestine (80%), with entero-hepatic

TABLE I

COMPARATIVE ANALYSIS OF MAINSTREAM SMOKE FROM MARIHUANA AND TOBACCO REFERENCE CIGARETTE.
(Average weight : 1.110 mg; Length : 85 mm)

Measurements	Marihuana Cigarette	Tobacco Cigarette
Gas phase		
Carbon monoxide, vol.%	3.99	4.58
Carbon dioxide, vol.%	8.27	9.38
Ammonia, μg	228	199
HCN, μg	532	498
Cyanogen $(CN)_2$, μg	19	20
Isoprene, μg	83	310
Acetaldehyde, μg	1,200	980
Acetone, μg	443	578
Acrolein, μg	92	85
Acetonitrile, μg	132	123
Benzene, μg	76	67
Toluene, μg	112	108
Vinyl chloride, ng*	5.4	12.4
Dimethylnitrosamine, ng*	75	84
Methylethylnitrosamine, ng*	27	30
Particulate phase		
Total particulate matter, dry, mg	22.7	39
Phenol, μg	76.8	138.5
O-Cresol, μg	17.9	24
m- and *p*- Cresol, μg	54.4	65
Dimethylphenol, μg	6.8	14.4
Catechol, μg	188	328
Cannabidiol, μg	190	-
Delta-9-tetrahydrocannabinol, μg	820	-
Cannabinol, μg	400	-
Nicotine, μg	-	2,850
N-Nitrosonornicotine, ng*	-	390
Naphthalene, μg	3	1.2
1-Methylnaphthalene, μg	6.1	3.65
2-Methylnaphthalene, μg	3.6	1.4
Benz(*a*)anthracene, ng*	75	43
Benzo(*a*)pyrene, ng*	31	21.1

* indicates known carcinogens.

From Marihuana and Health, National Academy of Sciences, Institute of Medicine Report, Washington, D.C., 1982.

recirculation which delays their elimination; urinary excretion does not exceed 20%. In addition, THC and its metabolites cross the placental barrier and are transferred to maternal milk.

The acute and chronic physicopathological effects of marihuana smoking may be related to three of its following properties.

- The property of THC, its psychoactive ingredient to attach, in nanomolar concentration to specific receptors in hippocampus, cerebellum and frontal lobes.

- The property of all psychoactive and nonpsychoactive cannabinoids and their metabolites to inhibit in micromolar concentration macromolecular synthesis.

- The properties of toxic xenobiotics contained in the gas and particulate phase of cannabis smoke.

EFFECTS OF THC ON THE BRAIN :
an impairment of information processing

The THC receptor and its distribution in the brain

THC interacts in nanomelecular concentration with stereospecific receptors first described in 1990 by Herkenham et al (7). They used a tritiated synthetic cannabinoid CP 54940, much more active than THC and presenting marked enantiomer selectivity. They were able to localize a specific receptor on brain slices sampled from rat, rhesus monkey and man. Herkenham reported that the density of the THC receptors in every animal species studied, followed the same distribution. It was especially marked in the limbic area, (globus pallidus and substantia nigra), hippocampus, cerebellum and frontal lobes.

The THC receptor was cloned by Matsuda (8) et al and THC cannabinoid was the only molecule which attached to the G protein coupled receptor, and exhibited adenylyl cyclase inhibition in transfected cells.

Observations of Herkenham extend and confirm those of McIsaac et al (9) who studied the distribution of delta 9 THC and its metabolites in the brain of squirrel monkeys after i.v. administration of 2-30 mg/kg of tritiated material. The dose response relationship noted with the increasing dosages was similar to that observed in man. Low doses had an euphoric quiet effect with disruption of perception; medium doses produced stimulation, lack of coordination. Higher dosages were accompanied by severe psychomotor incapacitation. A correlation was established between behavioral alterations and concomitant distribution of delta 9 THC in certain areas of the brain, the same as those mapped by Herkenham in his studies of the distribution of the THC receptors : frontal cortex, cerebellum, hippocampus.

Both groups of investigators noted the frontal cortex was the site of the processing of incoming information and initiated voluntary somatosensory stimuli required for equilibrium and motor coordination, that the hippocampus had a unique role in memory transfer and consolidation.

Mc Isaac's and Herkenham's studies on the distribution of THC in certain brain areas were also observed by Volkow et al (10) who used PET (positron emission tomography) to measure brain glucose utilization in man. They reported that marihuana produces immediate and long term changes in cerebellar metabolism which corresponds to the high density of THC receptor localized in that area. Chronic marihuana users displayed a decreased cerebellar activity, which could translate in "disruptions of functions associated with the cerebellum such as motor coordination, proprioception and learning."

Because of the limited spatial resolution of the PET instrument used, metabolic activities in the hippocampus and caudate nucleus could not be measured.

Other experimental studies have described morphological alterations in the hippocampus of animals exposed to marihuana (Table II). Heath (11) reported neuropathological alterations on electron microscopic sections of the hippocampus of monkeys exposed to marihuana smoke. Scallet (12), using morphometric techniques, reported a 44% reduction in the synapse and a significant decrease in the neuronal volume of the hippocampus of rats administered THC per os for 90 days. Landfield (13) also observed quantitative changes in hippocampal structure (decreased neuronal density) of rats administered THC 8 mg/kg subcutaneously for 8 months. Others have reported that THC suppresses hippocampal electrical activity (14). and interferes with cholinergic limbic system (15).

Because of the necessity of rapid fixation of the brain, such experimental procedures may not be performed in man.

Marihuana and memory

The hippocampus gates information during memory consolidation (16) and codes temporal and spatial relations between stimuli and responses (17).

Anecdotal accounts of cannabis use in man reports fragmentation of thought and confusion on attempting to remember recent occurrence (18). Most recent observations indicate that the most consistent effect of THC is a disruption of selective features of short memory recall, similar to that found in monkeys and patients with damage to limbic central areas (19).

Soueif (20) was the first to report, in 1951, significant cognitive impairments in a large cohort of hashish users, as compared with a control group. By contrast, a study by Fletcher (21), reported in 1973 that heavy marihuana users scored as well as non-smokers on several tests of learning and memory. However, in 1988, a follow-up study was performed by the same group (22) on the same cohort of Latin American marihuana users and on non-smoking controls. Selective cannabis impairment of short-term memory skills was then observed, contradicting the results obtained ten years earlier. In 1988, a study by Varma (23) also reported short-term memory impairment in heavy marihuana smokers studied in India. In 1989, Schwartz (24) reported the results of an exceptionally well-controlled study of persistent short-term memory impairment in a group of American, middle-class adolescents. Their median age was sixteen, and they had at least eight years of formal education. Their performance was compared with that of a group of controls matched for age and I.Q. When initially tested, the cannabis-dependent boys and girls did much worse on short-term memory tests than the control group, and after six weeks of supervised abstention from intoxicants they still presented short-term memory deficits. The Schwartz study proves the specific lasting property of marihuana to impair memory storage, an essential part of the learning process, and to adversely affect psychomotor performance.

The most striking evidence of the lingering disruption of marihuana on memory and coordination was reported by von Leirer *et al.* (25). These investigators recruited 10 experienced private pilots in order to conduct a double-blind experiment. The subjects were trained for 8 hours on a computerized flight simulator. The test started one morning with a control

TABLE II

Experimental studies describing morphological alterations in the hippocampus of animal exposed to marihuana.

	ANIMAL	Dose THC	Hippocampus
Heath (1979)	Monkey	5 mg/kg/day 2 months	Altered synaptic width Alteration endoplasmic reticulum Nuclear inclusions
Scallet (1987)	Rat	10-60 mg/kg/day 60 days	Decrease density neuronal cells 44% reduction in synapse number Reduction in dendritic lenght (pyramidal neurons)
Landfield (1988)	Rat	8 mg/kg/day 5 x week 8 months	Decreased neuronal density Increased cytoplasmic inclusion

"flight," after which each subject smoked either a marihuana cigarette containing 19 milligrams of THC or a placebo cigarette from which the THC had been removed. The simulated landing was repeated one, four and twenty-four hours later. In every case, the worst performance occurred one hour after THC inhalation. But twenty-four hours later, those pilots who had smoked marihuana still experienced significant difficulty in aligning the computerized landing simulator and in landing the "plane" at the center of the runway. The pilots, however, reported no awareness of any marihuana after-effects on their performance, mood or alertness. It is not known exactly how long beyond twenty-four hours a single marihuana cigarette will disrupt the fine brain mechanisms controlling memory banks, but it is known that traces of THC are still present in the brain twenty-four hours after administration (9). More down-to-earth tasks, such as operating complicated equipment or railroad trains, may also be susceptible to "day-after" marihuana effects.

Marihuana and road accidents

Chesher (6) has reported that THC was 4000 times more potent than alcohol in producing decrements in performance of subjects studied in controlled conditions. Since, then, several major railroad accidents have illustrated the impairing effects of marihuana on the performance of complex tasks. In January 1987, a freight train rammed into the Metroliner traveling from Washington to New York at full speed, resulting in sixteen dead and forty-eight injured passengers. The conductor of the train had ignored three red signals before the crash. Cannabinoids were detected in his body fluids. A year later, marihuana was detected in a thirty-year-old switchman who had fled his post in a control tower after a train derailment in Chester, Pennsylvania. In the ensuing crash, twenty-five people were injured.

A study issued in February, 1990 by the U.S. National Transportation Safety Board (26) has provided the most extensive evidence linking fatal accidents among truck drivers to cannabis. The study covered 182 accidents involving 86 trucks in which 210 people were killed. One-third of the victims whose bodies were examined had recently used alcohol or drugs. The highest percentage (12.8%) had used marihuana; next came alcohol (12.5%), then cocaine (8,5%).

Soderstrom (27) reported prior marihuana use prospectively in 1023 trauma victims resulting from vehicular and non vehicular accidents. Marihuana was detected in 34.7% of the subjects, more than alcohol (33.5%).

All of the clinical reports confirm experimental studies performed by Paule (28) who observed the disruptive effects of chronic marihuana smoke on the complex behavior of rhesus monkeys exposed daily or twice a week for one year to the smoke of one marihuana cigarette. This dose corresponds to the consumption of 52 to 104 cigarettes of cannabis a year. Both experiences resulted in impairment of their response to standard tests of acquired or conditioned behavior, which were adequately performed by control animals exposed to smoke without THC.

Marihuana and schizophrenia

It is well established that marihuana smoking will trigger an acute psychotic episode in schizophrenics (29). However, Andreasson *et al* (30), reporting on a fifteen-year prospective study of 55,000 Swedish military conscripts, showed that the relative risk for developing schizophrenia among those who were heavy consumers of cannabis (use on more than fifty occasions) was six times greater than in non-users.

The property of cannabis to induce long-lasting mental disturbances in western man, now epidemiologically documented, would confirm older anecdotal reports (31) from medieval Islam (1396), India (1878-1972), Egypt (1843-1925), Brazil (1955), the Bahamas (1970) and Jamaica (1976). If finally confirmed, when biochemical markers will be available, cannabis-induced psychosis would provide evidence that the repetitive disturbance of brain neurotransmission by THC carries the most serious risk of lastingly impairing the basic biochemical neural mechanisms which control coherent behavior.

EFFECTS OF CANNABINOIDS ON CELLULAR GROWTH AND FOETAL DEVELOPMENT

In addition to its stereospecific effects, THC and all other natural non psychoactive cannabinoids inhibit in micromolecular concentration macromolecular synthesis of DNA, RNA and proteins in cell culture (32). In similar concentrations, these same cannabinoids inhibit acetylation and phosphorylation of chromosomal proteins, altering thereby the transcription of DNA (33).

These effects of the cannabinoids might account for their depressant effect on cellular immunity, macrophage function and fetal development.

In the 1970's, Rosenkrantz (35) and others reported that marihuana products were toxic to fetal development in all species studied: fish, birds, rodents, hamsters, rabbits, dogs and monkeys. Offspring also displayed retarded development and behavioral anomalies (36).

In the 1980's, anomalies in newborn babies exposed to marihuana during gestation were reported by several investigators. Quasi (37) and Hingson (38) described deficits in babies born to marihuana-smoking mothers. These studies illustrating the damaging effects of cannabis on the developing human fetus were later confirmed by three independent groups of investigators: in 1986 by Hatch (39), in 1987 by Lester (40) and in 1989 by Zuckerman (41). Lester used high-speed computer voice analysis to assess the maturity of newborn infants. The cries of infants born to marihuana-smoking mothers in Jamaica showed a much higher incidence of voice anomalies than did cries of infants from non-smokers, suggesting possible impairment of fetal brain development. Zuckerman and colleagues reported a long-term study of 1226 mothers who were followed during pregnancy. Marihuana use was documented by urinalysis in 16% of the prospective mothers. Infants born to these marihuana-smoking mothers were shorter, weighed less and had smaller head circumferences at birth. Long-term developmental and behavioral effects which result from intra-uterine cannabis exposure were studied by Freid et al who reported that in 48 months of age children significant lower scores in verbal and memory were associated with maternal marihuana use (42).

EXPERIMENTAL AND CLINICAL REPORTS OF THE CARCINOGENICITY OF MARIHUANA SMOKE

In addition to the inherent properties of THC on the brain, and the inhibitory effect of all cannabinoids on cellular development one must consider the toxicity of the gas and particulate phase of marihuana smoke. Cannabis extracts and smoke condensate are mutagenic in the Ames test (43) Lung explants of human fibroblast culture exposed to the gas vapor phase of fresh marihuana smoke displayed abnormalities in DNA and chromosomal complement (44).

In rats, prolonged and repeated exposure to cannabis smoke results in lung tissue damage (45) and decreased activity of pulmonary macrophage (46).

Symptoms of airway obstruction (increased resistance to airflow) have been clinically documented in controlled experiments performed in young daily marihuana smokers (47). The symptoms are the more striking because acute exposure to THC dilates the bronchi. Microscopic examinations of bronchial biopsies sampled from heavy hashish users (20 to 26 years old) showed squamous cell hyperplasia (48). Finding of carcinogens in marihuana smoke, experimental evidence of the immunodepressive effects of THC and the presence of epithelial abnormalities known to be precursors of lung cancer in tobacco smokers, were good predictors of the development of cancer of the upper aerodigestive tract in chronic marihuana smokers, which have now been documented.

Donald (49) reported twelve cases of advanced head and neck cancer in young patients of average age twenty-six. One was nineteen. All had been daily marihuana or hashish smokers since high school, but did not smoke tobacco or use much alcohol. They were mostly squamous cell tumors of the tongue or jaw with local lymph node involvement, which Donald had only seen before among subjects sixty years of age or older who had been heavy drinkers and tobacco smokers for decades. "Such cases, says Donald, are unprecedented in this young age group." Taylor (50) reported that among 10 patients under age forty who were diagnosed with cancer of the respiratory tract, seven had a history of daily marihuana use. Taylor concluded that "regular marihuana use appears to be an additional significant risk factor for the development of cancer of the upper airways." Ferguson (51) reported the case of a twenty-seven-year-old man who died of metastic lung carcinoma. He had smoked marihuana heavily and steadily since age eleven. Cases of cancer of the tongue have also been reported in marihuana smokers (52).

MARIHUANA AND LEUKEMIA

The mutagenic and carcinogenic potential of marihuana is corroborated by the study of Robison (53). This "multicentric" investigation was undertaken in order to assess *in- utero* exposure to different medications in children who had developed non-lymphoblastic leukemia. Analyses were performed for reported maternal use of medications and drugs in the year preceding pregnancy as well as during their gestation of 204 case-control pairs of children. A tenfold increase in risk was found for the development of leukemia in the offspring of mothers who had smoked marihuana just before or during pregnancy. No other drug use during pregnancy (including tobacco, alcohol and painkillers) could be associated with such a risk. Children exposed to marihuana developed the disease earlier (at 19 instead of 93 months) and showed clonal abnormalities.

CONCLUSION

Results of standard *in vitro* and *in vivo* toxicological tests performed in the seventies on animal preparations to which marihuana extracts were administered were good predictors of the long-term pathophysiological manifestations observed twenty years later in chronic marihuana smokers. These manifestations also confirm anecdotal accounts of the damaging effects of cannabis reported throughout history and may be accounted for the properties of its biologically active cannabinoids, and the toxicity of marihuana smoke.

BIBLIOGRAPHY

1. Hollister L.E. *Health Aspects of Cannabis*. Pharmacol. Rev., 1987, 38:1-32.
2. Nahas G, Latour C. *First International Colloquium on Illicit Drugs*. Advances in the Biosciences, Oxford: Pergamon Press, 1991: 3-126.
3. Nahas G.G. Cannabis: toxicological properties and epidemiological aspects. *The Medical J. of Australia*, 1986;145:82-87.
4. Turner CE, Elsohly MA, Boeren EG. Constituents of *Cannabis sativa* L..XVII: A review of the natural constituents. *J Nat Prod.* 1980;43:169-234.
5. Busch FW, Seid DA, Wei ET. Mutagenic activity of marihuana smoke condensates. *Cancer Lett* 1979;6:319-324.
6. Chesher GB, Bird KD, Stramarcos A, Nikias M. *A comparative study of the dose response relationship of alcohol and cannabis on human skills performance*. In: Harvey DJ, ed. Marihuana 1984. Oxford: IRL Press, 1985:621-62.
7. Herkenham M, Lynn AB, Little MD et al. Cannabinoid receptor localization in brain. *Proc Natl Acad Sci* 1990;87:1932-1990.
8. Matsuda LA, Volait CJ, Brownstein MJ, Young AC and Bonner TI, *Nature*, 1990;346:561-64.
9. McIsaac WM, Fretchie GC, Idanpaan Heikkela JF, Ho BT and Englert LT. Distribution of marihuna in monkey brain and concomittant behavioral effecs. *Nature*, 1971;230:590-94.
10. Volkow ND, Gillepsi H, Mullani N, et al. Use of positron emission tomography to investigate the action of marihuana in the human brain. In *"First International Colloquium on Illicit Drugs"*. Advances in the Biosciences, Nahas G and Latour C., ed., Oxford: Pergamon Press, 1991:3-11.
11. Heath RG, Fitzjarrel AT, Fontana CJ and Garey RE. Cannabis sativa: effects on brain function and ultra-structure of Rhesus monkey. *Biol. Psych.*, 1980;15:657-691.
12. Scallet AC et al. Morphometric studies of the rat hippocampus following chronic delta-9-tetra-hydrocannabinol intoxication, *Brain Research*, 1987;436:193-198.

13. Landfield PW et al. Quantitative change in hippocampal structure following long-term exposure to delta-9-tetrahydrocannabinol: possible mediation by glucocorticoid systems. *Brain Research*, 1988;443:47-62.

14. Campbell KA, Foster TC, Hampson RE and Deadwyler SA. Delta-9-THC differentially affects sensory evoked potentials in the rat dentate gyrus. *J. Pharmacol. Exp. Ther.* 1986;239:936-940.

15. Miller LL and Branconnier. Cannabis effects on memory and the cholinergic limbic system. *Psychol. Bullet.* 1983;93:441-456.

16. Rawlins JND. Association across time: The hippocampus as a temporary memory store. *Behav. Brain Sci.* 1985;8:479-528.

17. Eichenbaum H, Cohen NI. Tends in neurosciences. 1988;11:244-248.

18. Moreau J. Hashish and mental illness. (1845). Translated by H. Peters, Raven Press, New York, 1973.

19. Miller LL. Cannabis and the Brain with special reference to the limbic system. In: *Marihuana Biological effects*, p. 539.

20. Soueif MI. Differential association between chronic cannabis use and brain function deficits. *Ann N Y Acad Sci* 1976;282:323-343.

21. Satz P, Fletcher J, Sutker L. Neuropsychologic, intellectual and personality correlates of chronic marjuana use in native Costa Ricans. In "Chronic cannabis use", *Ann N Y Acad Sci* 1976;282:266-306.

22. Page JB, Fletcher J, True WR. Psychosociocultural perspectives on chronic cannabis use: The Costa Rican follow-up". *J Psychoact Drugs* 1984;20:57-65.

23. Varma VK, Malhotra AK, Dang R et al. Cannabis and cognitive functions: A prospective study. *Drug and Alcohol dependency* 1988;21:147-152.

24. Schwartz RH, Gruenewald PJ, Klitzner M et al. Short-term memory impairment in cannabis-dependent adolescents. *Am J Dis Child.* 1989;143:1214-19.

25. Leirer VO, Yesavage JA. Marihuana carry-over effects on aircraft pilot performance. *Aviation, Space & Environmental Medicine*, 1991;62:221-227.

26. Department of Transportation, Washington. National Transportation Safety Board report. *Washington: Department of Transportation*:1990.

27. Soderstrom CA, Trifilis AL, Shankar BS, Clark WE and Cowley AR. Marihuana and alcohol use among 1023 patients. *Arch. Surgery*, 1988;123:733-737.

28. Paule MG et al. The effect of chronic marihuana smoke exposure on complex behavior in the Rhesus Monkeys. *The Pharmacologist*, 1989;A115.

29. Negrete J. *Cannabis and Schizophrenia*, Br J. Addiction, 34, 349-351.

30. Andreasson S, Allebeck P, Engstrom A et al. Cannabis and Schizophrenia; A longitudinal study of Swedish conscripts. *The Lancet* 1987;2:1483-85.

31. Brill M, Nahas GG. Cannabis intoxication and mental illness. In: GG. Nahas editor. *Marihana in science and medicine.*New York, Raven Press, 1985;228-279.

32. Nahas GG, Mokoshima A and Desoize B. Effects of cannabinoids on macreomolecular synthesis. Fed. Proc. 1977;36:1748-52.

33. Stein G et al. Cannabinoids : the influence of cell proliferation and macromolecular biosynthesis. In : *Marihuana Biological Effects,* p. 171.

34. Cabral GA, Vasquez R. Marijuana decreases macrophage antiviral and antitumor activities. In *"First International Colloquium on Illicit Drugs".* Advances in the Biosciences, Nahas GG and Latour C., ed., Oxford: Pergamon Press, 1991:93-105.

35. Rosenkrantz H. Effects of cannabis on fetal development of rodents. In *"Marihuana, Biological effects"*, Nahas GG. and Paton WDM, ed., Oxford: Pergamon Press, 1979:479-499.

36. Dalterio S, Bartke A, Mayfield D. Prenatal exposure to cannabinoids and sexual behavior of offspring. *Science* 1981;216:581-582.

37. Quasi QH, Mariano E, Milman DH et al. Abnormalities in offspring associated with prenatal marihuana exposure. *Dev Pharmacol Ther*, 1985;8:141-148.

38. Hingson R, Alpert JJ, Day N, Dooling E et al. Effects of maternal drinking and marihuana use on fetal growth and development. *Pediatrics* 1982;70:539-546.

39. Hatch EE, Bracken MB. Effect of marihuana use in pregnancy on fetal growth. *Am J Epidemiol* 1986;124:986.

40. Lester BM, Dreher MC. Effects of marihuana smoking in newborn cry analysis behavior. *Pediatrics* 1987;5.

41. Zuckerman B, Frank DA, Hingson R et al. Effects of maternal marihuana and cocaine use on fetal growth". *New Engl J Med* 1989;762-768.

42. Fried PA, Watkinson B. 36 and 48 month neurobehavioral follow up of children prenatally exposed to marihuana cigarettes and alcohol. *Developmental and Behavioral Pediatrics*, 1990;11:49-58.

43. Werner FC, Van Resburg SJ and Thiel PG. Mutagenicity of marijuana and transkei tobacco smoke condensates in the salmonella/microsome assay. *Mutat. Res.* 1980;77:135-142.

44. Leuchtenberger C, Leuchtenberger R, Ritter U, Inui N. Effects of marihuana and tobacco smoke on DNA and chromosomal complement in human lung explants. *Nature* 1973;242:403-404

45. Huber GL, Shea JW, Hinds WE et al. The gas phase of marijuana smoke and intrapulmonary bactericidal defenses. *Bull. Eur. Physiopatol. Resp.*, 1979;15:491-503.

46. Rosenkrantz H and Braude MC. Comparative chronic toxicities of delta-9-tetrahydrocannabinol administered by inhalation or orally in the rat. In : *Pharmacology of Marihuana*, edited by MC Braude and S. Szara, 1976;2:571-584, Raven Press, New York.

47. Tashkin DP, Calvarese BM, Simmons MS, Shapiro BJ. Respiratory status of seventy-four habitual marihuana

smokers. *Chest* 1980;78:699-706.

48. Tennant FS, Guerry RL, Henderson RL. Histopathologic and clinical abnormalities of the respiratory system in chronic hashish smokers. *Subst Alcohol Actions Misuse* 1980;1:93-100.

49. Donald PJ. Marijuana and upper aerodigestive tract malignancy in young patients. In *"First International Colloqium on Illicit Drugs"*. Advances in the Biosciences, Nahas GG and Latour C., ed., Oxford: Pergamon Press, 1991:39-54.

50. Taylor FM. Marijuana as a potential respiratory tract carcinogen: A retrospective analysis of a Community hospital population. *Southern Med J* 1988;81:1213-1216.

51. Ferguson RP, Hasson J, Walker S. Metastasic lung cancer in a young marijuana smoker. *J A M A* 1989;261:41-42.

52. Caplan GA, Brigham BA. Marijuana smoking and carcinoma of the tongue., *Cancer*, 1990;66:1005-1006.

53. Robison LL, Buckley JD, Daigle AE et al. Maternal drug use and risk of childhood nonlymphoblastic leukemia among offspring. *Cancer* 1989;63:1904-1910.

2. Visualization of brain alterations

USE OF POSITRON EMISSION TOMOGRAPHY TO STUDY DRUGS OF ABUSE.

Nora D. Volkow and Joanna S. Fowler.

Brookhaven National Laboratory, Upton, NY 11973

ABSTRACT

Positron Emission Tomography (P.E.T.) has been used in man to study the mode of action of illicit drugs, mainly cocaine, alcohol and marihuana on brain metabolism and circulation.

Acute cocaine administration decreases brain metabolic activity in cortical and subcortical regions. After chronic cocaine administration, a significant and persistent decrease in cerebral blood flow is observed. An increased metabolic activity is recorded in orbito-frontal cortex and basal ganglia during early cocaine withdrawal.

Alcohol intoxication is associated with a decrease in whole metabolic activity of the brain, mostly in the frontal and parietal lobes.

I.V. administration of THC (2 mg) was associated with a significant increase in cerebellar metabolic activity (glucose utilization). Chronic marihuana smokers also presented increased metabolic activity in the frontal lobe. However the increase cerebellar metabolism produced by THC administration was smaller in the chronic cannabis user than in the non smoking control. Deregulated cerebellar activity could translate in disruptions in functions associated with the cerebellum such as motor coordination, proprioception and learning.

KEY WORDS

Marijuana; cocaine; alcohol; PET; glucose metabolism; cerebellum; cerebral circulation.

Introduction

Positron Emission Tomography (PET) is a nuclear medicine imaging technique for measuring the regional and temporal concentration of positron emitters within a volume element of tissue (Phelps and Mazziotta 1986; Andreasen, 1988). Since carbon, oxygen, and nitrogen all have positron- emitting isotopes, they can, in principle, be used to label compounds without affecting their pharmacological behavior (Fowler et al. 1990). The relatively short half lives of the most frequently used positron emitters allows the performance of repeated studies on the same subject without jeopardizing the well-being of the individual.

From the perspective of the investigation of drugs of abuse, PET can be used in conjunction with various types of tracers to examine drug distribution and pharmacokinetics, mechanisms of action and toxicity of drugs, and the neurochemical events accompanying processes such as addiction and withdrawal.

For didactic purposes, we can divide the PET experimental strategies into those that assess the effects of acute and chronic drug administration on cerebral function, those that investigate the effects of acute and chronic drug administration on neurotransmitters, and those that directly investigate the behavior in the brain of the drugs of abuse labeled with a positron emitter. In this chapter, we will review the basic principles behind PET technology and its applications in the area of substance abuse. Even though we will limit our discussion mainly to the application of positron emission tomography (PET) in the investigation of substances of abuse, many of the strategies employed can also be utilized with single photon emission computed tomography (SPECT) and, when appropriate, we will illustrate with examples obtained from this technique.

Instrument

Imaging with PET is feasible because of the decay properties of positron emitters (For a more detailed description refer to Hoffman and Phelps 1986, Volkow et al. 1988a). Positron emitters are neutron-deficient isotopes that decay by liberating positrons. The positrons travel a short distance (mean free path) before losing their energy and colliding with an electron. Out of this collision two high-energy photons (511 keV) are liberated simultaneously at an angle of almost 180⁰. These photons are recorded by the PET camera if they arrive simultaneously into two interacting detectors. Events are recorded as simultaneous if they are detected within a preset time window period which usually averages 5-20 nsecs. The generation of these two photons liberated in opposite directions allows the location of the decay event within the volume delineated by the line of view of the two interacting detectors. This enables spatial discrimination without the need of collimators as required with SPECT. Because coincident detection of positron emitters by individual pairs of detectors is inefficient, PET cameras are designed in such a way that each detector operates in coincidence with other detectors simultaneously. In a typical PET camera, detectors are placed in a ring, or multiple rings, to obtain multiple images of the whole object at the same time. Events detected coincidentally by pairs of detectors acting in coincidence will represent the sum of activity along the line of view joining these detectors. There is a coincident line of detection for each possible combination of coincident detectors.

Detectors are crystals which fluoresce when exposed to ionizing radiation. Several types of crystals have been used in PET. The most commonly used has been bismuth germanate (BGO). The advantage of BGO is its high stopping power which permits the use of relatively small crystals. Since spatial resolution in the PET camera is related in part to the size of the crystal, BGO enables to achieve a higher spatial resolution than with other crystals. Cesium fluoride (CsF) is another detector material which has been used for PET cameras that require fast-counting statistics. Although CsF has a lower stopping power than BGO, it has a smaller decay constant. This translates into a better ability to handle high count rates without saturating the crystal. Barium fluoride (BaF) has a fast decay constant and also has a high stopping power, but it is very expensive.

In the detectors, crystals are coupled with a photomultipler. The photomultipler intensifies the light signals coming out of the crystal and passes this information into the computer which then applies mathematical algorithms that filter and back-project this information to obtain an image. These algorithms are not dissimilar to those used for CT scanning.. Mathematical models are then applied to translate the information in the image to information that is of physiological, pharmacological, or neurochemical relevance.

Isotopes and Radiopharmaceuticals

The positron emitters most commonly utilized in PET are carbon 11, oxygen 15, fluorine 18, and nitrogen 13. The half lives for these compounds are (F-18 ($t_{1/2}$: 110 min) C-11 ($t_{1/2}$:20.4 min); oxygen-15 ($t_{1/2}$:2 min); nitrogen-13 ($t_{1/2}$:10 min)) (Fowler et al. 1988). Because of the short half life of these isotopes, a cyclotron is required in the nearby vicinity to produce them. The radionuclides can be utilized in the chemical form in which they are produced, as in the case of O-15-labeled oxygen gas, to measure oxygen metabolism, or fluorine 18 in the chemical form of fluoride to measure bone density, or they can be incorporated into organic molecules to measure specific metabolic or chemical pathways providing that rapid synthetic routes can be developed. Most of the studies with PET have been done using an analog of glucose (deoxyglucose) labeled either with carbon 11 (CDG) (Reivich et al. 1982) or fluorine 18 (FDG) (Reivich et al. 1979). Deoxyglucose or 2-deoxy-2- fluoro-D-glucose enters the cell and is phosphorylated at a rate determined by the energy requirements of the cells (Sokoloff et al. 1977). Once phosphorylated, it is trapped within the cell and remains there without being further metabolized. This permits the measurement of the concentration of CDG or FDG in a steady-state condition. The strategy is similar to the one utilized with autoradiography using ^{14}C-deoxyglucose (Sokoloff et al. 1977). Once injected, 30 to 35 min are allowed to elapse for uptake of the tracer to occur. The PET measurements are made 35-55 min after injection. The disadvantage of the use of deoxyglucose is that information reflects the functional activity of the brain over a period of time of relatively long duration (the 30-min uptake for deoxyglucose). The advantage is that one is measuring directly the substrate that is required for energy formation (Greenberg et al. 1981). Because in the brain energy consumption and brain activity are tightly coupled, this information can then be utilized to derive information about regional brain function (Silver 1979). Under normal conditions, energy consumption and cerebral blood flow (CBF) are tightly coupled so that the measurement of CBF can also be utilized to assess regional brain function (Silver 1978). However, under pathological conditions, the normal coupling between CBF and brain energy metabolism can be lost (Wide et al. 1983). Furthermore, many of the substances of abuse have direct effects on cerebral blood flow that are separate from their effects on nervous tissue (Altura & Altura 1983). Because of this, it is important to delineate whether changes in CBF after drug administration are secondary to its effect on brain function, or whether they represent its vasoactive properties.

Oxygen 15 has also been utilized to measure energy metabolism in the human brain (Lammertsma et al. 1981). Surprisingly, studies done with cerebral activation and oxygen 15 have been unable to demonstrate regional increases in oxygen metabolism secondary to brain activation tasks (Fox and Raichle 1985) as observed when measuring glucose metabolism (Reivich et al. 1985) and/or CBF (Volkow et al. 1991a). These findings are controversial since heretofore it was believed that cerebral activation led to increases in glucose as well as oxygen metabolism. More work needs to be done to evaluate if in human during cerebral activation there is an uncoupling between glucose and oxygen metabolism.

Several tracers have been developed to measure different receptor systems (Schlyer et al. 1991). Most of the work, has focused on the dopamine system. Tracers have been developed to label sub-types of dopamine receptors. The most commonly used tracers to label D_2 receptors have been carbon 11 (Wong et al. 1986) and fluorine-18-labeled N-methylspiroperidol (NMS) (Arnett et al. 1986) and carbon-11-labeled raclopride (Farde et al. 1986). Both of these tracers differ with respect to their activity and specificity for the D_2 receptors. Methylspiroperidol binds to both dopamine and serotonin receptors (Wong et al. 1984). However, dissociation from the serotonin receptor is faster than that of the dopamine receptor. Thus, after 80 min of injection, most of the activity remaining in the brain represents binding into the D_2 receptor (Schlyer et al. 1991). In contrast, carbon-11 raclopride binds mainly to the D_2 receptors. The affinity of NMS for the dopamine receptor is higher than that of raclopride. This has the advantage of making it a much more insensitive tracer to concentration of dopamine (Seeman et al. 1990). This is relevant since the rate of uptake of raclopride is going to be affected by the intrasynaptic concentration of dopamine at the time of the study. This can lead to miscalculation in the quantitation of dopamine receptors. On the other hand, the relatively lower affinity of raclopride for the dopamine receptor allows the performance of displacement studies which are not feasible with N-methylspiroperidol. Displacement studies are carried out in the following way: the tracer is injected and allowed to reach peak concentration and then pharmacologic doses of a compound that affects synaptic dopamine concentration or interacts with the receptors can be given to displace ^{11}C-raclopride (Dewey et al. in preparation). Using this strategy, different parameters can be measured including the speed at which the drug enters the brain, the relative changes in dopamine concentration from drugs that directly or indirectly release dopamine, and/or the relative potency of drugs that bind to the dopamine receptors when compared with raclopride.

Tracers have also been developed to label the D_1 receptors. The most commonly used tracer has been C-11 SCH-23390 (Halldin et al. 1986). In the case of the investigation of drugs of abuse, interest has also focused on other receptor systems such as the benzodiazepine (Pappata et al. 1988), serotonin (Wong et al. 1987), and the cannabinoid receptor (Herkenham et al. 1990).

Psychoactive drugs can also be labeled directly with a positron emitter and their behavior can be monitored with respect to its distribution in the brain, its kinetics, its binding characteristics, its availability in plasma, and its distribution in other organs of the body.

Cocaine

PET has been utilized to investigate both acute and chronic effects of cocaine. These studies have been carried out to investigate mechanisms of action of cocaine, mechanisms of toxicity from cocaine, and processes involved in the reinforcing and addicting properties of cocaine.

a) Acute: The effects of acute cocaine administration in human subjects were measured using PET and FDG in a group of chronic cocaine abusers who were given 40 mg of cocaine i.v. (London et al. 1990). This study reports widespread decreases in brain metabolic activity both in cortical and subcortical structures. In this study, the rate of metabolism correlated with the subjective sense of intoxication. For this group of subjects brain metabolic response to cocaine was affected by the degree of ventricular enlargement. Subjects with enlarged ventricles were less sensitive to the effects of cocaine on brain metabolism than those that had less ventricular enlargement.

b) Chronic: Most of the studies done on cocaine have focused on the effects of chronic cocaine abuse on brain metabolism, cerebral blood flow, and neurotransmitter activity.

A study that investigated the long-term effects of cocaine in cerebral blood flow was done in a group of 20 chronic cocaine abusers and 24 normal controls (Volkow et al. 1988b). Oxygen-15-

labeled water was used to determine regional cerebral blood flow (Raichle et al. 1983). Measurements of cerebral blood flow were done on two separate occasions on the patients: within three days of last cocaine use and then 10 days after cocaine detoxification. Upon the first evaluation, most of the patients showed widespread disturbances in cerebral blood flow. Fourteen of the 20 cocaine abusers showed patchy areas of uptake for oxygen-15-labeled water at various regions of the brain cortex. These effects were predominantly in the frontal cortex and in the left hemisphere. The repeated study done after ten days of the last use of cocaine continued to show disturbances of CBF. The decreases in CBF were interpreted as representing vascular pathology secondary to the use of cocaine. These findings have been replicated using SPECT to measure changes in CBF of cocaine abusers (Mena et al. 1989). These studies have documented defects in perfusion in both moderate as well as heavy cocaine abusers. Cocaine could produce vascular pathology secondary to its known vasoconstricting actions (Altura, Altura 1985, Isnert and Chokoshi 1989). In fact, clinical studies have documented the occurrence of cerebral strokes and hemorrhages associated with cocaine consumption (Levine 1987, Lichtenfeld 1984). Vasoconstriction from cocaine, if prolonged, could produce vascular damage, tissue ischemia, and necrosis. Also, the increases in blood pressure secondary to cocaine could favor the occurrence of cerebral hemorrhages. These findings are of pertinence in understanding the mechanisms of toxicity in cocaine. This is relevant inasmuch as cocaine accounts for a considerable amount of morbidity and mortality. Although most of the mortality associated with cocaine stems from its cardiotoxic properties, cerebral vascular accidents lead to a considerable amount of morbidity and are increasingly being reported.

The effects of chronic cocaine have also been investigated using FDG to measure regional glucose metabolism. These studies examined the pattern of regional metabolism at the different phases of cocaine detoxification. A study that investigated the effects of acute withdrawal of cocaine was performed on a group of 15 chronic cocaine abusers and 10 normal controls (Volkow et al. 1991b). Ten of the cocaine abusers were tested within one week of last cocaine use, whereas 5 were tested 2 weeks to 4 weeks after last detoxification. When compared with the normal controls, the cocaine abusers tested within one week of last use of cocaine showed higher metabolic activity in the orbito-frontal cortex and in the basal ganglia. Metabolism in these regions was negatively correlated with the time period since last use of cocaine. In fact, increased metabolic activity in the orbito-frontal cortex and basal ganglia were not seen when patients were tested 2-5 weeks after last use of cocaine, thus suggesting that the increase in metabolic activity into the basal ganglia and orbito-frontal cortex was related to cocaine withdrawal. Metabolic activity in these regions was also positively correlated with the intensity of the craving. The fact that these regions, orbito-frontal cortex and basal ganglia, have been implicated in a circuit that involves repetitive and compulsive patterns of behavior (Kolb 1977), suggests that one of the mechanisms by which cocaine induces pathological self administration may involve its ability to disrupt this circuit.

A second study was carried out to determine if brain metabolic defects persisted after the early withdrawal from cocaine had subsided (Volkow et al. submitted/a). For this study, a total of 20 chronic cocaine abusers were investigated 10 days to 2 months after last use of cocaine. Seven of these subjects were retested 3-4 months after a drug free period. Patients were hospitalized and maintained on a rehabilitation program for the entire length of the study. The normal controls constituted 20 healthy volunteers who were screened for the absence of psychopathology, neurological or medical disease, and past or present history of cocaine abuse. The study documented significant reductions in frontal metabolic activity in cocaine abusers when compared with normal controls. The decreases were larger for the left frontal cortex than for the right frontal area and persisted when the subjects were retested 3-4 months after detoxification. The rate of decrease in metabolic activity in the left frontal cortex was correlated with the intensity of depressive symtoms as assessed by the Beck Depression Inventory.

These studies show differences in brain metabolism between normal controls and cocaine abusers and documents differential brain metabolic changes for the early and late phases of cocaine detoxification. Cocaine abusers tested during the early phase of cocaine withdrawal, showed increased metabolic activity into areas connected with the dopamine system whereas cocaine abusers tested in late withdrawal showed decrease metabolic activity in the frontal cortex. Persistence of metabolic abnormalities even after 4 months of detoxification points to the long term deleterious actions of cocaine in brain metabolism.

Positron emission tomography has also been utilized to determine if there are changes in dopamine receptors secondary to chronic cocaine use. This is relevant inasmuch as it has been hypothesized that cocaine addiction may result from a decrease of dopamine activity secondary to chronic use of cocaine (Dachis et al. 1985). Studies measuring changes in dopamine receptors with chronic cocaine are controversial, some authors report increases, decreases, or no changes (For review Post et al. 1987). In order to test this hypothesis, a study was done to measure the availability of brain dopamine receptors with PET and F-18 N-methylspiroperidol. The study was done in 10 chronic cocaine abusers, (7 were tested within one week of last use of cocaine, and 3 were tested 2 to 4 weeks after cocaine withdrawal), and in a group of 10 normal controls (Volkow et al. 1990a). This study documented decreases in dopamine receptor availability in subjects who were tested within one week of last cocaine use. In contrast, there was no evidence of changes in dopamine receptors on the patients tested 2 to 4 weeks of last use of cocaine. This study showed that chronic use of cocaine leads to changes in dopamine receptors. The decrease in dopamine receptor availability during the initial withdrawal may represent an adaptation to the dopaminergic overstimulation secondary to chronic use of cocaine. With withdrawal, the receptor availability may revert to values similar to those seen in normal controls. The extent to which this occurs, however, can not be inferred from this investigation due to the very small sample size of subjects tested after the initial cocaine withdrawal.

c) Pharmacokinetics: Carbon-11-labeled cocaine ([N-^{11}C-methyl] cocaine) has been used in conjunction with PET to measure the behavior of cocaine in the human and baboon brain <u>in vivo</u> (Fowler et al. 1989). The pattern of distribution of cocaine in the human brain was heterogenous with maximal concentration occurring in the basal ganglia which is the area with the highest density of dopamine receptors in the human brain. In principle, the PET image of the brain after C-11 cocaine could represent both the parent drug and labeled metabolites. However, it is known that the principal labeled metabolites of cocaine, ^{11}C-benzoylecgonine and ^{11}C-ecgonine methyl ester, do not cross the blood-brain barrier and, therefore, the PET image of the brain is cocaine itself.

Animal experiments carried out in baboon brain with PET to characterize the pattern of binding of cocaine showed that cocaine binds predominantly to the dopamine transporters since pretreatment with nomifensine, a drug that inhibits the dopamine transporter selectively reduced cocaine binding into the basal ganglia. In contrast, desipramine, a drug that binds to norepinephrine and serotonin transporters, did not affect its binding. Even though the highest concentration of cocaine was observed in the basal ganglia, there were other brain areas that showed a significant accumulation of cocaine. In particular, the thalamus showed an initial accumulation of cocaine as high as that in the basal ganglia. However, its clearance from this region was faster than in the basal ganglia.

With PET, one can measure both the regional distribution of a compound on the brain and its changes with time, resulting in a direct measurement of the pharmacokinetics of a drug in the brain. Studies measuring the pharmacokinetics of ^{11}C-cocaine showed a very fast uptake of cocaine by the human brain. Peak uptake occurred 4-6 minutes after injection. This study also showed that cocaine clears from the brain very rapidly and less than 50 percent of the activity remains 20 minutes after injection. The pattern of uptake and clearance of cocaine parallels the time changes in the intensity of the euphoria experienced after intravenous cocaine. This parallelism suggests that the kinetics of cocaine are important in understanding its reinforcing effects.

A PET study designed to measure changes in distribution and kinetics of cocaine in the brain secondary to chronic use of cocaine was done in a group of 10 chronic cocaine abusers and 10 normal controls (Volkow et al. unpublished data). This study showed that chronic cocaine abusers had decreased uptake of carbon-11 cocaine in the brain. These decreases were significantly lower in the basal ganglia and in thalamus. Whether these changes represent decreases in the dopamine transporters secondary to chronic use of cocaine or to presynaptic dopamine neuron degeneration or whether they reflect decreases in cerebral blood flow needs to be further investigated.

The whole body distribution of cocaine was investigated with carbon-11 methylcocaine and PET to try to understand the widespread toxicity of cocaine to the various organs of the body (Volkow et al. submitted b)). Of particular interest are the cardiotoxic properties of cocaine since they account for a considerable amount of the morbidity and mortality seen in cocaine abusers (Huester 1987). The mechanisms of cardiotoxicity from cocaine have been related to both its central (Wilkerson 1988, Jones Tackett 1990) and peripheral effects (Pitts 1989). Peripheral mechanisms involve the release of catecholamines from the adrenals (Chiueh Kopin, 1978; Nahas et al. 1991) as well as possible direct interactions of cocaine with the cardiac tissue (Peng et al. 1989). The extent to which cocaine accumulates in the human heart has not been investigated in vivo. Postmortem studies have demonstrated binding of cocaine to muscarinic receptors in myocardial tissue (Sharkey et al. 1988). A PET study done on 10 normal controls measured the rate of uptake of carbon-11 cocaine in the human heart. This study showed a similar absolute uptake of cocaine in the heart to that in the brain (.01% of the injected dose per cc of tissue). The kinetics of cocaine into the heart were faster than those seen in the brain. Maximal uptake was achieved 3 to 4 minutes after injection; half-peak clearance was achieved by 10 minutes after injection. Neither preadministration of desipramine nor of benztropine affected the binding of cocaine in the heart. More work is still required to characterize the pattern of cocaine binding in myocardium. However, the demonstration of significant accumulation of cocaine into the human heart is of relevance since it validates the possibility that cardiotoxicity from cocaine could be related to the direct interaction of cocaine with the myocardial cells. In addition, cocaine's local anesthetic properties could result in depressant effects on excitable tissue and thus could account for arrythmias and asystole (Przywara Dambach 1989).

Cocaine did not bind into the lungs and in this organ the pattern of activity paralleled that of plasma. There was significant accumulation of cocaine in the liver where the kinetics were much slower than in any other organ. Peak uptake occurred 30 minutes after injection and at that point it plateaued and remained stable for the remaining of the study (45 minutes). The uptake of radioactivity by the liver probably represents not only carbon-11 cocaine but also its metabolites since at 30 minutes 60% of the radioactivity represents cocaine metabolites and only 40% represents the parent compound. Accumulation of cocaine was also seen into the kidneys where the kinetics paralleled those seen in the heart. Maximal uptake of cocaine in the human body was observed in the adrenals. For this organ, the uptake per unit weight of tissue was 5 times higher than that seen in the brain. The kinetics of uptake, however, were similar to those of the brain and maximal peak was achieved 6 to 8 minutes after injection and half-peak clearance occurred by 20 minutes. The extremely high accumulation of cocaine into the adrenals may be of relevance in accounting for the massive release of catecholamines observed after acute cocaine administration (Nahas et al. 1991). Even though the catecholamine release from adrenals was related to its central effect, there is now direct evidence that cocaine can interact directly with a chromaffin cell to release norepinephrine (Chieuh Kopin, 1978). This, in turn, is also of relevance in understanding the cardiotoxicity from cocaine.

In summary, these PET studies have been able to document cerebral vascular toxicity secondary to cocaine, increases in metabolic activity in basal ganglia and orbito-frontal cortex during

early cocaine withdrawal,
long-term decreases in regional brain metabolism after chronic use of cocaine, as well as very fast kinetics for cocaine in brain that may account for some of its reinforcing properties and high accumulation of cocaine into various organs that could account for its cardiotoxic and hepatotoxic properties.

Alcohol

Most of the work done with alcohol has involved the investigation of changes in the regional brain glucose metabolism secondary to chronic use of alcohol. Of the four studies reported, (Samson et al. 1986, Sachs et al. 1987, Wik et al. 1988, Volkow et al. 1991c) three of these have documented evidence of decrease whole brain metabolic activity in these patients. Although regional patterns of defects reported by various investigators vary, the most consistently reported as abnormal is the frontal cortex. In addition, left hemispheric abnormalities have been reported with more frequency than those of the right hemisphere. Differences among the investigators could reflect differences in population investigated as well as differences in the period after alcohol withdrawal at which the subjects were investigated. These issues were addressed in a study done on 20 alcoholics that were investigated 6-42 days after alcohol withdrawal (Volkow et al. 1991c). This study showed that the decrease in whole brain metabolic activity was correlated with the days since last alcohol use. When normalizing for the effects of whole brain metabolism to examine if there were areas of the brain more affected than others, alcoholics were found to have significant lower frontal and left parietal metabolism. These normalized changes in metabolism were independent of the time period after alcohol withdrawal at which subjects were investigated. Thus, these studies document the existence of effects not only of alcohol withdrawal, but also long-term changes in brain metabolism that may translate in dysfunction of those regions affected.

Studies have also been done with PET to investigate the effects of acute alcohol administration on regional brain glucose metabolism. A study was done in 6 normal controls and 6 chronic alcoholics who were tested under two conditions: one done under no pharmacological intervention, and the second done after acute administration of 1 g/kg of ethanol (Volkow et al. 1990b). This study documented decreases in brain glucose metabolism after acute alcohol administration. The decreases in metabolism were heterogeneous, and the areas with the highest response to alcohol were the areas which are known to have the highest density of benzodiazepine receptors in the human brain. This study interpreted the decrease in regional brain metabolism after ethanol as reflecting an interaction of ethanol with the benzodiazepine-GABA receptor system. For this study, alcoholics showed greater response to the effects of ethanol relative to normal controls. Despite the larger reductions in brain glucose metabolism upon ethanol administration, the alcoholics showed less subjective sense of intoxication than that reported by the normal controls. This paradoxical higher metabolic responsivity and decreased subjective responsivity to the effects of ethanol may reflect tolerance of the brain to decreased metabolism secondary to alcohol. Since the regional metabolic pattern of response to acute alcohol administration correlated with the known concentration of benzodiazepine receptors, the increased metabolic responsivity by the alcoholics was interpreted as reflecting changes in the benzodiazepine-GABA receptor system in chronic alcoholics. These changes could be a result of chronic alcohol administration, of alcohol withdrawal, or they could relate to genetic differences that may predispose these individuals to alcoholism. Animal work has demonstrated that chronic alcohol induces increased sensitivity in the benzodiazepine GABA receptor system (Ticku 1983). Since the mechanisms by which alcohol interacts with the benzodiazepine GABA receptor system have not been delineated, the nature of possible changes in the benzodiazepine-GABA receptor system could not be clarified by this study. A second study was done in order to address specifically the extent to which the benzodiazepine receptor was abnormal in chronic alcoholics. This study was done in a group of 15 normal controls and 10 chronic alcoholics (Volkow et al. 1991d). Subjects were tested with FDG to measure brain glucose metabolism under two conditions. The first scan was done with no pharmacological intervention, whereas the second scan was done 40 minutes after intravenous administration of the benzodiazepine agonist lorazepam. Five of the normal controls received 1 mg of lorazepam and the rest of the subjects received 30 μ/kg of lorazepam (2 mg). Both doses of lorazepam decreased regional brain glucose metabolism. The magnitude of response was dose-related and represented a 10% decrease for 1 mg, and 20% for the 2-mg dose. The low dose of lorazepam produced a pattern of response similar to that observed with alcohol. The maximal response was observed in the occipital cortex and in the pre-frontal cortex. The high dose of lorazepam induced a marked decrease in metabolic activity in the thalamus, which was not observed with alcohol nor with the low-dose of lorazepam. Surprisingly, the alcoholics showed much less response than the normal controls to lorazepam administration. The seeming inconsistency in the results observed in the studies done in alcohol and those with lorazepam could be explained if, the site of interaction of alcohol and benzodiazepine in the GABA receptor complex is different. This possibility is supported by animal work which has shown that for some paradigms the effects of alcohol and benzodiazepines are not only different but sometimes even antagonistic (Najim and Karim 1991). Direct measurements of benzodiazepine receptor concentration in human brain using ligands to label the benzodiazepine receptors have been done with PET and SPECT. A preliminary study done with PET in three chronic alcoholics reported decreases in benzodiazepine receptors in these patients (Litton et al. 1991). The results were too preliminary to draw any specific conclusions.

Marijuana

The effects of acute marijuana administration have been investigated, both in normal controls and in chronic marijuana abusers (Volkow et al. 1991e, Volkow unpublished data). These studies were done using FDG to measure regional brain glucose metabolism under two conditions: the first scan was done with no pharmacological intervention, and the second scan was done 24 hrs later, after the subjects have been given 2 mg i.v. of THC (the main psychoactive component of marijuana). There was a very variable response to THC, both with respect to the subjective effects of the drug as well as the regional brain metabolic changes. Eleven of the 18 subjects reported the experience as pleasant, 3 became extremely anxious and paranoid, and 4 reported minimal effect. Seven of the subjects showed an increase of more than 10% in whole brain metabolism, 5 showed more than a 10% decrease, and 6 showed changes of less than 10% in whole brain metabolism. Despite this wide variability in metabolic response, there was a consistent increase in cerebellar metabolism observed both in the normal controls as well as the chronic marijuana users. The increased metabolism in cerebellum correlated with the plasma concentration of THC; individuals with the highest THC plasma levels showed the highest cerebellar metabolic values. The intensity of the subjective sense of intoxication after THC was also correlated with the cerebellar metabolic activity. Comparisons between the normal and marijuana abusers revealed that the marijuana abusers showed less of a cerebellar response upon THC administration than the normal controls. In addition, marijuana abusers also showed increased metabolic activity in the prefrontal cortex after THC. Prefrontal cortex activation upon THC administration was not observed in the normal controls. Comparisons of the metabolic images obtained during the baseline scan showed significantly lower cerebellar metabolism in the marijuana abusers when compared with normal controls. Thus, this study demonstrates that for THC the area of the brain which is the most sensitive to the acute effects of marijuana is also the one that is mostly affected under chronic conditions. Marijuana appears to produce both immediate, as well as long-term changes in cerebellar metabolism. The well- localized cerebellar effects of marijuana correspond to the high density of cannabinoid receptors reported in this area (Herkenham et al. 1990) and are consistent with a direct interaction of THC with the cannabinoid receptors. Other brain areas have been shown to contain cannabinoid receptors such as the hippocampus, substantia nigra pars reticulata, and the caudate nucleus. Because of the limited spatial resolution of the PET instrument used in these studies, these regions could not be measured in the present investigation. This PET study supports the notion that the effect of marijuana in brain glucose metabolism is mediated through its interaction with receptors, rather than through diffuse non-specific effects by alteration of membrane function. Similarly, this study documents deleterious effects of chronic marijuana in brain function. More specifically, marijuana abusers showed evidence of decreased cerebellar activity. This could translate in disruptions in functions that are associated with the cerebellum, such as motor-coordination, proprioception, and learning. All the above cognitive processes have been shown to be disrupted during acute marijuana intoxication (Roth et al. 1973, Yesavage et al. 1985, Dornbush et al. 1971) and to be abnormal in chronic marijuana users (Wig and Varma 1977, Mendhiratta et al. 1978, Varma et al. 1988). Cerebellar disruption by THC could explain these functional disturbances.

Attempts have also been made to label a THC analog to measure its regional concentration in the baboon brain (Macrianik et al. 1990). Unfortunately, because of the high lipophilicity of marijuana, this study did not yield any particular regional pattern of localization. The tracer was widely distributed in the brain with a relatively higher concentration in the cerebellum. However, preadministration of pharmacological doses of THC failed to change the regional pattern of binding, thus suggesting that the binding corresponded to nonspecific binding of the tracer. Development of

THC ligands with higher affinity for the cannabinoid receptors may make it feasible to measure the regional concentration of cannabinoid receptors <u>in vivo</u> in the human brain.

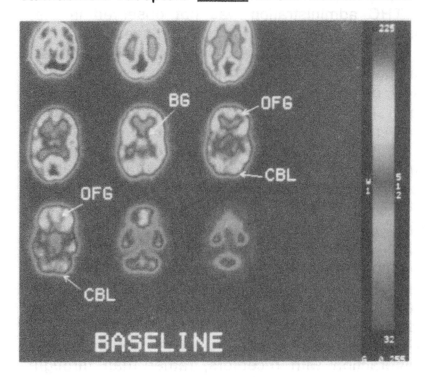

Figure 1. Image of brain glucose metabolism obtained with 18F-deoxyglucose (FDG) and PET in a normal control tested under baseline conditions (resting, eyes open, ears unplugged).

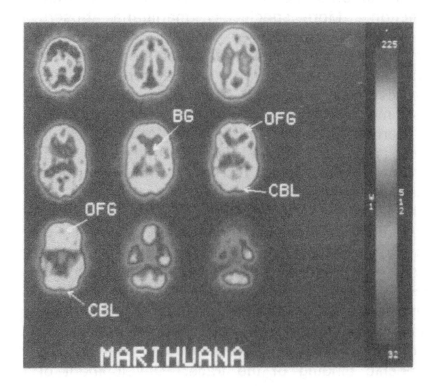

Figure 2. Images of brain glucose metabolism for the same subject as Figure 1 but 40 minutes after intravenous administration of 2 mg of tetrahydrocannabinol. Notice the relative increases in metabolism in orbital frontal cortex and cerebellum during marihuana intoxication.

BIBLIOGRAPHY

Altura BM, Altura BT, and Gebrewold A. Cocaine induces spasms of cerebral blood vessels: Relation to cerebral vascular accidents, strokes and hypertension. Fed Proc 44:1637, 1985.

Andreasen NC. Brain imaging: Applications in Psychiatry. Science 239:1381- 1388, 1988.

Arnett CD, Wolf AP, Shive CY, et al. Improved delineation of human dopamine receptors using ^{18}F-N-methylsperoperidol and PET. J Nucl Med 27:1878-1882, 1986.

Chiueh CC, Kopin IJ. Endogenous epinephrine and norepinephrine from the sympathoadrenal medullary system of unanesthetized rats. J Pharmacol Exp Ther 205:148-154, 1978.

Dackis CA, Gold M. New concepts in cocaine addiction: the dopamine depletion hypothesis. Neuroscience Behavioral Reviews 9: 464-477, 1985.

Dewey SL, Smith G, Wolf AP, et al. GABA ergic inhibition of endogenous dopamine released measured in vivo with ^{11}C raclopride and positron emission tomography. Submitted.

Dornbush RL, Fink M, and Freedman AM. Marijuana, memory, and perception. Amer J Psychiatry, 128(2):194, 1971.

Farde L, Hall H, Ehrin E, et al. Quantitative analyses of D_2 dopamine receptor binding in the living human brain by PET. Science 231:258-261, 1986.

Fowler JS, Wolf AP, and Volkow ND. New directions in positron emission tomography part II. In Annual report in medicinal chemistry, Vol. 24, Allen RC (ed) New Jersey Academic Press, pp. 261-269, 1990.

Fowler JS, Volkow ND, and Wolf AP, et al. Mapping cocaine binding sites in human and baboon brain in vivo. Synapse 4:371-377, 1989.

Fowler J, Wolf AP. Cyclotrons and radiopharmaceuticals in positron emission tomography. JAMA 259:1854-1860, 1988.

Fox PF, Raichle ME. Regional uncoupling of cerebral blood flow and oxygen metabolism during focal physiological activation: a positron emission tomography study. Proceedings of the XII International Symposium on Cerebral Blood Flow and Metabolism. Lund/Ronneby 1985, p. 125.

Greenberg JH, Reivich M, Alavi A, et al. Metabolic mapping of functional activity in human subjects with the ^{18}F-fluoro-deoxyglucose technique. Science 212:678-680, 1981.

Halldin C, Sone-Elander S, Farde L, Ehrin E, Fasth K-J, Langstrom B, and Sedvall G. Preparation of ^{11}C-labeled SCH 23390 for the in vivo study of dopamine D-1 receptors using positron emission tomography. Int J Radiat Appl Instrum 37(A):1039, 1986.

Herkenham M, Lynn AB, Little MB, Johnson RM, Melvin LS, de Costa BN, Rice KC. Cannabinoid receptor localization in brain. Proceedings of the National Academy of Sciences of the United States of America 87: 1932-1936, 1990.

Hoffman E, Phelps M. Positron emission tomography in Positron Emission Tomography and Autoradiography: Principles and Applications for the Brain and Heart. Edited by Phelps ME, Mazziotta J, Schelbert HR. New York, Raven Press, 1986.

Huester DC. Cardiovascular effects of cocaine. JAMA 257: 979-980, 1987.

Isnert JM, and Chokoshi S. Cocaine and vasospasm. New Engl J Med 23:1604- 1606, 1989.

Jones LF, Tackett RL. Central mechanisms of action involved in cocaine- induced tachycardia. Life Sciences 46:723-728, 1990.

Kolb B. Studies on the caudate-putamen and the dorsomedial thalamic nucleus of the rat: Implications for the mammalian frontal lobe functions. Physiol Behav 18:237-244, 1977.

Lammertsma AA, Jones T, Frackowiak RSJ, et al. A theoretical study of the steady state model for measuring regional cerebral blood flow and oxygen utilization using oxygen-15. J Comput Assist Tomogr

5:544-550, 1981.

Levine SR, and Welch KMA. Cocaine and stroke: Current concepts of cerebrovascular disease. Stroke 22:25-30, 1987.

Lichtenfeld PJ, Rubin DB, and Feldman RS. Subarachnoid hemorrhage precipitated by cocaine snorting. Arch Neurol 41:223-224, 1984.

Litton JE, Farde L, Neiman J, Halldin C and Sedvall G. Positron emission tomography of benzodiazepine receptors in alcoholics - equilibrium quantification. J Cereb Blood Flow Metabolism 11: S617, 1991.

London ED, Cascella NG, Wong DF, Phillips RL, Dannals RF, Links JM, Herning R, Grayson R, Jaffe JH, Wagner HN. Cocaine-induced reduction of glucose utilization in human brain. A study using positron emission tomography and [fluorine-18]-fluorodeoxyglucose. Archives of General Psychiatry 47: 567-576, 1990.

Macrianik G, Charalambous A, Schiue CY, Dewey SL, Schlyer DJ, Makrigannis A, and Wolf AP. ^{18}F labelled tetrahydrocannabinol: synthesis, distribution in mice and PET studies in a baboon. J Nucl Med 31:902, 1990.

Mena I, Miller B, and Garrett K, et al. Neurospect in cocaine abuse: rCBF and HMPAO findings. Clinical Nuclear Medicine 14:12, 1989.

Mendhiratta SS, Wig NN, and Varma VK. Some psychological correlates of long- term heavy cannabis users. Br J Psychiatry, 132:482, 1978.

Nahas G, Trouve R, Manger W, Latour C. Cocaine and sympathoadrenal system. In Nahas GG, Latour G, eds. Physiopathlogy of illicit drugs: Cannabis, Cocaine Opiates. Oxford, Pergamon Press 1991; 151-164.

Najim RA, Karim KH. Protection against ethanol-induced gastric damage by drugs acting at the GABA-benzodiazepine receptor complex. Psychopharmacology 103:110-114, 1991.

Pappata S, Samson Y, Chavoix C, Prenant C, Maziere M, and Baron JC. Regional specific binding of [^{11}C]RO 15 1788 to central type benzodiazepine receptors in human brain: quantitative evaluation by

PET. J Cereb Blood Flow Metab 8:304, 1988.

Peng S-K, French WJ, Pelikan PCD. Direct cocaine cardiotoxicity demonstrated by endomyocardial biopsy. Arch Pathol Lab Med 113:842-845, 1989.

Phelps M, Mazziotta J (eds) Positron Emission Tomography and Autoradiography Principles and Application for the Brain and Heart. New York:Raven Press, 1986.

Pitts OK, Marwah J. Autonomic actions of cocaine. Can J Physiol Pharmacol 67:1168-1176, 1989.

Post R, Weiss SR, Pert A, and Uhde T. Chronic cocaine administration: sensitization and kindling effects. In: Cocaine: Clinical and Biobehavioral Aspects. Fischer S, Maskin A (eds) New York, Oxford 1987, pp. 109-173.

Przywara DA, Dambach GE. Direct actions of cocaine on cardiac cellular activity. Circulation Res 65:185-192, 1989.

Raichle ME, Martin WRW, and Herscovitch P, et al. Brain blood flow measured with intravenous $H_2^{15}0$. II. Implementation and validation. J Nucl Med 24:790-798, 1983.

Reivich M, Kuhl D, and Wolf A, et al. The ^{18}F Fluorodeoxyglucose method for the measurement of local cerebral glucose utilization in man. Circ Res 44:127- 137, 1979.

Reivich M, Alavi A, Wolf AP, Greenberg JH, Fowler J, Christman D, MacGregor R, Jones SC, London J, Shiue C, and Yonekura Y. Use of 2-deoxy-D[^{11}C]glucose for the determination of local cerebral glucose metabolism in humans: Variation within and between subject. J Cereb Blood Flow Metab 2:307-319, 1982.

Reivich M, Alavi A, Gur NC, et al. Determination of local cerebral glucose metabolism in humans: methodology and applications to the study of sensory and cognitive stimuli, in Brain Imaging and Brain Function. Edited by Sokoloff L. New York, Raven Press, 1985.

Roth WT, Tinklenberg JR, Whitaker CH, Darley CF, Kopell BS, and Hollister LE. The effect of marihuana on tracking task performance.

Psychopharmacologia, 33:259, 1973.

Sachs H, Russell JAG, Christman DR, and Cook B. Alterations in regional cerebral glucose metabolic rate in non-Korsakoff chronic alcoholism. Arch Neurol 44:1242-1251, 1987.

Samson Y, Baron JC, Feline A, Bories J, and Crouzel Ch. Local cerebral glucose utilization in chronic alcoholics, a positron tomography study. J Neurology Neurosurgery and Psychiatry 49:1165-1170, 1986.

Schlyer D. The use of positron emission tomography in identifying and quantitating receptors involved in schizophrenia. In: Positron Emission Tomography in Schizophrenia Research, Edited by Volkow ND, Wolf AP. Washington DC, American Psychiatric Press, 1991, p. 75.

Seeman D, Niznik HB, Guan HC. Elevation of dopamine D_2 receptors in schizophrenia is underestimated by radioactive raclopride. Arch Gen Psychiatry 47: 1170-1172, 1990.

Sharkey J, Ritz M, Schenden JA, Hanson RC, Kuhal MJ. Cocaine inhibits muscarinic cholinergic receptors in heart and brain. J Pharmacol Exp Ther 246:1048-1052, 1988.

Silver IA. Cellular microenvironment in relation to local blood flow in Cerebral vascular smooth muscle and its control. New York; Elsevier Excerpta Medica, 1979.

Sokoloff L, Reivich M, Kennedy C, et al. The ^{14}C-deoxyglucose method for the measurement of local cerebral glucose utilization: theory, procedure and normal values in the conscious and anesthetized albino rat. J Neurochem 28:897-916, 1977.

Ticku MK. Ethanol and the benzodiazepine-GABA receptor-ionophore complex. Experientia 45: 413-418, 1989.

Varma VK, Malhorta AK, Dang R, Das K, and Nehra R. Cannabis and cognitive functions: a prospective study. Drug and Alcohol Dependence, 21:147, 1988.

Volkow ND, Mullani NA, Bendriem B. Positron emission tomography instrumentation: an overview.   Am J Physiological Imaging 3:142-153, 1988a.

Volkow ND, Mullani N, Gould L, Adler S, and Krajewski K.   Cerebral blood flow in chronic cocaine users: a study with positron emission tomography. Brit J Psych   152:641-648, 1988b.

Volkow ND, Fowler JS, Wolf AP et al.   Measurement of postsynaptic dopamine receptors in cocaine abusers.   Am J Psychiatry   147:719-724, 1990a.

Volkow ND, Hitzemann R, Wolf AP, Logan J, Fowler JS, Christman D, Dewey SL, Schlyer D, Burr G, Vitkun S, Hirschowitz J.   Acute effects of ethanol on regional brain glucose metabolism and transport. Psychiatry Res 35:39-48, 1990b.

Volkow ND, Mullani N, Gould L, Gatley J.   Sensitivity of measurements of regional brain activation with ^{15}O-water and PET to time of stimulation and period of image reconstruction.   J Nucl Med 32:58-61, 1991a.

Volkow ND, Fowler JS, Wolf AP, Hitzemann R, Dewey S, Bendriem B, Alpert R, Hoff A.   Changes in brain glucose metabolism in cocaine dependence and withdrawal.   Am J Psych   148:621-626, 1991b.

Volkow ND, Hitzemann RH, Wolf AP, Fowler JS, Dewey SL, Wang GJ. Decreased metabolic activity in the brain of chronic alcoholics.   J Cereb Blood Flow Metab   11 (Suppl.)   S876, 1991c.

Volkow ND, Hitzemann R, Fowler J, Wolf AP, Wang GJ, Pappas N, Burr G, Dewey S, and Piscani K.   Brain metabolic responses to benzodiazepines in normals and alcoholics.   J Nucl Med   32:1022, 1991d.

Volkow ND, Gillespie H, Mullani N, Tancredi L, Grant L, Ivanovic M, and Hollister L.   Cerebellar metabolic activation by Delta-9-tetrahydrocannabinol in human brain: A study with Positron Emission Tomography and F-18-2 Fluoro- 2-deoxyglucose. Psychiatry Res   40:69-70, 1991e.

Volkow ND, Hitzemann R, Wang GJ, Fowler JS, Wolf AP, Dewey SL. Long-term frontal brain metabolic changes in cocaine abusers. (submitted/a)

Volkow ND, Fowler JS, Wolf AP, et al. Distribution of ^{11}C cocaine in human heart, lungs, liver and adrenals. A dynamic PET study. J Nucl Med (submitted/b).

Wide RJ, Bernardi S, and Frackowiak R, et al. Serial observation on the pathophysiology of acute stroke; the transition from ischemia to infarction as reflected in regional oxygen extraction. Brain 106:197-222, 1983.

Wig, NN and Varma VK. Pattern of heavy long-term cannabis use in North India and its effects on cognitive functions: a preliminary report; Drug and Alcohol Dependence, 2:211, 1977.

Wik G, Borg S, and Sjörgen I, et al. PET determination of regional cerebral glucose metabolism in alcohol-dependent men and healthy controls using ^{11}C- glucose. Acta Psychiatr Scand 78:234-241, 1988.

Wilkerson RD. Cardiovascular effects of cocaine in conscious dogs: Importance of fully functional autonomic and central nervous system. J Pharmacol Exp Ther 246:466-471, 1988.

Wong DF, Lever JR, Harting PR, Dannals RF, Villemagne V, Hoffman BJ, Wilson AA, Ravert HT, Links JM, Scheffel U, and Wagner HN, Jr. Localization of serotonin 5-HT$_2$ receptors in living human brain by positron emission tomography using N1-([^{11}C]-methyl2-Br-LSD. Synapse 1:393, 1987.

Wong DF, Wagner HN, Dannals RF, et al. Effects of age on dopamine and serotinin receptors measured by positron tomography in the living human brain. Science 226:1393-1396, 1984.

Wong DF, Gjedde A, Wagner HN. Quantification of neuroreceptors in the living human brain. J Cereb Blood Flow Metab 6:137-146, 1986.

Yesavage JA, Leirer VO, Denari M, and Hollister LE. Carry-over effects of marijuana intoxication on aircraft pilot performance: A preliminary report. Am J Psychiatry, 142(11):1325, 1985.

3. Effects on psychomotor performance

Marijuana Carry-Over Effects on Psychomotor Performance: A Chronicle of Research.

Von O. Leirer, Ph.D.
Decision Systems, P.O. Box 6489, Stanford, California 94305

Jerome A. Yesavage, M.D.
Stanford University School of Medicine
Veterans Administration Medical Center

Daniel G. Morrow, Ph.D.
Decision Systems, Stanford CA

Abstract

The purpose of this research was to develop a cognitive factors research laboratory that could measure the subtle, carry-over effects of various drugs such as marijuana and other factors such as aging on psychomotor performance found in complex human-machine interaction. The three experiments reviewed are a chronology of our efforts. The findings reported indicate that marijuana impairs complex psychomotor performance for up to 24 hours after smoking a low to moderate social dose. The results also indicate that marijuana and other factors such as task difficulty and aging act to cumulatively impair performance. These findings suggest that there may be conditions under which real world complex human-machine performance is significantly impaired with out the subjective awareness of the marijuana user.

Acknowledgments: Research was supported by the National Institute on Drug Abuse, Grant No: DA 03593 and conducted at the Stanford University School of Medicine and the Palo Alto Veterans Administration Medical Center. Manuscript preparation was supported by, National Institute on Aging, Grant No. 2 R44 AGO6957-02.

Key words: psychomotor performance, working memory, cognitive factors, human-machine interface, marijuana.

Marijuana Carry-Over Effects on Psychomotor Performance:
A Chronicle of Research.

This research had two purposes. First, we were interested in the question of how long low to moderate social doses of marijuana affects complex psychomotor performance. Second, we were interested in developing a more sensitive methodology to measure subtle effects of marijuana and other drugs such as alcohol.

Our conceptual framework for this research included two central ideas. First, complex psychomotor performance found in most human/machine interaction is largely a human information processing task (1). Second, in any unit of time, humans have a limited capacity to consciously process the encoded and recalled information used when performing such tasks (2,3). In the terminology of cognitive psychology, when there is an abundance of environmental input, complex human/machine performance is limited by cognitive capacity rather than the availability of information (4,5). Finally, within this framework, previous empirical findings show that one of marijuana's acute effects is to further reduce humans' capacity to process information (6,7).

We began this research believing there were several problems with existing laboratory measures of complex psychomotor performance. First they did not collect enough data at a fast enough rate to capture transient, subtle effects of marijuana that might occur several hours after ingestion. Human information processing is occurring at the millisecond level with important human/machine interactions occurring more frequently than one time per second (1). Thus, we reasoned that a sensitive measure of psychomotor performance should be sampled more frequently than once per second.

A second suspected problem with existing approaches was that subjects could effectively compensate for impairment by strategically allocating their cognitive resources (2,8) to just those two or three simultaneous tasks that the experiment required them to perform (4). I   n order to prevent compensation, we thought that the psychomotor task should be complex and require the subject to simultaneously perform as many tasks as possible and that these tasks should include different sensory inputs, i.e., visual, auditory, and tactile. In other words, we wanted to insure that the subject's would not hold any extra processing capacity in reserve. Under this condition, if marijuana impairs their processing capacity, it will be expressed as measurable performance decline rather than an unmeasurable increase in the pilot's level of effort.

We believed a third problem was that most laboratory measures lacked face validity and public understanding. The laboratory findings that existed were not easily communicated to even sophisticated lay audiences. For example, the public has a hard time understanding the relationship

THC ligands with higher affinity for the cannabinoid receptors may make it feasible to measure the regional concentration of cannabinoid receptors <u>in vivo</u> in the human brain.

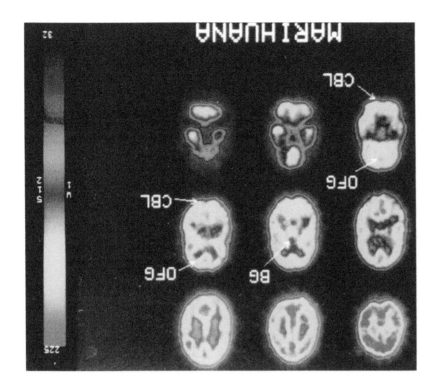

Figure 1. Image of brain glucose metabolism obtained with 18F-deoxyglucose (FDG) and PET in a normal control tested under baseline conditions (resting, eyes open, ears unplugged).

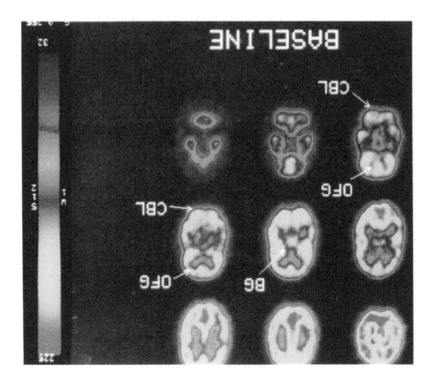

Figure 2. Images of brain glucose metabolism for the same subject as Figure 1 but 40 minutes after intravenous administration of 2 mg of tetrahydrocannabinol. Notice the relative increases in metabolism in orbital frontal cortex and cerebellum during marihuana intoxication.

between a divided attention task such as target tracking on a CRT while monitoring for other events such as blinking lights (9,10,11) and the possible psychomotor impairment of a heavy equipment operator who smokes marijuana after work. Thus, to effectively communicate basic research findings to the public it would be ideal to use a psychomotor task that simulates non-trivial, real world tasks.

A fourth related problem was that the existing psychomotor tests that did have high face validity were considerably less sensitive than typical laboratory measures and, therefore, we believed they failed to show significant impairment except immediately after smoking large doses of marijuana. Examples include an early aviation piloting experiment (12) employed a paper and penile checklist rating of performance. These tasks were collecting only one or two data points every several seconds or minutes.

We believed that these measurement problems lead to many non-significant effects and in turn lead many scientists and laymen alike to suspect that moderate social doses of marijuana had little or no impact on psychomotor performance with only a one or two hour delay after smoking.

To address these concerns and shortcomings, we selected simulated aircraft piloting as a psychomotor task for investigating marijuana carry-over effects. Piloting is a non-trivial, real world task. It is also a complex psychomotor task that can be systematically manipulated in experiments using modern, computer-controlled, realistic aircraft simulators. These aircraft simulators allow for the collection of a large number of data points at sampling rates greater than one time per second. For example, the simulator we used in our final two experiments collects data on 52 variables at about 22 times per second. Finally, while piloting is a complex psychomotor task requiring considerable skills and experience, there is an abundance of well trained pilots available to act as subjects in experiments.

The three experiments reviewed here are a chronicle of our efforts to develop a research laboratory and an experimental paradigm that can be used to measure and study both immediate as well as longer term carry-over effects of marijuana and other drugs such as alcohol.

Experiment One. Carry-Over Effects of Marijuana Intoxication on Aircraft Pilot Performance: A Preliminary Report

In 1985 Jerome Yesavage, Leo Hollister, Mark Denari, and I (13) published our first findings using what would be considered by our current standards, a very primitive computerized aircraft simulator. This simulator consisted of a control yoke similar to those found in small aircraft, a black and white CRT used to display the simulated environment in which the pilot flies, and a computer keyboard to allow the pilots to regulate simulated aircraft's power settings and flaps (see Figure 1). This simulator was controlled by an Apple II (6502 cpu) computer which allowed us to collect

Figure 1

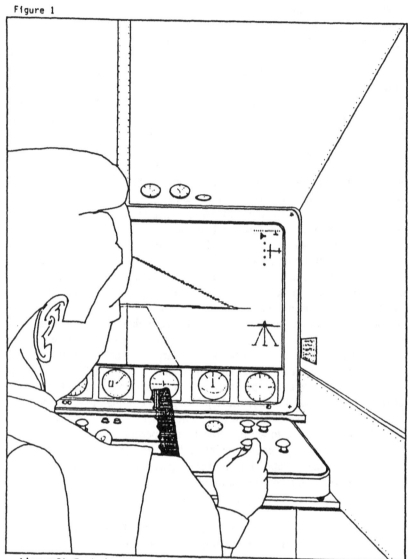

Aircraft Cockpit Simulator

data on eight variables at about once every 500 msec. One important feature of this primitive simulator which we were to recognize later was that there was a lag time of about one second between the time the pilot made a flight correction (i.e., operated the yoke or keyboard) and the simulator's response to the correction. While pilots complained that this made the simulator very difficult to fly, the limits of the technology made it impossible for us to correct the problem.

In this first experiment, pilots first practiced for eight hours on the piloting task that would be used in the actual experiment. This piloting task consisted of a simple simulated flight. Pilots first took off at one airport, climbed to 700 feet, made a small course correction, and then descended and landed at a second airport. Because of restrictions on our data collection capabilities, performance data was gathered primarily on the last part of the flight which includes descent and landing. On the day of the experiment, pilots first flew two practice flights and then a test flight to establish a base rate of non-drug performance. They then smoked a 19mg dose of THC delta 9 and after a one, four, and 24 hour delay, flew the same route.

Table 1.

Mean performance decrements after smoking 19mg delta 9 THC
Baseline (before) and 24 hours after smoking.

Time Delay

Dependent Measure	Baseline	24 hrs.	t-test value
Distance of center on landing	12	24	-3.52, p<.05
Mean lateral deviation	19	34	-3.25, p<.05
Mean vertical deviation	25	40	-1.90, NS
Aileron, number of changes	60	76	-3.66, p<.05
Aileron, mean size	53	65	-2.61, p<.05
Elevator, number of changes	264	285	-0.83, NS
Elevator, mean size	54	74	-2.46, p<.05
Number of throttle changes	22	24	-1.83, NS

From Yesavage, Leirer, Denari, and Hollister, 1985

As shown in Table 1 above, an analysis of these findings revealed that at 24 hours after smoking, on five of the eight variables measured, pilots were still not performing at the level they had previous to their smoking the 19mg of THC delta 9.

We classified this study as a preliminary report because it lacked many of the experimental controls to permit confidence in the findings. The study also failed to give us any idea of when we would no longer detect an effect of the drug. With these reservations in mind, we considered this first experiment suggestive evidence that the carry-over effects of moderate to low social doses of marijuana may last many hours longer than the previous body of research suggested. The findings also suggested that this new methodology might indeed provide a much more sensitive measure of drug impairment and that we would be able to study the psychomotor impairment of different drugs even when these effects are subtle and small in magnitude.

Experiment Two: Marijuana, Aging, and Task Difficulty Effects on Pilot Performance.

With the experience gained from our first experiment and with the advent of new technology in aircraft simulators, Jerome Yesavage, Daniel Morrow, and I (14) were able to construct a much more sophisticated and realistic flight simulator and research laboratory. This new simulator used a Silicon Graphics Computer to generate color visual displays of the simulated environment and a Frasca 141 computer-controlled aircraft simulator. The Frasca 141 simulator included a realistic small aircraft cockpit with a full compliment of flight instrumentation (see Figure 2 on next page). With this simulator we were able to collect data on 52 different variables at approximately 22 times per second. Our pilots liked the simulator and thought it was realistic in most aspects of flight. With this new laboratory, we began a more systematic and multifaceted study of the effects of marijuana on complex psychomotor performance.

In addition to studying the time course of the drug effects, we wanted to study the possibility that other factors combine with marijuana to jointly influence performance. Specifically, we wanted to investigate any possible cumulative effects of marijuana, aging, and task difficulty on psychomotor performance. If limited processing capacity is the source of impairment, we should find best performance in the condition of a placebo dose of marijuana smoked by young pilots, performing our easiest piloting tasks. Impairment should increase as we add factors that either reduce processing capacity (increased drug dose or increased pilot age) or increase demands on processing capacity by increasing task difficulty (increasing weather turbulence). Thus, in the condition of highest task difficulty, highest drug dose, and older pilots we should see the greatest performance decrement.

To investigate these possibilities we trained nine young pilots (Mean age=25.5) and nine old pilots (Mean age 37.6) on a simulated flight that included the following set of standard maneuvers while flying a rectangular flight pattern around the runway: a) take-off, b) climbing turn, departure stall, c) slow flight, d) approach stall, e) standard rate turn, d) base leg approach to landing that involved straight and level flying, e) final leg approach to landing, f) descent to landing, and g) landing. Each of these components included specific and realistic instructions for air speed, heading, altitude, and rate of turns. Thus, pilots were asked to perform as closely to the ideal of

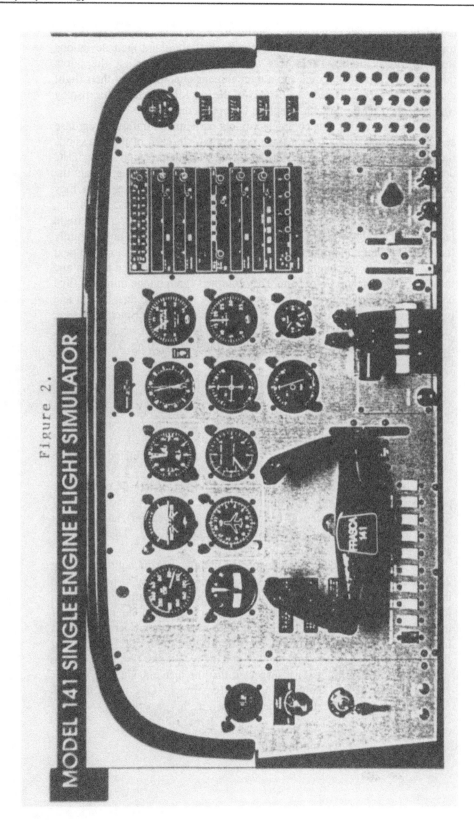

Figure 2. MODEL 141 SINGLE ENGINE FLIGHT SIMULATOR

the flight instructions as possible.   During the eight hours of pre-experiment training, pilots were given feedback after each flight about their deviations from the ideal flight.   This was accomplished by showing the pilots numerical performance scores and a three dimensional display of their flight path that was superimposed on a display of the ideal flight path. This display was generated on the Silicon Graphics Computer monitor.

After training was completed, pilots participated in three drug dose sessions, each occurring over a three day period. In each of these sessions the pilots smoked either a) Placebo, b) 10 mg THC delta 9, and 20 mg THC delta 9 cigarette.   On the first day of each session, pilots began the experiment by completing two practice flights as described above. Next, they smoked a cigarette containing one of the three drug dose levels. Then one, four, and eight hours after smoking, they flew the above described flight pattern under calm weather simulation and a second flight under mildly turbulent weather conditions. They then returned to the laboratory the next day (24 hours after smoking) to fly two practice flights and then a calm and turbulent experimental flight pair. They returned again on the third and final day (48 hours after smoking) to fly two practice flights and then a calm and turbulent experimental flight pair. To summarize the experimental design, young and old pilots flew in calm and moderately turbulent weather, at 1, 4, 8, 24, and 48 after smoking a placebo, 10mg, and 20mg THC delta 9 cigarette.

This experiment investigated two related questions: a) how long are the drug's carry-over effects, and b) are there cumulative effects of marijuana use, task difficulty, and age of drug user. To measure "performance", we created a single dependent measure by standardizing difference-from-ideal-performance scores for each variable across subjects and experimental conditions. This allowed us to create a single, equally weighted aggregate score for each subject in each experimental condition.

As expected the results indicated an effect of age of pilot (older pilots performed less well than younger pilots), an effect of task difficulty (calm weather performance was better than turbulent weather), and an effect of time after smoking (the drug effect decreased over time). These findings are shown in Table 2 below.

The results included an unexpected finding. We did not replicate the 24 hour carry-over effects of our first experiment. Looking at performance in the high dose condition, we found no evidence of impairment beyond a four hour delay.   In the low dose condition we found no evidence of impairment even one hour after smoking. After careful consideration of this failure to replicate, we could only guess that it was the result of improving the realism of the simulator and thereby making it an easier task for the pilots to perform. Related to this, we suspected that the unavoidable delays in the responsiveness of the simulator used in Experiment One made the task significantly more difficult than the task used in this experiment.

Table 2.
Mean performance decrements of Young and Old
Pilots in the three Drug Dose and the five Time
Delay conditions. Significant impairment occurred
at the 1 and 4 hour period in the 20mg condition only.

Age	Dose		
	20 Mg	10 Mg	0 Mg
Young			
1 hour	102.14	95.75	92.83
4 hour	101.89	95.36	92.85
8 hour	87.32	88.77	88.05
24 hour	87.80	91.72	91.83
48 hour	85.46	90.18	88.86
Old			
1 hour	123.68	101.38	104.41
4 hour	112.50	100.90	106.95
8 hour	113.15	100.55	106.35
24 hour	106.60	95.39	107.71
48 hour	106.23	97.18	109.92

From Leirer, Yesavage, and Morrow (1988)

Findings about cumulative impairment were more encouraging. Looking only at the one hour time point after smoking, we found the predicted pattern of cumulative effects. Young pilots, smoking placebo dose, in calm weather showed the least performance decrements. As shown in Figure 3 below, performance decrements continued to increase until we get to the condition of older pilots in turbulent weather under the high dose condition, who show the greatest performance decrements.

These cumulative effects predict the following type of real world scenario. When operating complex equipment under routine conditions, marijuana smokers may find that they perform adequately only a few hours after smoking. This in turn leads them to mistakenly believe that they are acting responsibly by leaving a few hours between smoking marijuana and performing some complex psychomotor task at work. Then because of some unusual circumstance or emergency, they are required to perform an unusually difficult psychomotor task. Our findings suggest that at certain levels of increased task difficulty, they would have been able to deal with the emergency had it not been for the cumulative impact of both increased task difficulty and marijuana carry-over effects. This sort of scenario might be expected to occur in complex divided attention tasks such as automobile driving, aircraft flying, boat navigation, or heavy equipment operation when unusually severe conditions such as accident avoidance, or adverse weather occur.

Figure 3.
Mean summary performance decrements
for the five categories of flights
predicted to show progressively higher
decrements.

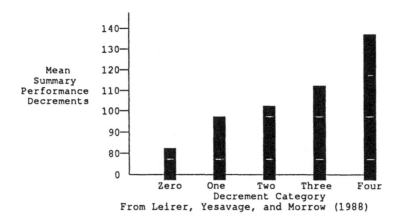

From Leirer, Yesavage, and Morrow (1988)

Experiment Three: Marijuana Carry-Over Effects on Aircraft Pilot Performance.

With our failure in Experiment Two to replicate the 24 carry-over effects of Experiment One, Jerome Yesavage, Daniel Morrow, and I (15) designed and conducted a third experiment specifically to investigate the question of 24 hour carry-over effects using the new, more realistic aircraft simulator. As suggested above, we suspected that the new more realistic simulator inadvertently made the task too easy to detect impairment more than a few hours after smoking a moderate social dose of marijuana. If this was the case, then by keeping the realistic simulator but making the piloting task more difficult, we should be able to again detect 24 hour carry-over effects.

To test this hypothesis, we used the same rectangular flight pattern as before with the same flight maneuvers. We added to this task radio communication with a simulated air traffic controller, visual detection and avoidance of approaching aircraft, and engine failure and correction. We also increased the weather turbulence. And finally, we reduced the amount of practice time so that the task would not become over learned. This created a very difficult though realistic flying task. (It very much reminded me of routine flying in the San Francisco Bay area.) We reduced the number of conditions in this experiment to include only a placebo and a 20mg THC delta 9 dose conditions, and a single high turbulent weather condition.

Table 3.
Pilot Performance Decrement Scores
1/4, 4, 8, 24, and 48 Hours After Smoking
Either a 20 mg or Placebo Dose of THC delta 9
(Probability Values Are Based on 1-Tailed T-Tests)

Time Delay	Mean Performance			Statistical Analysis	
	20 mg	Placebo	Dif. Score(S.D.)	t-value	p-value
Before Dose	95.48	90.83	4.65 (26.47)	0.53	0.61
1/4 Hour	151.88	94.58	57.30 (67.78)	2.54	0.02
Four Hours	114.00	86.35	27.65 (31.05)	2.67	0.02
Eight Hours	112.55	88.35	24.20 (34.12)	2.13	0.04
24 Hours	104.55	83.90	20.65 (31.06)	1.99	0.04
48 Hours	91.20	84.32	6.88 (21.23)	0.92	0.19

From Leirer, Yesavage, and Morrow (1991)

On the first day of the experiment, pilots flew two practice flights and then flew two experimental flights to establish their base rate performance level. Next, they smoked either a placebo cigarette or a 20mg dose cigarette. Fifteen minutes, four hours, and eight hours, 24 hours, and 48 hours after smoking they flew two flights using the same procedure as described in Experiment 2. Thus, each pilot flew these six sessions under a placebo and a 20mg dose condition. We compared each pilot's performance in the 20mg dose condition to this placebo dose condition at each of the six time points.

As reported in Table 3 above and graphically shown in Figure 4 on the following page, we found significant increases in impairment at 15 minutes, 4 hours, 8 hours, and 24 hours after smoking the 20mg dose. Thus, the findings replicated the 24 hour carry over effects first reported in Experiment One. At 48 hours after smoking, we found no indication of impairment.

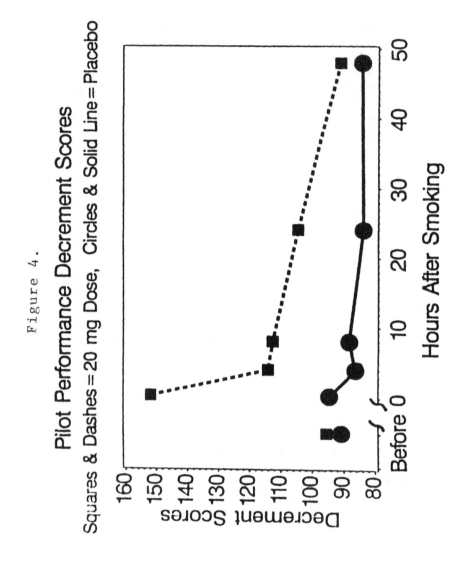

Figure 4.

Pilot Performance Decrement Scores

Squares & Dashes=20 mg Dose, Circles & Solid Line=Placebo

Discussion

The findings from these three experiments suggest a number of points about the carry-over effects of a low to moderate social dose of marijuana on complex psychomotor performance. The first point is that after about four hours, a low to moderate dose produces effects that are relatively subtle and difficult to measure. However, I do believe that our findings indicate that impairment does exist as long as 24 hours after smoking a moderate social dose of marijuana.

The second point is that real world psychomotor tasks most likely affected by this long term carry-over will be those requiring divided attention extending over several minutes or hours and that require all of the person's available cognitive capacity. For example, while I do not think there would be any measurable impairment to a street sweeper's performance 24 hours after smoking certain types of heavy equipment operations, some medical procedures, and so forth may very well be impaired.

The third point is that the effects of a low to moderate social dose at several hours after smoking are so subtle that users do not realize they exist and therefore cannot depend their own self judgment to know they are not performing as well a normal. This point is supported by our finding that only one of the pilots in Experiment Three indicated that he felt any awareness of the drug effect at 24 hours after smoking.

The final related point is that effects of one low to moderate social dose of marijuana after 4 or so hours will show up as a statistical increase in error or accident rates. It is unlikely that the individual users at risk for such accidents will ever realize the impact of marijuana on their performance. This lack of awareness is the result of little or no subjective experience of the drugs effect and the fact that the increase in the individual's base rate of recognizable errors or accidents will be too low to detect.

I would like to conclude by discussing the time investment required to investigate these more subtle effects. Cognitive factors research such as this requires first developing and validating a computerized laboratory. It is reasonable to expect to spend three to five years developing effective apparatus and procedures. Prior research demonstrates that standard divided attention tasks are not sensitive enough to detect these effects. However, once the time investment is made, a laboratory such as ours can be used for a number of different research topics. For example, we have used the same equipment and procedures to investigate other drugs such as alcohol as well as impact of aging on complex psychomotor performance.

LITERATURE

1. Card, S. K. , Moran, T. P. , and Newell, A.   The psychology of humancomputer interaction. Hillsdale, N. J. : Erlbaum. 1983.

2. Baddeley, A. D. , Hitch G.   Working memory.   In: Bower G. ed. The psychology of learning and motivation. San Diego, CA: Academic Press. 1974.

3. Roscoe, S.   Aviation Psychology.   Ames, IA: Iowa State University Press, 1980.

4.   Norman, D. A. and Bobrow, D. G.   On data-limited and resource limited processes, Cognitive Psychology 7, 44-64, 1975.

5. Waugh, N. C. and Norman, D. A.   Primary memory. Psychological Review, 72, 89-104, 1965.

6. Darley, C. F. , Tinklenberg, J. R. , Roth, W. T. , and Atkinson, R. C. The nature of storage deficits and state-dependent retrieval under marihuana. Psychopharmacologia, 37, 139-149, 1974.

7. Melges, F. T. , Tinklenberg, J. R. , Hollister, L. E. , and Gillespie, H. K. Marihuana and temporal disintegration. Science, 168 1118-1120, 1970.

8.   Salthouse, T. A.   Influence of experience on age differences in cognitive functioning.   Human Factors, 17, 551-569, 1990.

9. Roth, W. T. , Tinklenberg, J. R. , Whitaker, C. A. , Darley, C. F. , et al. The effect of marihuana on tracking task performance. Psychopharmacology, 33, 259-265, 1973.

10. Tinklenberg, J. R, Meleges, F. T, Hollister, L. E. , and Gillespie, H. K. Marijuana and immediate memory.   Nature, 226, 1171-1172, 1970.

11. Moskowitz, H. S. , Sharma, S. , McGlothlin, W.   Effect of marihuana upon peripheral vision as a function of the information processing demands in central vision. Perceptual Motor Skills, 35, 875-882, 1972.

12. Janowsky, D. S. , Meacham, M. P. , Blaine, J. D. , Schoor, M. , Bozzetti, L. P.   Simulated flying performance after marijuana intoxication.   Aviation, Space, and Environmental Medicine, 47, 124-128, 1976.

13. Yesavage, J. A. , Leirer, V. O. , Denari, M. , and Hollister, L. E. Carry-Over effects of marijuana intoxication on aircraft pilot performance: A preliminary report.   American Journal of Psychiatry, 142, 1325-1329, 1985.

14. Leirer, V. O. , Yesavage, J. A. , and Morrow, D. G.   Marijuana, aging, and task difficulty effects on pilot performance.   Aviation, Space, and Environmental Medicine, 60, 1145-1152, 1989.

15. Leirer, V. O. , Yesavage, J. A. , Morrow, D. G.   Marijuana carry-over effects on aircraft pilot performance.   Aviation, Space, and Environmental Medicine, 62, 221-227, 1991.

CHRONIC MARIHUANA SMOKING AND SHORT-TERM MEMORY IMPAIRMENT

Richard H. Schwartz

Department of Pediatrics, Georgetown University, Medical School, Washington D.C., USA

ABSTRACT

We evaluated The auditory/verbal and visual/spatial memory of 10 cannabis-dependent adolescents and compared the results with performance of 17 subjects in two control groups. The control groups included 8 adolescent drug abusers who had not been long-term users of cannabis and another 9 adolescents who had never abused any drug. All three groups were matched age, IQ, and absence of previous learning disabilities. Adolescents with a history of frequent alcohol or phencyclidine abuse were excluded from entering the study. A battery of seven neuropsychological tests was administered initially to all subjects and a parallel test battery was administered 6 weeks thereafter. Significant differences between the cannabis-dependent group and the two control groups were obtained initially on the Benton Visual Retention Test ($F_{[2.24]}=6.04$) and the Wechsler Memory Scale Prose Passages ($F_{[2.23]}=7.04$). After 6 weeks of supervised abstention from intoxicants, subjects in the cannabis-dependent group showed some significant improvement on the Wechsler Memory Scale Prose Passages and on the Benton Visual Retention Test; however, the improvement failed to achieve statistical significance. We concluded that cannabis-dependent adolescents have selective short-term memory deficits that continue for at least 6 weeks after the last use of marijuana.

KEYWORDS

Marijuana, delta-9-THC, adolescents, memory, Benten visual retention test, Wechsler memory scale.

Marihuana is the dried leafy part of the Cannabis plant, which is the preferred form in the United States and in most of Central and South America and in Canada. Europeans and the Middle East people seem to prefer hashish the most. Hashish in Switzerland probably has 7% delta-9-THC. In the United States it is in the neighborhood of 3-4% delta-9-THC. The dried plant seems to deteriorate in the ocean voyage. Aficionados of marihuana smoking disdain the ordinary "grass" and go to sinsemilla which has 7-8% average delta-9-THC concentration or double that of hashish. Hash oil is a distillation product of the dried leafy form of usually inferior grades of marihuana. Distillation makes it into a very concentrated product containing in the neighborhood of 40% delta-9-THC.

Summary of Results of Published Studies on Memory

A. Abel (1970)
- 8 cannabis smokers and no controls
- cross-over study with test-retest design
- recall of story during period of intoxication compared to recall in sober state
- results: under intoxicated state less recall of exact words, two and four-word sequences, and no. of idea units

B. Dornbush (1974)
- 20 cannabis-smoking medical students
- placebo-controlled study
- replicated findings of Abel: intoxication from marihuana interrupted the storage phase of short-term memory.

The studies that looked at short-term memory impairment began around 1970. One of the very first studies was done by AA. Abel. In a cross-over study design, Abel tested 8 marihuana smokers but no controls, who were asked to recall a story during a period of intoxication. The subjects compared listening the story and how much information was recalled in a sober state. The results showed that under the intoxicated state there was definitely less recall of the exact words of two to four word sequences and then of the number of idea units. Somewhat later, in 1974, Dornbush studied 20 cannabis smoking medical students in a placebo controlled study. He found that intoxication from marihuana interrupted the storage phase of short-term memory. Basically Dornbush replicated the findings of Abel.

Author	Year	Journal	Result
Tinklenberg	1970	Nature	Decrement in memory
Melges et al.	1970	Science	Decrement in memory
Abel	1971	Nature	Decrement in memory
Abel	1971	Science	Decrement in memory
Darley et al.	1973	Memory and cognition	Decrement in memory
Dittrich	1973	Psychopharmacologia	Decrement in memory
Dornbusch et al.	1974	Am. J. Psychiatry	Decrement in memory

Early Literature on Adverse Effects of Marihuana on Memory during Period of Intoxication.

In this figure are summarized all the early studies from 1970-1974 including several by Abel and the study by Dornbush in 1974. I must remind you that the average concentration of delta-9-THC for all of these studies was between 1 and 1.5% compared to today's average of 5-7% in Switzerland and 7-8% (for the much sought - after sinsemilla type of marihuana) in the United States. All these studies came to the same conclusion: when a person smokes marihuana, the short-term memory is affected during the 2 or 3 hours of intoxication.

In the German journal *Psychopharmacologia*, Dittrich stated in 1973: "... that the adverse effects of cannabis on attention varied according to the task. Short-term memory was forgotten to a greater degree after use of marihuana compared to use of placebo drug. The greater the dose and the longer the interval from presentation of newly-learn material to its retrieval the greater the impairment of short-term memory."

Marihuana as a psychoactive drug alters certain brain functions of short-term memory. But the major question to be resolved is whether the impaired neurological functioning persists beyond the period of acute intoxication. This means, if somebody is stoned and forgets things for 2-3 hours, that is one thing, but it is another thing if he continues to have a problem with short-memory even after it is impossible to know that he is high. If they talk to you they are normal, if you ask them how they feel, they say perfectly normal. But will their memory still be impaired? This is really important if students smoke marihuana in the morning on their way to school. And it is also important on the workplace, especially when skilled workmanship is necessary.

Literature Review: Cannabis Does Not Impair Short-term Memory

	No. of subjects
1. Bowman & Pihl, Psychopharmacologia, 1973	16
2. Satz et al.,* Annals N.Y. Acad. Sci., 1976	41
3. Mendhiratta,* British J. Psychiatry, 1978	<u>50</u>
	107

* Re-test, one decade later found that there was significant impairment of short-term memory when better tests were used!

The initial studies that were done in the early and mid seventies showed that marihuana did not seem to have an effect on short-term memory after the high wore off. There were three such studies. An Irish one by Bowman and Pihl, a Costa Rican study by Satz et al. and one from India by Mendhiratta. They tested 107 patients in the total. But two of these authors, Satz and Mendhiratta, repeated their study ten years later. They found 70% of the same exact subjects as in the initial study. All these studies were improved controlled studies. They retested the patients with more sophisticated tests of short-term memory and they reversed their initial findings. Then they said they were wrong with their initial study. Marihuana has a longer lasting effect on short-term memory than in the immediate two to three hour period when people realize that they are high and where they look high. These are the significant findings of the retests a decade later. Anybody who wants to read the Costa Rican re-study by Page and al. published in the *Journal of Psychoactive Drugs* (Jan/Mar. 1988, Vol. 20) is welcome to. There you can see exactly how they did reverse their findings with selective tests of short-memory on the Costa Rican marihuana smokers and the control group. You can also see it in the study of Mendhiratta published in 1988 in the *British Journal of Addiction* (Vol. 83, pp. 749-753). The bottom line is that is that this showed significant additional deterioration in the case of marihuana users on a number of psychological measures after the high wore off. This leaves only the study of 16 patients which showed no effect on short-term memory after the high wore-off, on digit span, on speed and accuracy tests.

The Swiss pride themselves on accurate machine tool making, on accurate locomotive making, on Sesselbahn (chair-lift) making. I would not ride on a Swiss machine or use a Swiss machine if they have pot smokers in the workplace.

Soueif: Conclusions from studies of effects of chronic cannabis smoking on 850 Egyptian prisoners:

Chronic cannabis use, particularly by educated men, was associated with diminished attention to detail and seriously impaired short-term memory.

To date, a lot of controlled studies with a large number of patients have shown a definite impairment of short-term memory after the high wore off, e.g. 850 persons in the study of Soueif, who tested Egyptian prisoners. There was a control group of non-marihuana smokers and a large group of marihuana smokers. Smoking of hashish in Egypt is endemic and is a debilitating part of the Egyptian workforce. Soueif concluded that chronic cannabis use, particularly by educated men, was associated with diminished attention to detail and seriously impaired short-term memory. This means, if somebody is a subsistence farmer and so has very simple chores, it is unlikely that the use of hashish will impair that person's workplace tasks. The more sophisticated the task, the more educated the person, the greater the impairment will become obvious if carefully measured.

Wig and Varma, in 1977, studied 23 cannabis smokers and 11 controls in India. They were matched for age and occupation. They studied them during the Wechsler Memory Scale. They found impairment of the short-term memory. They studied them using the Bender-Gestalt test, they found impairment. This was the only study, before 1986, that reported differences that could not be really explained by uncontrolled confounding variables. The studies are becoming more carefully done, but the conclusions remain the same.

Composition of Study Groups

Group 1. Cannabis-dependent subjects (N=10)
- amount of marihuana smoked = 18 gm/week, mean
- frequency of use = 5.9 days/week, mean
- duration of use = 7.6 months, mean

Group 2. Control group = infrequent use of marihuana
- maximum lifetime use = 35 times
- no use in 96-hour period prior to testing
- matched by age, I.Q., and socio-economic status

Group 3. Non-drug using control group
- matched by age, I.Q., and socio-economic status

The study that we did was reported in the "American Journal of Diseases of Children". It was a very, very carefully controlled study and that is the reason why the number of subjects was small. You will see that the care that went into the study was very important. We tested 10 teenage marihuana smokers. They did not abuse alcohol to any great extend, they did not abuse intravenous drugs, they were not cocaine takers, and you will see from some of the exclusions how rigidly we controlled the study. They were heavy marihuana smokers, they used it almost daily and 18 grams per week. They had to at least use it 4 days a week to get into the study and they had to smoke marihuana for a minimum of 4 continuous months before entering the study. The average amount of almost daily marihuana use was 7.6 months.

There were two control groups that we compared with these 10 heavy users. One control group had used marihuana infrequently (maximum amount was 35 times in their life). There was definitely no use of marihuana in the 96 hour period prior to testing and their urine tests were free of marijuana. Most of the other studies have a hard time quantitating how much marihuana was actually being used. We not only asked them but we tested the urine on all of the people in the study. Not only the first time, but for those that were in the treatment program, group one and group two, we tested the urine two times every week until the second retest was done. And groups two and three were matched by age, I.Q., and socio-economic status. The mean I.Q. was 112, so these were average performing students. Group three never used marihuana. They were brothers and sisters of additional marihuana users. They were carefully interviewed, their parents were interviewed and they also had urine tests for marihuana, not only initially but periodically throughout the study, every two weeks. They were also matched by age, I.Q., and socio-economic status. There were more girls in the two control groups but we controlled the results for that statistically.

Short-term Memory Study

1. Verification of school performance by examination of report cards
2. Interview with parents to verify absence of learning problems during childhood
3. All testing supervised by PhD neuropsychologist
4. Cannabis smokers (N=10) and control group (N=17) matched by age, I.Q., and socio-economic status
5. Double testing separated by six week interval, comprehensive I.Q. test and six tests of visual or auditory short-term memory

---

We looked at the sixth grade school record and we verified that there was no learning disability, no hyperactivity and that the students had good school performance before then began to use marihuana. All the testing was supervised by a PhD neurophysiologist. There were two testing periods: initially, and then after a period of supervised abstinence for the heavy marihuana smokers 6 weeks later.

Short-term Memory Study: Exclusions from Study

1. Frequent drunkenness
2. Learning disability
3. I.Q. below 90 or above 120
4. Serious head injury
5. Cannabis use during period between test and retest

To improve our results we used strict exclusions for the patients to come into the study. They could be not using alcohol more than once a week, not getting drunk. They had no learning disability, there were no I.Q.'s below 90 and none above 120. This is important because the more superior the I.Q., the more they can compensate for marihuana, except in the most sophisticated of tasks and then, although they may perform normally, the normal is average performance. They rarely will get superior performance commensurate with their I.Q. And of course there was no history by the parents or by the participants of any serious head injury with unconsciousness, and there was no cannabis use between the test and the retest period.

IQ-Test
Wechsler Intelligence Test for Children
Neuropsychological Tests for Short-term Memory
1. Wechsler Memory Scale for prose passages
2. Benton visual retention test
3. Petersen-Petersen short-term memory test
4. Wechsler Memory Scale for paired associate learning
5. Buschke selective reminding test
6. Complex figure drawing test (Ray-Osterrieth)

We allowed a long time for any marihuana "high" to go away. The first test was done at least 36 hours after the heavy users went into a drug treatment program. First we did a full scale i.Q. test. Thee was no difference between the control groups and marihuana using groups. So marihuana use does not impair the I.Q. Maybe it does if you are intoxicated but not after the intoxication has passed. There were five tests of neuropsychological functions. The Benton Visual Retention Test, Petersen-Petersen short-term memory test and several other test. The Benton test is a test of visual memory. There were ten cards containing simple designs. They had to look at the cards for a while, several seconds, and then the cards were hidden and they had to produce the designs some seconds to minutes later.

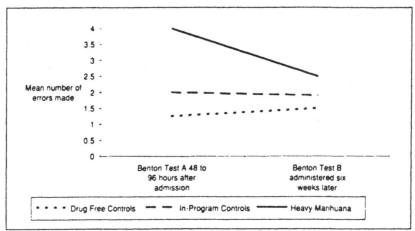

Fig. 1: *Comparison of Benton Visual Retention Test scores soon after admission and again six weeks later.*

In this figure you can see the results of the Benton test. The heavy marihuana smokers not only made significantly more errors than the two control groups initially, but such errors continued, even though there was substantial improvement between tests one and two. The improvement never reached statistical equality with the two control groups six week later.

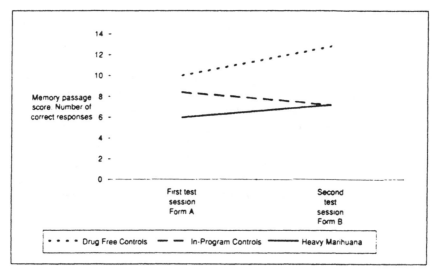

Fig. 2: *Comparison of Wechsler Memory Scale for prose passages soon after admission and again six weeks later.*

The Wechsler Memory Scale for prose passages is a test of auditory memory. There were 22-24 memory units and after 30 minutes they were asked to repeat word for word what was said. In the above figure you see the number of correct responses in the Wechsler test. In this test the heavy marihuana users made less correct responses initially. The infrequent marihuana users (one of the two control groups) did worse the second 6 weeks, I do not know why. But, even though there was improvement in the other two groups, you can see that there wa still a great big gap for the non-marihuana smokers.

Short-term Memory Study: Conclusions

1. Chronic cannabis use impairs short-term memory in teenagers
2. Short-term memory may remain intact in some individuals
3. Exceptionally gifted heavy marihuana smokers are less affected while the learning disabled and those with low I.Q. have major short-term memory problems from use of marihuana

---

The conclusion is that chronic marihuana use impairs short-term memory in teenagers. The memory remains intact in some individuals. Although a few heavy marihuana smokers are unaffected, the majority suffer from loss of short-term memory ability for weeks after the last puff of marihuana smoke. The exceptionally gifted, we believe, are less affected, while the learning disabled and those with low i.Q.'s have major and devasting effects on their short-term memory. If somebody has a learning disability and become a hashish smoker, their school performance is going to suffer greatly. Marihuana greatly adds to the burden of their learning.

Real-Life Correlates of Marihuana-induced Short-term Memory Impairment

1. Tart, C. T.: Marihuana intoxication: Nature 226, 701–704 (1979).
   In 1970, Tart surveyed 150 university students:
   59% were concerned about forgettung content of conversations.
   41% answered positive to: if I read while smoking cannabis,
   I remember less (than when I am sober).
2. Hendin, H. et al.: Living high. Human sciences press, New York, 1987.
   Surveyed 150 adults who smoked cannabis daily.
   A. 65% noted impaired memory as the most bothersome long-term effect of their use of cannabis.
   B. 45% noted impaired ability to:
      a. concentrate on complex tasks
      b. think clearly
      c. get things done

What does it means in real life? There are three different studies that show there is not just something in the experimental laboratory but has real life correlates. For example, a researcher named Tart reported in the British journal "Nature" that he surveyed 150 university students. Almost 60% were very concerned about the forgetting contents of conversations (these were the marihuana smokers). 41% answered, if they read while smoking cannabis, they remember less than when they are sober.

In a book called "Living high" you can read about a study of 150 adult American daily cannabis smokers. There appear to be some success stories, but if you read about the success stories, while they may appear superficially successful in their profession, their marriage, their relationship with their children and their social problems with devastating to them. 65% noted impairment of memory as the most bothersome long-term effect of their use of cannabis. 45% noticed impaired ability to concentrate on complex tasks, to think clearly and get things done.

A Telephone Survey of 110 Adults

(mean age = 32 years, mean annual income = $ 23 000) who smoked hashish every day and did not abuse alcohol, concluded:

    – 73% complained of serious memory loss.

Roffman, R.A. et al.: Treatment of marihuana dependence: preliminary results. J. Psychoactive Drugs 20, 129–137 (1988).

A recent study of a telephone survey was reported in the *Journal of Psychoactive Drugs* in 1988. They asked 110 adults with the mean age of 32 years, they were middle class, they smoked hashish every day and they did not abuse alcohol. 73% complained about memory loss.

In summary, we can say that the use of marihuana, even at a less frequent basis, impairs short-term memory well beyond the period of intoxication.

Freebase, Cocaine and Memory

Despite the seriousness of acute medical and psychological consequences of cocaine abuse, little knowledge exists about the chronic effects of the drug. Investigation of a sample of abstinent freebase (crack) abusers in the Bahamas provides the first research evidence that prolonged cocaine abuse may result in persistent short-term memory disturbances.

Manschreck, T. C. et al.: Compr. Psychiatry 31, 369–375 (1990).

Crack smoking like marihuana smoking seems to also impair short-term memory. Why should we permit our sons and daughters to develop a disease of forgetfulness like Alzheimers Disease of the young?

REFERENCES

Abel, E.L. (1970). Marijuana and memory. <u>Nature</u>, <u>1227</u>, 1151-1152.

Abel, E.L. (1971). Marijuana and memory : Acquisition or retrieval? <u>Science</u>, <u>173</u>, 1038-1040.

Bowman, M., R.O. Pihl. (1973). Cannabis : Psychological effects of chronic heavy use : A controlled study of intellectual functioning in chronic users of high-potency cannabis. <u>Psychopharmacologia</u>, <u>29</u>, 159-170.

Darley, C.F., J.R. Tinklenberg, L.E. Hollister, R.C. Atkinson. (1973). Influence of marijuana on storage and retrieval processes in memory. <u>Mem. Cognit.</u>, <u>1</u>, 196-200.

Dittrich, A., K. Battig, I.V. Zeppelin. (1973). Effects of (-) delta-9-tetrahydrocannabinol (delta-9-THC) on memory, attention and subjective state. <u>Psychopharmacologia</u>, <u>33</u>, 369-376.

Dornbush, R.L., M. Fink, A.M. Freedman. (1971). Marijuana, memory, and perception. <u>Am. J. Psychiatry</u>, <u>128</u>, 194-197.

ElSohly, M.A., C.T. Abel. (1990). Quarterly Report : <u>Potency Monitoring Project Report N° 32</u>, Oct-Dec. 1989. University City, MS. Research Institute of Pharmaceutical Sciences.

Hendin H., A.P. Haas, P. Singer, et al. (1987). Living High. <u>Daily Marijuana Use among Adults</u>. Human Sciences Press, Inc. New York.

Melges, F.T., J.R. Tinklenberg, L.E. Hollister, H.K. Gillespie. (1970). Marijuana and temporal disintegration. <u>Science</u>, <u>168</u>, 1118-1120.

Mendhiratta, S.S., N.N. Wig, V.K. Varma. (1978). Some psychological correlates of long-term heavy cannabis users. <u>Br. J. Psychiatry</u>, <u>13-2</u>, 482-486.

Page J.B., J. Fletcher, W.R. True. (1988). Psychosociocultural perspectives on chronic cannabis use : The Costa Rican follow-up. <u>J. Psychoactive Drugs</u>, <u>20</u>, 57-65.

Satz, P., J.M. Fletcher, L.L. Sutker. (1976). Neuropsychologic, intellectual and personality correlates of chronic marijuana use in native Costa Ricans. <u>Ann. NY. Acad. Sci.</u>, <u>282</u>, 266-306.

Schwartz, R.H., P.J. Gruenewald, M. Klitzner , P. Fedio. (1989). Short-term memory impairment in cannabis-dependent adolescents. <u>Am. J. Dis. Child.</u>, <u>143</u>, 1214-1219.

Souief, M.I. (1976). Differential association between chronic cannabis use and brain function deficits. <u>Ann. NY. Acad. Sci.</u>, <u>282</u>, 323-343.

Tart, C.T. (1979). Marijuana intoxication : Common experiences. <u>Nature</u>, <u>226</u>, 701-704.

Tinklenberg, J.R., F.T. Melges, L.E. Hollister, H.K. Gillespie. (1970). <u>Nature</u>, <u>226</u>, 1171-1172.

Varma, V.K., A.K. Malhotra, R. Dang, et al. (1988). Cannabis and cognitive functions : A prospective study. <u>Drug Alcohol Depend.</u>, <u>21</u>, 147-152.

Wig, N.N., V.K. Varma. (1977). Patterns of a long-term heavy cannabis use in North India and its effects on cognitive functions : A preliminary report. <u>Drug Alcohol Depend</u>, 2, 211-219.

DETECTION OF CANNABIS AND OTHER DRUGS IN 120 VICTIMS OF ROAD ACCIDENTS

Michel Aussédat et S. Niziolek-Reinhardt.

Department of Surgical Intensive Care
Medical Center Metz-Thionville, France

ABSTRACT

Analysis of blood or urine to detect substances which impair psychomotor performance were performed in 120 victims of road accidents, admitted as emergencies in the medical center of Metz-Thionville. 34% of samples were negative, 36% contained alcohol, 14% cannabis, 10% benzodiazepines and barbiturates and 1% opiates. Several drivers were polydrug users. This study illustrates the high incidence of previous drug consumption in drivers victims of road accidents. It calls for measures of information and prevention, as well as systematic detection among subjects victims of vehicular accident.

KEY WORDS

Driving, vehicular accidents, alcohol, cannabis, opiates, benzodiazepines.

INTRODUCTION

Circulation accidents are the third cause of mortality in France after cardiovascular accidents and cancer. Statistic and epidemiologic data allow better understanding of road security problems but do not permit an analytic approach in the causes of accidents. Forensic medicine, will alienate these uncertainties by the use of methods which permit to detect presence of substances which impair driving.

It appears that road accidents result from many factors that are difficult to dissociate from one an other. In the majority of accidents, abnormal behavior of the driver is involved and human error is a primary cause. Multiple causes, like physiological and psychological impairment may result from intake of alcohol, or other legal drug intake. In addition, illicit drug consumption impairs psychomotor performance.

Drug consumption, especially of cannabis has markedly increased during the past two decades and has been associated with vehicular accidents (1). In the present study, to detect that incidence, a systematic detection of licit and illicit drugs in blood and/or in urine was performed on every driver victim of a vehicular accident and admitted in a regional medical center.

MATERIALS AND METHODS

Our study (1,2) took place from April 1986 to January 1987 and included 120 individuals who were victims of vehicular accidents. 119 samples of blood and 64 urine samples were analyzed. It was noted that during a week, there is a sharp increase in accidents from Saturday to Monday. These week end accidents account for 65% of all traffic highways accidents with 20% on Saturday, 21% on Sunday and 20% on Monday. During a period of 24 hours, there is a greater incidence of accidents between 7 p.m. and 7 a.m.

Our study comprised 84 car accidents (70%), one truck and 35 motorcycles casualties. 104 men and 16 women were involved. The average age was 30 years with a peak in 20-29 years old or 47,5% of the population studied. Others were in groups: 15-19 years and 30-39 years of age. Seven (7,5%) were killed.

The severely injured who stayed in the hospital at least 6 days accounted for 35% or 42 victims, and 16 of them required intensive care. 59% or 71 victims were less severely wounded: among them 3 had left the hospital against medical advice, 20 had uncomplicated head injury and 40 presented head injuries with loss of consciousness.

Samples of blood and urine, taken from patients victims of vehicular accident, were analyzed for ethanol, benzodiazepines, barbiturates, tricyclic drugs, opiates, cocaine, amphetamines and cannabis. Immuno enzymatic (EMIT) and mass spectrometric methods of analysis were used.

The EMIT method was used to identify illicit substances in body fluids and results were quantified by Gas Chromatography-Mass Spectrometry.

RESULTS

In our study, 34% of the samples were negative, 36% contained alcohol, 14% cannabis, 7% opiates and 10% benzodiazepines and barbiturates.

At entrance, neurologic examinations of all patients were normal and there were no signs of consumption of illicit drugs. The most detailed examination by a neurologist reported trembling, disturbance of coordination and disturbance of speech in cannabis intoxication.

In 35.3% of cases plasma concentration of alcohol ranged between 0.80 and 3.6 g/l. In 60% of cases, no alcohol was found. In one third of the cases blood alcohol concentration was 0.80 g/l or greater.

Cannabis metabolites were identified by the EMIT method in 17 urine samples, and 9 of them were confirmed by mass spectrometry. 7 samples contained cannabis with alcohol (2 g/l) and 1 sample contained cannabis with barbiturates and opiates.

Opiates were detected in 6 subjects: a 25 years old woman addicted to heroin and in whom opiates and barbiturates were also detected; two 30 years old men addicted to heroin (one of them denied his addiction), and two subjects of 65 and 58 years old who had consumed a syrup with codein for chronic bronchitis.

In 10% of the samples (12 subjects) benzodiazepines were detected in blood or urine: in 3 cases, only benzodiazepines were present, in 4 cases benzodiazepines were associated with alcohol.

Our result reflect the present poly drug usage of many consumers of recreational drugs, licit and illicit.

DISCUSSION

In this study 14% of the injured had consumed cannabis prior to their accident. This figure is in the lower range of that reported by others. In the USA: 16% by Sterling Smith (4), 37% by Williams (5), 34,7% by Soderstrom (1) in Sweden, 12% by Holmgren (6), , in Australia 20% by Hendtlass (7). Cannabis consumption impairs motor coordination, reaction time and sensory perception, glare recovery as described in a series of studies by Adams (8) and Chesher (9). Studies on a driving simulator by Rafaelson (10) or in real situation by Hansteen (11) have illustrated serious driving impairment after cannabis consumption, rebutting the initial misleading study of Crancer (12) and the report of Mason and Mac Bay (13).

Furthermore, the residual effects of cannabis on motor coordination, persists for 24 hours following acute exposure (Von Leirer, this monograph). Cremona (14) finds that the risk to have one accident is 3 to 5 times greater after cannabis consumption. Warren and Simpson (15) have determined a guilt index which they define as self reported responsability of the driver who provokes an accident under the influence of a drug. The guilt index of a person who does not take a drug is 1, and the guilt index of a subject who had cannabis detected in the urine is 1.7, identical to that of a person who drives in the state of alcohol intoxication.

CONCLUSION

Detection of drugs which impair information processing by the brain indicates that one in 3 drivers involved in vehicular accidents had consumed alcohol, 1 in 7, cannabis, 1 in 10 barbiturates and tranquilizers and 1 in 11, opiates. Furthermore this study reflected the present poly drug use of a significant portion of the population.This report should be followed up by a national survey. Pharmacological effect on psychomotorperformance of prescription of psychotropic drugs (Benzodiazepine) taken separately and in association with other drugs should be investigated. Non invasive techniques should be developed (detection in saliva) and cost of analysis should be reduced. In France, a legislation aimed at penalizing drivers who drive under the influence of drugs which impair coordination and psychomotor performance should be formulated and enforced. This legislation exists only for alcohol.

REFERENCES

1. Soderstrom C.A., Trifillis A.L., Shankar B.S., Clark W.E., Coley R.A. Marijuana and Alcohol use among 1023 trauma patients. *Arch. Surg.* Vol 23, 733-737, 1988.

2. Nizioloek-Reinhardt S. Influence de la consommation de cannabis sur les accidents de la voie publique. A propos d'une étude réalisée au CHR de Metz-Thionville. *Thèse de Médecine - Nancy*, 1989.

3. Coudane H., Blanquart D., Aussédat M., Niziolek S., Peton P. Influence de la consommation de cannabis sur les accidents de la voie publique. *Journal de Médecine Légale, Droit Médical* Tome 33 n° 4, 287-290, 1990.

4. Sterling Smith R.S. Alcohol, drugs and traffic safety. *Israelstam and Lambert Ed.* Toronto, 93-105, 1975.

5. Williams A.F., Peat M.A., Crouch D.J., Wells J.K., Finckle B.S. Drugs in fatally injured young male drivers. *Public Health Rep.*, 100 (1): 19-25, Jan-Feb. 1985.

6. Holmgren P., Loch E., Schuberth J. Drugs in motorists travelling mads: on the road detection of intoxicated drivers and screening for drugs in these offenders. *Forensic Sci. Int.*, 27, 57-65, 1985.

7. Hendtlass J. Drugs and driving: the road to action. *Aust. Fam. Physician*, 16, 25-31, 1987.

8. Adams A.J., Brown B., Haegerstrom-Portnoy G., Flomm M.C. Marijuana, alcohol and combined drug effects on time of glare recovery. *Psychopharmacology* 56, 81-86, 1978.

9. Chesher G.B., Bird K.D., Stramarcos A., Nikias M. A comparative study of dose response relationship of alcohol and cannabis on human skills performance. In: Harvey D.J. editor. Marihuana nineteen eighty-four: proceedings of the Oxford Symposium on Cannabis. Third International Symposium, Ninth International Congress of Pharmacology, *Oxford: IRL Press*, 621-627, 1985.

10. Rafaelson O.L., Bech P., Christiansen J., Christrup H., Nyobe J. Cannabis and alcohol: effects on simulated car driving. *Science*, 179: 920-923, 1973.

11. Hansteen R.W., Miller R.D., Lonero L., Jones B. Effects of cannabis and alcohol on automobile driving and psychomotor tracking. *Annals of the New York Academy of Sciences*, 240-246, 1976.

12. Crancer A.J., Dille J.M., Delay J.C., Wallace J.E., Hakin M. Comparaison of the effects of marijuana and alcohol on simulated driving performance. *Science*, 164, 851-854, 1969.
13. Mason A.P., Mac Bay B.S. Ethanol, marijuana and other drug use 600 600 drivers killed in single vehicle crashes in North California 1978-1981. *J. Forensic Sci.*, 29 (4): 172-174, 1984.
14. Cremona A. Mad drivers: psychiatric illness and driving performances. *Br. J. Hosp. Med.*, 35, 193-195, 1986.
15. Warren R.A., Simpson H.M. Cannabis and driving: implication of moderation in penalties. *Traffic Injury Research Foundations of Canada*, Ottawa, 1980.

MARIJUANA AND ALCOHOL USE AMONG 1023 TRAUMA PATIENTS[1]

Carl A. Soderstrom[1], Anna L. Trifillis[2], Belavadi S. Shankar[3], William E. Clark[4] and Adams Cowley[1].

[1]Department of Surgery-Traumatology, Maryland Institute for Emergency Medical Services System, Baltimore, [3]Operations and Systems Analysis, Maryland Institute for Emergency Medical Services System, Baltimore, [4]Emergency Medical Services Field Operation Program, Maryland Institute for Emergency Medical Services System, Baltimore and [2]Department of Pathology, University of Maryland, Baltimore, U.S.A.

ABSTRACT

Marijuana use prior to injury was determined prospectively in 1023 patients injured as the result of vehicular (67.6%) and nonvehicular (32.4%) trauma. Most were men (72.8%); most were 30 years of age or younger (58.4%). All were admitted directly from the scene of injury. Serum ∂-9-tetrahydrocannabinol activity was ascertained using a radioimmunoassay. Activity of 2 ng/ml or more was detected in 34.7% of subjects. Blood alcohol determinations were made in 1006 patients; 33.5% were positive. Marijuana use among vehicular and nonvehicular trauma victims was not significantly different. Marijuana use was higher among those 30 years of age or younger and among men. Vehicular crash victims consumed alcohol more frequently. Use of marijuana and alcohol in combination (16.5%) was highly significant compared with marijuana alone (18.3%), alcohol alone (16.1%), or neither drug (49.1%).

KEY WORDS
Marijuana; Alcohol detection; vehicular accidents; prospective study.

---

[1] *Article reproduced with permission from Arch. Surg., 1988;123:733-737 (Copyright 1988, American Medical Association).*

While alcohol abuse among victims of traumatic insults has been well documented,[1-8] little is known about marijuana use by trauma victims proximate to the time of injury. In the few large series reported, confirmation of such use involved victims of vehicular crashes. These studies did not reflect the overall use of marijuana because they limited their scope to studying only those who died [9-11] or those agreed to be tested.[12]

No previous prospective study has measured marijuana use in a large population of vehicular and nonvehicular trauma victims. The purpose of this prospective study was to determine the use of marijuana proximal to the time of injury in approximately 1000 trauma victims and to relate that use to a number of other factors.

Radioimmunoassay (RIA) tests for ∂-9-tetrahydrocannabinol (∂-9-THC), the major psychoactive ingredient in marijuana, allow for documentation of marijuana use proximal to the time of injury. Urine tests for cannabinoid metabolites cannot be used to ascertain such a relationship, because such tests yield positive results four to ten days after smoking marijuana and for as many as 36 days after cessation in chronic users.[13,14] In contrast, the RIA test for serum ∂-9-THC activity is not positive for such activity four hours after stopping marijuana use.[15]

MATERIALS AND METHODS

All trauma victims admitted from the scene of injury to the Shock Trauma Center of the Maryland Institute for Emergency Medical Services Systems, Baltimore, between July 29, 1985 and May 8, 1986, were candidates for the study. This demographic study was designed to document marijuana and alcohol use among trauma victims.

Prior to testing for marihuana use, the following data were collected: age ($\leq$ 30 years; >30 years), sex, mechanism of injury (vehicular and nonvehicular), Injury Severity Score (ISS),[16] and blood alcohol level (BAL) (0, $\leq$99, $\geq$100 mg/dl).

The ISS was designed to measure and compare degrees of injury among blunt trauma victims. For ISS, the body is divided into six regions. Each injury is assigned a number from 1 ("minor") to 6 ("maximum injury virtually unsurvivable") that reflects degree of injury from the abbreviated injury scale.[17] To derive an ISS, the highest score from the three most severely injured regions are squared and then added. (One investigator (C.A.S.) determined all ISSs).

After data collection, a serum sample was tested for ∂-9-THC. The serum sample was a leftover (usually discarded) from venous blood samples for routine admission diagnostic studies. In no case was serum drawn for the sole purpose of testing for ∂-9-THC. before testing, specimen tubes were assigned new identifiers in a random fashion; all previous identifiers were removed and discarded. Approximately 200 specimens were submitted for testing at one time to one of the investigators (A.L.T.).

Patient confidentiality was protected by (1) selecting broad demographic and clinical data set points and (2) "blinding" patient identifiers to ∂-9-

THC results. With the patient confidentiality assured and the use of normally discarded serum samples, the need for patient consent was obviated.

This study was approved by the University of Maryland's Human Volunteers Research Committee and the Shock Trauma Center's Clinical Research Committee.

TESTING FOR ∂-9-THC

Serum samples were stored at -70°C. Analysis for ∂-9-THC was made using an RIA kit prepared by the Research Triangle Institute (Research Triangle Park, NC). A series of standards (2.5, 5.0, 10.0, 25.0 and 50.0 ng/ml) and control serum samples (8 and 30 ng/ml) were assayed with each series of unknowns. Standard, control, and unknown serum samples were extracted with methyl-alcohol and centrifuged. Aliquots of supernatant were incubated overnight at 4°C with ∂-9-THC-specific antiserum and the radioligand, tritiated ∂-8-THC. The unlabeled ∂-9-THC reduced the fraction of radioactivity bound to the antibody. Subsequent charcoal treatment removed free radioligand, and bound radioactivity was counted in a liquid scintillation counter (Beckman 7800).

The ∂-9-THC levels for unknowns were interpolated from the standard curve in the following manner. Counts per minute values of standards, controls, and unknowns were corrected for nonspecific binding, divided by the total counts per minute of the radioligand, and multiplied by 100 to obtain the percent radioligand bound. On logit-log paper, the percent radioligand bound for each of the standards was plotted vs the ∂-9-THC concentration, and the concentrations in controls and unknowns were interpolated from the standard curve.

The RIA technique is less specific than gas chromatography, high-pressure liquid chromatography, or mass spectrophotometry. In the forensic community, gas chromatography/mass spectrophotometry (GC/MS) is the most accepted method to confirm cannabinoid activity detected by RIA.[14] As a check, 50 serum samples, 39 of which were positive for ∂-9-THC, were submitted for GC/MS analysis. All 39 RIA-positive serum samples were also positive by GC/MS, and all RIA-negative serum samples were negative by GC/MS. In general, RIA yielded values that were approximately 2.5 times those found by GC/MS. Because immunologic reagents may react to THC metabolites other than ∂-9-THC, these tests generally give higher values than chromatographic methods specific for ∂-9-THC.

Unless otherwise indicated, data were analyzed with the X^2 test.

RESULTS

Between July 29, 1985 and May 8, 1986, a total of 1109 patients were candidates for the study (Table I). Serum was available from 1023 (92.2%). Most patients were admitted within one hour of injury; in no case did that time exceed three hours.

It was postulated that blood might not be drawn as frequently from

TABLE 1. - Profile of Patients.*

	No. (%) of Patients	No. (%) Men
Vehicular crash victims		
Automobile occupants		
Drivers	398	261 (65.6)
Passengers	145	76 (52.4)
Total	543 (53.1)	337
Motorcycle riders		
Drivers	72	69 (95.8)
Passengers	4	3 (75.0)
Total	76 (7.4)	72
Pedestrians struck	73 (7.1)	52 (71.2)
Nonvehicular victims	331 (32.4)	274 (82.8)
Total	1023 (100)	735 (71.8)

* Patients aged 30 years or less, 597 (58.4%); Injury Severity Scores of crash victims 13 or higher, 294 (42.5%).

patients who were essentially "dead on arrival" or in whom heroic resuscitation efforts failed. This hypothesis is supported by an analysis of survival and mortality data (including time of death) from independent lists of patients who had serum available for testing compared with those from whom serum was not available. Of the 1023 patients, 91 died (8.9%), 13 (14.3%) of whom were dead on arrival or pronounced dead within one hour of admission. In contrast, 36 (41.9%) of 86 patients in whom serum was not available for testing died. of those 36 patients, 29 (80.6%) were dead on arrival or pronounced dead within one hour of admission.

MARIJUANA USE

The use of marijuana among 34.7% of the 1023 patients is reflected in Table 2. The serum samples of those 355 patients were positive for ∂-9-THC at a level of 2 ng/ml or higher. (Levels below 2 ng/ml were considered negative for ∂-9-THC activity). The results, ranging from 2 to 75 ng/ml, were subdivided arbitrarily into three category of ∂-9-THC activity as follows : 2 to 4.9 ng/ml (n=141, 13.8%), 5.0 to 9.9 ng/ml (n=91, 8.9%), and 10 ng/ml or higher (n=123, 12.0%). Marijuana use was significantly greater in those 30 years of age or younger and among men. There was no statistical difference in marijuana use among vehicular and nonvehicular trauma victims.

ALCOHOL USE

Admitting BALs were determined on 1006 (98.3%) of the 1023 patients (Table 2); 328 (32.6%) were positive. Over three quarters of the BALs were 100 mg/dl or higher. Alcohol use was significantly linked to being 30 years of age or younger, a man, and injured in a vehicular crash.

COMBINED MARIJUANA AND ALCOHOL USE

Almost half of the 1006 patients tested for both marijuana and alcohol showed no evidence of having used either drug (Table 3). Of the remainder, 18,3% were positive for marijuana only, 16.1% had consumed only alcohol, and 16.5% had used both drugs. The number of patients using both marijuana and alcohol (166) is significantly larger (p<.001) than the expected number (114) if the use of marijuana and the use of alcohol were unassociated.

MARIJUANA AND ALCOHOL USE IN VESICULAR AND NONVEHICULAR TRAUMA

Marijuana use was significantly greater among victims of both vehicular and among men injured in vehicular crashes (Table 4). While alcohol consumption was greater among those 30 years of age or younger

TABLE 2. - Use of Marijuana and Alcohol Among Victims of Trauma.

Variable	n	No.(%) Using Drug	p*
Marijuana			
Age			
≤ 30 y	597	248 (41.5)	
> 30 y	426	107 (25.1)	<.001
Sex			
M	735	274 (37.3)	
F	288	81 (28.1)	<.02
Cause			
Vehicular	692	234 (33.8)	
Nonvehicular	331	121 (36.6)	NS
Total	1023	355 (34.7)	
Alcohol			
Age			
≤ 30 y	585	215 (36.8)	
> 30 y	421	113 (26.8)	<.001
Sex			
M	722	268 (37.1)	
F	284	60 (21.1)	<.001
Cause			
Vehicular	680	249 (36.6)	
Nonvehicular	326	79 (24.2)	<.001
Total	1006+	328 (32.6)	

* *NS indicates not significant; X^2 values were calculated from observed and expected numbers from 2x2 contingency tables.*
+ *Blood alcohol levels were not determined in 17 of the 1023 patients tested for marijuana use.*

TABLE 3. - Marijuana and Alcohol Use Among Victims of Trauma.

	No Drug	Marijuana Only	Alcohol Only	Both Drugs
Automobile occupants				
Drivers (n=393)	196	61	72	64
Passengers (n=139)	68	24	19	28
Motorcycle riders				
Drivers (n=70)	25	10	18	17
Passengers (n=4)	3	1	0	0
Pedestrians (n=71)	30	10	17	14
Other	172	78	36	43
Total patients	494	184	162	166*
% of total patients(1006)+	49.1	18.3	16.1	16.5

* p<.001
+ Blood alcohol levels were not determined in 17 of the 1023 patients tested for marijuana use.

TABLE 4.- Marijuana and Alcohol Use Among Vehicular and Nonvehicular Trauma Victims Relative to Age and Sex.

Variable	n	No.(%) Using Drug	p*
Marijuana			
<u>Vehicular</u>			
Aged ≤ 30 y	408	166 (40.7)	
Aged > 30 y	284	68 (24.0)	<.001
<u>Nonvehicular</u>			
Aged ≤ 30 y	189	82 (43.4)	
Aged > 30 y	142	39 (27.5)	<.001
<u>Vehicular</u>			
Men	274	103 (37.6)	
Women	57	18 (31.6)	<.03
<u>Nonvehicular</u>			
Men	461	171 (37.1)	
Women	231	63 (27.3)	NS
Alcohol			
<u>Vehicular</u>			
Aged ≤ 30 y	396	170 (42.9)	
Aged > 30 y	284	79 (27.8)	<.001
<u>Nonvehicular</u>			
Aged ≤ 30 y	184	45 (24.4)	
Aged > 30 y	142	34 (23.9)	NS
<u>Vehicular</u>			
Men	448	197 (44.0)	
Women	228	52 (22.8)	<.001
<u>Nonvehicular</u>			
Men	274	71 (25.9)	
Women	56	8 (14.3)	NS

* NS indicates not significant; X^2 values were calculated from observed and expected numbers from 2x2 contingency tables.

and among men injured in vehicular crashes, age and sex were not factors in alcohol consumption among nonvehicular trauma victims.

MARIJUANA/ALCOHOL USE IN VEHICULAR TRAUMA

Among automobile drivers, 31.7% used marijuana, with no difference among men and women drivers. Almost twice as many drivers 30 years of age or younger (n=225) used marijuana, compared with older drivers (n=173)(40.0% and 20.8%, respectively)(p<.001). Fifty percent of male passengers (n=76) used marijuana, compared with 23.2% of female passengers (n=69)(p<.001). Age was not a factor in the use of marijuana.

Overall, 28.9% of drivers had BALs of 100 mg/dl or higher. Such levels were found more in male drivers than in female drivers (36.8% and 13.9%, respectively)(p<.001). Blood alcohol levels of 100 mg/dl or higher were noted more frequently in the 225 drivers 30 years of age or younger than in those older than 30 years of age: 32.9% and 23.7%, respectively (p<.04). Among 145 passengers, 31 (21.4%) had BALs of 100 mg/dl or higher. Age and gender were not factors among passengers with such levels.

Marijuana or alcohol use or both among motorcycle occupants and pedestrians is noted in Table 3.

FACTOR AFFECTING ISS

The effects of the use of alcohol or marijuana or both, the sex and age of the patient, and the status of vehicular crash victims (passenger/driver, pedestrian) on injury severity were evaluated. With the use of an analysis of variance, only pedestrian status correlated with an increase in ISSs (p<.02).

COMMENT

This is the first prospective study of marijuana use in a large number of consecutive patients injured as the result of vehicular and nonvehicular causes. Marijuana use was detected in 34.7% of the 1023 patients.

After smoking marijuana, blood levels of ∂-9-THC peak within one hour and then rapidly decline.[18-19]∂-9-Tetrahydrocannabinol RIA activity is usually negative four hours after marijuana use. On the basis of RIA analysis of serum samples of subjects who smoked up to three marijuana cigarettes consecutively, Soares and Gross[20] concluded that "a plasma sample negative for ∂-9-THC would seem to be unambiguous evidence that Marihuana had not been smoked in the preceding hour." Their study did not include long-term users.

One must consider the possibility of false-positive results in long-term heavy users of marijuana. Being highly lipophilic, ∂-9-THC is rapidly taken up and stored in fatty tissues.[14,19,30] Chesher[19] noted that long-term users may have considerable stores; which may be steadily released into the blood. He cited studies indicating that levels of 2 to 22 ng/ml have been measured in heavy users of marijuana a day after they reported not using marijuana. Patterns of use among our patients were not ascertained.

One must consider the possibility that ∂-9-THC levels may be detected in passive nonusers in the vicinity of marijuana smokers. Experimental data do not support this concept. Mason and colleagues[21] exposed a passive subject to marijuana smokers in a confined space. Radioimmunoassay of the subject's serum indicated ∂-9-THC activity was between 2.0 and 2.2 ng/ml, but "the passive exposure conditions used were extreme." Analysis of the room's air indicated that the subject was exposed to ∂-9-THC levels 60 to 600 times greater than that of nicotine in public places.[22] In another closed-space study in which six smokers developed ∂-9-THC levels of 7.5 ng/ml, no activity was detected in four passive subjects.[22]

OVERALL MARIJUANA USE

Marijuana use in trauma patients has been assessed only in those injured in vehicular crashes, including several large groups of subjects who were fatally injured. Mason and McBay[10] noted marijuana use among 47 (7.8%) of 600 North Carolina drivers who died within an hour of being injured in single-vehicle crashes. in those patients, ∂-9-THC activity was measured with an RIA technique in which values 3.0 ng/ml or higher were considered positive. The mean age of the victims was 31.2 years; 86.7% were men. Cimbura and colleagues[9] noted a 3.3% incidence of marijuana use among 401 fatally injured drivers and 83 pedestrians. Radioimmunoassay of ∂-9-THC (≥2.0 ng/ml was considered positive) was used in these victims dying within one hour of injury. The highest incidence of marijuana use among fatally injured drivers was reported by Williams and colleagues.[11] Their subjects were a select group, ie, young male California drivers. Among 440 such individuals, marijuana use varied from 32% to 39% among age groups between 15 and 34 years. Williams and colleagues used a GC/MS method to measure whole blood ∂-9-THC activity. Measuring levels of activity as low as 0.2 ng/ml using this method, they noted that serum levels would be twice as high. Zimmerman et al[23] detected ∂-9-THC levels of 5.5 ng/ml or higher among 14.4 % of 1792 "erratic" drivers stopped by California police. Finally, a study[12] of nonfatally injured drivers in Monroe County, New York, found an incidence of marijuana use similar to the values reported by others.[9,10,23] In that study, Terhune and Fell[12] detected ∂-9-THC activity in the serum of 47 (9.5%) of 497 drivers. That figure was probably conservative, considering that 29.3% of candidates for study refused to participate despite being "assured that their identities would be protected."

MARIJUANA USERS

The 34.7% incidence of marijuana use in this report is similar to the 36.8% rate reported by Williams and colleagues,[11] who studied "a population that has high drug use and crash rates: young California males." Although marijuana use in both of these studies is higher than that reported by others,[9,10,12] the predilection for that use among younger individuals, particularly men, has been noted by them. Among the 16 fatally injured drivers who used marijuana studied by Cimbura and colleagues,[9] 15 were men, and all were younger than 30 years of age. In the group of nonfatally injured drivers reported by Terhune and Fell,[12] those who had used marijuana compared with those who had not were more frequently under the age of 31 (84.8% vs 55.6%) and men (80.9% vs 58.2%). Among the 600 fatally injured drivers described by Mason and McBay,[10] 37 (78.7%) of the 47 marijuana users were younger than the mean age of 31.2 years of all of the drivers (p<.02); sex was not a factor in marijuana use. The pattern of marijuana use among injured young men is a reflection of the overall use among the population at large.[24]

COMBINED MARIJUANA AND ALCOHOL USE

Among our patients, the use of marijuana correlated significantly with the use of alcohol and vice versa. The combined reports of Cimbura and colleagues[9] and Terhune and Fell,[12] in which the incidence of marijuana use among vehicular crash victims was 3.3% and 9.5%, respectively, illustrate the use of one drug with the other. The linkage of use of the two drugs is accentuated in the report of Williams et al[11] in which marijuana was detected in 36.8% of subjects. The combined use of marijuana and alcohol reported in the national survey of 1982 is similar to the 16.5% rate observed in the current study. In that survey,[24] 16% of young adults (18 to 25 years of age) indicated that when they used marijuana, they usually consumed alcohol.

ETIOLOGIC ROLES OF MARIHUANA AND ALCOHOL

A recent consensus report by the Council on Scientific Affairs of the American Medical Association[25] indicated that (1) clinical, epidemiologic, and experimental data accumulated over 50 years have demonstrated an etiologic link between increasing BALs in drivers involved in vehicular crashes and (2) deterioration of driving skills begins at a BAL of 50 mg/dl or less. No definitive studies have demonstrated a significant link between the use of marijuana and vehicular crashes. Due to the complex pharmacokinetics of marijuana, blood and serum levels of ∂-9-THC activity compatible with "intoxication" or "impairment" have yet to be defined.[18,20,26]

INJURY SEVERITY

An analysis of variance was performed to assess which of several factors may be associated with an increase in severity of injury. Considering the vulnerability of pedestrians, it is not surprising that they had significantly higher ISSs (p<.02) than automobile and motorcycle riders.

Alcohol consumption did not correlate with injury severity. This finding is consistent with the findings in two other studies, in which it was noted that the ISSs of drinking trauma patients were not significantly different from those of patients who had not been drinking.[27,28] In fact, in another study, Ward and colleagues[29] found that among surviving trauma patients, those who drank alcohol had significantly higher ISSs than those who did not (p<.001). A greater percentage of patients who did not consume alcohol died (p<.02). In a study of blunt and penetrating trauma victims, in which the trauma score[30] was used as a measure of severity of injury, Thal and coworkers[6] found no difference in the degree of injury or mortality among those who had consumed alcohol compared with those who had not. The possible influence of marijuana use on injury severity was examined for the first time in this study. No correlation was found between injury severity and the presence of ∂-9-THC activity in a patient's serum.

CONCLUSIONS

A method that protected patient confidentiality was developed to prospectively evaluate the use of marijuana and alcohol among trauma patients proximal to the time of injury. The following observations were made :

1. One third of victims used marijuana or alcohol or both, with the use of one being associated with use of the other.

2. Marijuana and alcohol use was more common in men and in those 30 years of age or younger.

3. The incidence of marijuana use was similar among victims of vehicular and nonvehicular trauma, but alcohol was used more by victims of vehicular trauma.

4. Injury severity did not correlate with either drug.

This study was supported by US Department of Transportation/ National Highway Safety Traffic Administration contract DTNH 22-85-05124.

Radioimmunoassay kits were supplied by the National Institute on Drug Abuse (Rockville, Md) by authority of Richard L. Hawks, PhD. Roger Foltz, PhD, assisted in confirming RIA results by mass spectroscopy at the University of Utah Center for Human Toxicology, Salt Lake City. Colleen Rohrbeck and others collected clinical and demographic data. The manuscript was edited by Linda J.Kesselring.

REFERENCES

1. Perrine MW: Alcohol involvement in highway crashes: A review of the epidemiologic evidence. *Clin Plast Surg* 1975;2:11-34.
2. Soderstrom CA, DuPriest RW Jr, Benner C, et al: Alcohol and roadway trauma: Problems of diagnosis and management. *Am Surg* 1979;45:129-136.
3. Zuska JJ: Alcohol and trauma: Wounds without cause. *Bull Am Coll Surg* 1981;66:5-10.
4. Baker SP, O'Neil B, Karpf RS: *The Injury Fact Book*. Lexington, Mass, DC Heath & Co, 1984.

5. National Commission Against Drunk Driving: *A Progress Report on the Implementation of Recommendations by the Presidential Commission on Drunk Driving.* US Dept of Transportation publication (HS 806 885). National Highway Traffic Safety Administration, 1985.

6. Thal ER, Bost RO, Anderson RJ: Effects of alcohol and other drugs on traumatized patients. *Arch Surg* 1985;120:708-712.

7. National Research Council and the Institute of Medicine, Committee on Trauma Research and Commission on Life Sciences: *Injury in America: A Continuing Public Health Problem.* Washington, DC, National Academy Press, 1985.

8. Richman A: Human risk factors in alcohol-related crashes. *J Stud Alcohol* 1985;10(suppl):21-39.

9. Cimbura G, Lucas DM, Bennett RC, et al: Incidence and toxicologic aspects of rugs detected in 484 fatally injured drivers and pedestrians in Ontario. *J Forensic Sci* 1982;27:855-867.

10. Mason AP, McBay AJ: Ethanol, marijuana, and other drug use in 600 drivers killed in single-vehicle crashes in North Carolina, 1978-1981. *J Forensic Sci* 1984;29:987-1026.

11. Williams AG, Peat MA, Crouch DJ, et al: Drugs in fatally injured young male drivers. *Public Health Rep* 1985;100:19-25.

12. Terhune KW, Fell JC: The role of alcohol, marijuana, and other drugs in the accidents of injured drivers. Read before the American Association for Automotive Medicine, San Francisco, Oct 1, 1981.

13. Mason AP, McBay AJ: Cannabis: Pharmacology and interpretation of effects. *J Forensic Sci* 1985;30:615-631.

14. Schwartz RH, Hawks RL: Laboratory detection of marijuana use. *JAMA* 1985;254:788-792.

15. Soares JR, Gross SJ: Separate radioimmune measurements of body fluid ∂-9-THC and 11-nor-9-carboxy-∂-9-THC. *Life Sci* 1976;19:1711-1718.

16. Baker SP, O'Neill B, Haddon W Jr, et al: The injury severity score: A method for describing patients with multiple injuries and evaluating emergency care. *J Trauma* 1974;14:187-196.

17. Petrucelli E (ed): *An International Bibliography of Abbreviated Injury Scale Usage.* Morton Grove, Ill, American Association for Automotive Medicine, 1982.

18. Gross SJ, Soares JR: Validated direct blood ∂-9-THC radioimmune quantitation. *J Anal Toxicol* 1978;2:98-100.

19. Chesher GB: Alcohol and other drugs in road crashes: What does pharmacokinetics have to do with it? *Alcohol, Drugs, and Driving* 1985;1:1-20.

20. Nahas GG: Pharmacologic and epidemiologic aspects of alcohol and cannabis. *N Y State J Med* 1984;84:599-604.

21. Mason AP, Perez-Reyes M, McBay AJ, et al: Cannabinoid concentrations in plasma after passive inhalation of marijuana smoke. *J Anal Toxicol* 1983;7:172-174.

22. Law B, Mason PA, Moffat AC, et al: Passive inhalation of cannabis smoke. *J Pharm Pharmacol* 1984;36:578-581.

23. Zimmerman EG, Yeager EP, Soares JR, et al: Measurement of ∂-9-

tetra-hydrocannabinol (THC) in whole blood samples from impaired drivers. *J Forensic Sci* 1983;28:957-962.

24. Miller JD, Cisin IH, Gardner-Keaton H, et al: *National Survey on Drug Abuse: Main Findings 1982,* Department of Health and Human Services publication (ADM)83-1263. Rockville, Md, National Institute on Drug Abuse, 1983.

25. Council on Scientific Affairs: Alcohol and the driver. *JAMA* 1985;255:522-527.

26. Consensus Development Panel: Drug concentrations and driving impairment. *JAMA* 1985; 254:2618-2621.

27. Huth JF, Maier RV, Simonowitz DA, et al: Effect of acute alcoholism on the course and outcome of injured automobile drivers. *J Trauma* 1983;23:494-498.

28. Luna GK, Maier RV, Sowder L, et al: The influence of ethanol intoxication on outcome of injured motorcyclists. *J Trauma* 1984;24:695-700.

29. Ward RE, Flynn TC, Miller PW, et al: Effects of ethanol ingestion on the severity and outcome of trauma. *Am J Surg* 1982;144:153-157.

30. Champion HR, Sacco WJ, Carnazzo AJ, et al: Trauma score. *Crit Care Med* 1981;9:672-676.

4. Psychopathology and behavior

HISTORICAL OUTLOOK OF THE PSYCHOPATHOLOGY OF CANNABIS.

Gabriel Nahas.

New York University, School of Medicine, New York, USA &
Laboratoire de Pharmacologie et Toxicologie Cellulaires, Paris, France.

Throughout recorded history in the past 2,000 years, the use of *Cannabis* has been associated with mental disturbances ranging from distorted perceptions to hallucinations to dementia (Table 1). (By contrast, the use of opiates, including that of heroin, has been rarely associated in the literature with similar mental pathology).   It is clear from the observations of Moreau and of the French Romantics, and from the experiences of Taylor (1849) and Ludlow (1857), that ingestion of *Cannabis* extracts might precipitate episodes of acute mental confusion reminiscent of an acute brain syndrome.

Moreau mentions only one acute psychotic episode among his associates who had taken hashish; all the others kept their self-awareness and their consciousness while being transported in the fantastic world created by the drug. They did not choose to use *Cannabis* as a pleasure-inducing substance, but as a method to explore mental pathology. None of them absorbed the drug over a prolonged time and on a regular basis, and its chronic effect on mental function could not be assessed. However, Moreau mentions that "one of the determining causes of insanity among the Orientals is the excessive use of hashish" and that large amounts of the drug could readily induce a true psychotic episode "for a period of time which cannot be foreseen."

The widespread use of *Cannabis* extracts in the nineteenth century for therapeutic purposes, in Europe and the United States, was not accompanied by any report of adverse acute or chronic mental reactions.

Until recently, the only reports of the adverse effects of *Cannabis* on mental functions came from India, the Middle East, or Africa. Such adverse reactions have been described with increasing frequency over the past 25 years, as *Cannabis* intoxication has appeared on the American scene. They resemble those described earlier in the older literature.

Furthermore, the syndrome of mental and physical deterioration first reported by Moreau (1845) and attributed to chronic *Cannabis* intoxication, and confirmed by Chopra (1969), Miras (1969), and Benabud (1957), has now been reported in the United States (McGlothlin and West, 1968).

---

[1] *Reprint from "Marihuana in Science and Medicine", G. G. Nahas Ed, Raven Press, 1984*

TABLE 1. *Medical reports associating* Cannabis *use
with mental illness (until 1981)*

Country	Investigator	Date
China	Pen-Ts'ao Ching	2nd century, A.D.
Islam	Az Zarkashi } in (Rosenthal, 1971)	1396
	Al Badri	1464
India	Bengal Psychiatric Hospital	1878
	Dinshaw	1896
	Ewens	1904
	Robertson Milne	1906
	Peebles and Mann	1914
	Dhunjibhoy	1930
	J. D. Chopra, G. S. Chopra, R. N. Chopra	1942,1967,1969, 1971,1974,1976
	Grossman	1969
	Gaskill	1945
	Goel	1975
	Mendhiratta	1978
	Varma	1972
Afganistan	Gobar	1974
Egypt	Moreau	1843
	Warnock	1896,1903
	El Guindy	1925
Morocco	Benabud	1957
	Christosov	1965
	Defer and Diehl	1968
Nigeria	Lambo	1965
	Boroffka	1966
	Asuni	1964
Brazil	De Farias	1955
Panama	Siler et al.	1933
Bahamas	Spencer	1970
Jamaica	Knight	1976
So. Africa and Africa	Medical Staff of Pretoria Mental Hospital	1938
	Tonker	1966
	Conos	1925
USSR	Skliar and Iwanow	1932
	Landau	1927
	Anziferow	1929
	Kamajew	1931
	Sengal	1981
France	Baruk	1935
	Defer and Diehl	1968
	Deniker and Ginestet	1969
Sweden	Bejerot	1972
	Bernhardson and Gunne	1972
Denmark	Jorgensen	1968
Greece	Stringaris	1939
	Stefanis et al.	1977
	Bartolucci et al.	1969
Canada	Thurlow	1971
	Campbell	1976
	Blumenfield	1972
	Tylden	1967
	Dobell	1863
England	Baker-Bates	1935
	Dally	1967
	Baker and Lucas	1969
	George	1970
United States	Bromberg	1934,1939
	Curtis and Wolfe	1939
	Allentuck	1944
	Marcovitz and Myers	1944
	Charen and Perelman	1946
	Keeler	1967,1968
	Hekimian and Gershon	1968
	Keup	1969
	Klee	1969
	Marten	1969
	Perna	1969
	Talbott and Teague	1969
	Weil	1970
	Bialos	1970
	Colbach and Crowe	1970
	Smith and Mehl	1970
	Bey and Zecchinelli	1971
	Kolansky and Moore	1971
	Halikas	1972
	Kornhaber	1971
	Tennant	1972
	Kaplan	1971
	Milman	1969,1981
	Voth	1981
	Carranza	1981

REFERENCES

Allentuck, S. (1944): Medical aspects. In: *Marihuana Problem in the City of New York* (Mayor's Committee on Marihuana). Cattell Press, Lancaster, Pennsylvania.

Anziferow: Quoted in Walton (1938): Hashischmus in Turkestan, 1929.

Asuni, T. (1964): Socio-psychiatric problems of cannabis in Nigeria. *Bull. Narc.*, 16:17–28.

Baker, A. A., and Lucas, E. G. (1969): Some hospital admissions associated with *Cannabis. Lancet*, I:148.

Baker-Bates, E. T. (1935): A case of *Cannabis indica* intoxication. *Lancet*, I:811.

Bartolucci, G., Fryer, L., Perris, C., and Shagass, C. (1969): Marihuana psychosis: A case report. *Can. Psychiatry Assoc. J.*, 14:77–79.

Bejerot, N. (1972): *Addiction, an Artificially Induced Drive*, p. 23. Charles C Thomas, Springfield.

Benabud, A. (1957): Psychopathological aspects of the cannabis situation in Morocco: Statistical data for 1956. *Bull. Narcotics*, 9:1–16.

Bernhardson, G., and Gunne, L. M. (1972): Forty-six cases of psychosis in cannabis abusers. *Int. J. Addict.*, 7:9–16.

Bey, D. R., and Zecchinelli, V. A. (1971): Marijuana as a coping device in Vietnam. *Mil. Med.*, 136:448–450.

Bialos, D. S. (1970): Adverse marijuana reactions: A critical examination of the literature with selected case material. *Am. J. Psychiatry*, 127:819–823.

Blumenfield, M., Riester, A. E., Serrano, A. C., and Adams, R. L. (1972): Marijuana use in high school students. *Dis. Nerv. Syst.*, 33:603–610.

Boroffka, A. (1966): Mental illness and Indian hemp in Lagos. *E. Afr. Med. J.*, 43:377–384.

Bromberg, W. (1934–35): Marihuana intoxication. A clinical study of Cannabis sativa intoxication. *Am. J. Psychiatry*, 91:303–330.

Bromberg, W. (1939): Marihuana, a psychiatric study. *JAMA*, 113:4–12.

Campbell, I. (1976): The amotivational syndrome and cannabis use with emphasis on the Canadian scene. *Ann. NY Acad. Sci.*, 282:33–36.

Carranza, J. (1981): Marihuana-induced psychoses. In: *Drug Abuse in the Modern World: A Perspective for the Eighties*, edited by G. G. Nahas and H. C. Frick, pp. 57–61. Pergamon Press, New York.

Charen, S., and Perelman, L. (1946): Personality studies of marihuana addicts. *Am. J. Psychiatry*, 102:674–782.

Chopra, G. S. (1969): Man and marihuana. *Int. J. Addict.*, 4:215–247.

Chopra, G. S. (1971): Marihuana and adverse psychotic reactions. *Bull. Narcotics*, 28:15–22.

Chopra, G. S., and Smith, J. W. (1974): Psychotic reactions following cannabis use in East Indians. *Arch. Gen. Psychiatry*, 30:24–27.

Chopra, G. S., and Jandu, B. S. (1976): *Ann. NY Acad. Sci.*, 282:95–108.

Chopra, I. C., and Chopra, R. N. (1967): The use of cannabis drug in India. *Bull. Narcotics*, 9:4–29.

Chopra, R. N., Chopra, G. S., and Chopra, I. C. (1942): *Cannabis sativa* in relation to mental diseases and crime in India. *Indian J. Med. Res.*, 30:155–171.

Christozov, C. (1965): L'aspect marocain de l'intoxication cannabique d'apres des etudes sur des malades mentaux chroniques: 1ᵉʳᵉ partie et 2ᵉᵐᵉ partie. *Maroc. Med.*, 44:630–642.

Colbach, E. M., and Crowe, R. R. (1970): Marihuana associated psychosis in Vietnam. *Mil. Med.*, 135:571–573.

Conos, B. (1925): Trois cas de cannabisme avec psycose consecutive. *Bull. Soc. Path. Exot.*, 18:788–793.

Curtis, H. C., and Wolfe, J. R. (1939): Psychosis following the use of marihuana with report of cases. *J. Kansas Med. Soc.*, 40:515–517.

Dally, P. (1967): Undesirable effects of marijuana. *Br. Med. J.*, 3:367.

DeFarias, C. (1955): Use of maconha (*Cannabis sativa* L.) in Brazil. *Bull. Narcotics*, 7:5–19.

Defer, B., and Diehl, M. L. (1968): Les psychoses aigues: A propos de 560 observations. *Ann. Med. Psychol.*, 2:260–266.

Deniker, P., and Ginestet, B. (1969): Pharmacologie humaine de l'usage incontrole des drogues psychodysleptiques. *Laval. Med.*, 40:25–36.

Dhunjibhoy, J. E. (1930): A brief resume of the types of insanity commonly met with in India with a full description of "Indian hemp insanity" peculiar to the country. *J. Ment. Sci.*, 76:254–264.

Dinshaw, V. (1896): Complete aphonia after ganja-smoking recovery. *Indian Med. Rec.*, 11:14.

Dobell, H. (1863): On some effects of *Cannabis indica. Med. Times Gaz.*, 2:245–246.

El Guindy, M. (1925): Hashish: Proposal of the Egyptian delegation that hashish should be included in the list of narcotics with which the conference has to deal. In: *League of Nations*. Records of the Second Opium Conference. Geneva, November 17, 1924–February 19, 1925. Vol. I: Plenary Meetings: Text of Debates. Geneva, 132–138.

Ewens, G. F. W. (1904): Insanity following the use of Indian hemp. *Indian Med. Gaz.*, 39:401–413.

George, H. R. (1970): Two psychotic episodes associated with cannabis. *Br. J. Addict.*, 65:114–121.

Gobar, A. H. (1974): Drug Abuse in Afghanistan. In: *Intern. Ment. Health Res. Newslett.*, 16:4.

Goel, D. S. (1975): Cannabis: the habit and psychosis. *Indian J. Psychiatry*, X:238–243.

Grossman, W. (1969): Adverse reactions associated with cannabis products in India. *Ann. Intern. Med.*, 70:529–533.

Halikas, J. A. (1974): Marijuana use and psychiatric illness. In: *Marijuana: Effects on Human Behavior*, edited by L. L. Miller, pp. 265–302. Academic Press, New York.

Hekimian, L. J., and Gershon, S. (1968): Characteristics of drug abusers admitted to a psychiatric hospital. *JAMA*, 205:125–130.

Jorgensen, F. (1968): Abuse of psychotomimetics. *Acta. Psychiatr. Scand. Suppl.*, 203:205.

Kamajew (1931): Der Anaschismus (Russian).

Kaplan, H. B. (1980): *Deviant Behavior in Defense of Self*. Academic Press, New York.

Kaplan, H. S. (1971): Psychosis associated with marijuana. *NY State J. Med.*, 71:433–435.

Keeler, M. H., and Reifler, C. B. (1967): Grand mal convulsions subsequent to marihuana use. *Dis. Nerv. Syst.*, 28:474–475.

Keeler, M. H., Reifler, C. B., and Liptzin, M. B. (1968): Spontaneous recurrence of marihuana effect. *Am. J. Psychiatry*, 125:384–386.

Keup, W. (1969): Marihuana. *Science*, 163:1144.

Klee, G. D. (1969): Marihuana psychosis. A case study. *Psychoanal. Quart.*, 43:719–733.

Knight, F. (1976): Role of cannabis on psychiatric disturbance. *Ann. NY Acad. Sci.*, 282:64–71.

Kolansky, H., and Moore, W. T. (1971): Effects of marihuana on adolescents and young adults. *JAMA*, 216:486–492.

Kornhaber, A. (1971): Marihuana in an adolescent psychiatric outpateint population. *JAMA*, 215:1000.

Lambo, T. A. (1965): Medical and social problems of drug addiction in West Africa. *Bull. Narcotics*, 17:3–14.

Landau: Nasha intoxication. (1927): *Sovrem. Psychonevrol.*, No. 4.

Marcovitz, E., and Myers, H. J. (1944): The marihuana addict in the army. *War Med.*, 6:382–391.

Marten, G. W. (1969): Adverse reaction to the use of marijuana. *J. Tenn. Med. Assoc.*, 62:627–630.

McGlothin, W. H., and West, L. J. (1968): The marijuana problem: An overview. *Am. J. Psychiatry*, 125:370–378.

Medical Staff Pretoria Mental Hospital (1938): Mental symptoms associated with smoking of dagga. *S. Afr. Med. J.*, 12:85–88.

Mendhiratta, S. S., Wig, N. N., and Verma, S. K. (1978): Some psychological correlates of long-term heavy cannabis use. *Br. J. Psychiatry*, 132:482–486.

Milman, D. H. (1981): Effects on children and adolescents of mind altering drugs with special reference to cannabis. In: *Drug Abuse in the Modern World*, edited by G. G. Nahas and H. C. Frick, pp. 57–61. Pergamon Press, New York.

Milman, D. H. (1969): The role of marihuana in patterns of drug abuse by adolescents. *J. Pediatr.*, 74:283–290.

Miras, C. J. (1969): Experience with chronic hashish smokers. In: *Drugs and Youth*, edited by J. Wittenborn, H. Brill, J. P. Smith, and S. A. Wittenborn, pp. 191–198. Charles C Thomas, Springfield.

Moreau, J. J. (1845): *Du Hachisch et de l'Alienation Mentale*. Librarie de Fortin. Masson, Paris (English edition: Raven Press, New York, 1972).

Moreau, J. (1843): Recherches sur les alienes en Orient. *Ann. Med. Psychol.*, Paris.

Peebles, A. S. M., and Mann, H. W. (1914): Ganja as a cause of insanity and crime in Bengal. *Indian Med. Gaz.*, 49:395–396.

Perna, D. J. (1969): Psychotogenic effect of marihuana. *JAMA*, 209:1085–1086.

Robertson-Milne, C. J. (1906): Hemp and insanity. *Indian Med. Gaz.*, 41:129.

Rosenthal, F. (1971): *The Herb Hashish versus Medieval Muslin Society*. E. J. Brill, Leiden.

Segal, B. (1981): Hashish Use in Soviet Russia. In: *Drug Abuse in the Modern World*, edited by G. G. Nahas and H. C. Frick, p. 246. Pergamon Press, New York.

Siler, J. F., et al. (1933): Marihuana smoking in Panama. *Mil. Surg.*, 73:269–280.

Skliar, N., and Iwanow, A. (1932): Ueber den Anascha-Rausch. *Allg. Zeit. Psychiat*, 98:300–330.

Smith, D. E., and Mehl, C. (1970): An analysis of marihuana toxicity. In: *The New Social Drug*, edited by D. E. Smith. Prentice-Hall, Inc., Englewood Cliffs, New Jersey.

Spencer, D. J. (1970): Cannabis-induced psychosis. *Br. J. Addict.*, 65:369–372.

Stefanis, C., Boulougouris, J., and Liakos, A. (1977): Incidence of mental illness in hashish users and controls. In: *Hashish Studies of Long-Term Use*, edited by C. Stefanis, R. Dornbush, and M. Fink, pp. 49–53. Raven Press, New York.

Stringaris, M. G. (1939): *Die Haschischusucht*. Springer-Verlag, Berlin, reissued 1972.

Talbott, J. A., and Teague, J. W. (1969): Marihuana psychosis: Acute toxic psychosis associated with the use of cannabis derivatives. *JAMA*, 210:299–302.

Tennant, F. S., and Groesbeck, C. J. (1972): Psychiatric effects of hashish. *Arch. Gen. Psychiatry*, 27:133–136.

Thurlow, H. J. (1971): On drive state and cannabis. A clinical observation. *Can. Psychiatry Assoc. J.*, 16(2):181–182.

Toker, J. L. (1966): Mental illness in white and Bantu opulations of the Republic of South Africa. *Am. J. Psychiatry*, 123:53–65.

Tylden, E. (1967): A case for cannabis. *Br. Med. J.*, 2:556.

Varma, L. P. (1972): Cannabis psychosis. *Indian J. Psychiatry*, 14:241–255.

Voth, H. (1981): A psychiatrist's perspective on marijuana. In: *Proceedings: 1980–1981*, p. 33. National Parents Movement for Drug Free Youth. Georgia State University, Parent Resources and Information on Drug Education.

Warnoch, J. (1903): Insanity from hasheesh. *J. Ment. Sci.*, 49:96–110.

Weil, A. T. (1970): Adverse reactions to marihuana: Classification and suggested treatment. *N. Engl. J. Med.*, 282:997–1000.

Ludlow F. (1857)*The Hasheeschh Eater: Being passages from the Life of a Pythagorean.* Harper and Row, New Yok.

Taylor M. (1849), Flight from Reality. Duell, Sloan and Pearce, New York.

PROSPECTIVE STUDY OF 104 PSYCHIATRIC CASES ASSOCIATED WITH CANNABIS USE IN A MOROCCAN MEDICAL CENTER

Taieb Chkili and J.E. Ktiouet

Department of Psychiatry, Rabat-Salé Medical Center
Rabat, Morocco

ABSTRACT

104 male patients admitted to the psychiatric ward of the Rabat-Salé Medical Center had smoked cannabis before their hospitalization. They were mostly unemployed, between 20 and 30 years of age. They were all treated with phenothiazenes or butyrophenones. 44 patients presented a progressive and structured psychopathology, from schizophrenic-like to paranoiac psychosis, lasting from 14 days to 7 months. In these cases cannabis use might have triggered an underlying psychosis. 60 patients presented acute undifferentiated psychiatric symptoms, mostly "acute toxic psychosis" which lasted less than 2 weeks. This prospective study does not permit the definition of a specific "cannabis psychosis."

KEY WORDS

Cannabis, psychosis, schizophrenia, acute toxic psychosis

INTRODUCTION

The use of "kif", the local name for Cannabis Sativa is very ancient in Morocco, where it was brought by the Arab conquerors around the tenth century. Its consumption by the local population has been previously associated with admission of patients to the Berrechid Psychiatric Hospital for Psychiatric Disorders (Benabud 1957; Christozov 1965). The present study is based on the case studies of 104 patients admitted to the psychiatric unit of the Rabat-Sale Medical Center, and who had consumed cannabis before their psychiatric symptomatology developed.

PATIENT POPULATION

All patients are males 18 to 58 years old, mostly clustered between 21 and 30 years of age. Half of them are city dwellers (Rabat-Salé), and half of them come from the countryside. They belong to the poorer, destitute working class. The younger ones came from large families. The majority have little education beyond 6th grade, and a few have gone to college. Most of them are unemployed, or occupied at menial part-time tasks.

Most started consuming cannabis between 10 and 20 years of age, and were initiated into smoking by a group of their peers. Half of the patients cannot give any reason for starting smoking. Among reasons given by the other half of the sample are: a need to escape and forget family problems, a way of obtaining pleasure, well-being and euphoria, a remedy to anxiety, a way of dissipating loneliness.

Two-thirds of the sample smoked only cannabis on a daily basis, one third used other drugs, mostly alcohol or prescription drugs such as barbiturates. The duration of consumption for 3/4 of the sample did not exceed ten years.

RESULTS

All patients studied reported cannabis consumption before the appearance of their psychotic symptoms. The length of hospitalization is summarized in Table I. All patients were treated with phenothiazines or butyrophenones. In 67% of cases, hospitalization lasted less than a month. However, in 32% of cases, it lasted for 1 to 7 months in spite of the pharmacological treatment.

TABLE I

Length of hospitalization of 104 cannnabis users* admitted to the psychiatric ward of the Rabat-Salé MedicalCenter

14 days or less	14-28 days	1-3 months	3-7 months
44(42.3%)	26(25.0%)	26(25.0%0	8(7.7%)

*all of these patients were treated with phenothiazines or butyrophenones

The general symptoms presented by the patients fell into two distinct groups (TABLE II). This empirical classification was based upon a distinct difference in psychopathological manifestations and evolution between the two groups.

TABLE II

Symptoms displayed by 104 cannabis users hospitalized in the psychiatric ward of the Rabat-Salé Medical Center

1) Chronic Psychopathology
   44 patients prolonged psychotic reaction

   35 schizophrenic episodes
   2 paranoiac psychosis
   2 manic episodes
   5 depressive states

2) Acute Psychopathology
   60 patients acute psychiatric effects

   4 anxiety states with insomnia and headache
   20 aggressive and hyperactive behavior
   32 acute toxic psychosis with delirium
   4 confusion, disorientation

The first group of 44 patients, who all smoked cannabis for some length before presenting their psychiatic symptoms, present a progressive and "structured" psychopathology, from schizophrenic-like to paranoiac psychosis, displaying the classical progression described for these conditions. This has led us to believe that cannabis, as a toxic agent, has revealed a latent preexisting pathology. The drug appears to act as a precipitating factor, which triggers the underlying disease. But we do not believe that one can refer to a pathology which is specific to cannabis.

The second group of 60 patients is characterized by poorly defined psychiatric symptoms, except for the cases of "acute toxic psychosis" which closely follow episodes of cannabis consumption. These cases are indistinguishable from other acute toxic psychosis associated with ingestion of psychoactive drugs.

In conclusion, this prospective study does not permit us to define the existence of a "cannabis psychosis" presenting specific features.

REFERENCES

Benabud, A. Psychopathological aspects of the cannabis situation in Morocco: Statistical data for 1956. Bull.Narcotics, 9:1-16,1957.

Christozov, C. L'aspect marocain de l'intoxication cannabique d'apres des etudes sur des malades mentaux chroniques: 1ere partie et 2eme partie. Maroc.Med., 44:630-642, 1965.

EFFECTS OF CANNABIS ON SCHIZOPHRENIA.

Juan C. Negrete.

McGill University, Addictions Unit, Montreal General Hospital, Canada

ABSTRACT

A large number of schizophrenics abuse cannabis; many of them assert that the drug helps them cope with some symptoms and with unpleasant side-effects of medication. Others, however, report a worsening in their condition, particularly in symptoms of the "positive" type. Several studies confirmed these reports and elicited additional evidence of harmful effects. There appears to be agreement on the following findings: cannabis use is associated with an increased risk of developing schizophrenia; cannabis use precipitates a more sudden and earlier onset on the illness; cannabis use enhances the "positive" symptoms of schizophrenia (excessive dopaminergic activity, greater hallucinatory and delusional activity); cannabis worsen't the "negative" symptoms of schizophrenia (lethargy, autism, anhedonia, social withdrawal). These symptoms might result from cholinergic hyperactivity. These effects of cannabis in the symptoms of schizophrenia are likely to be due to an interactive of THC with central dopaminergic and cholinergic neurotransmission.

KEY WORDS

Cannabis; Schizophrenia; adverse effects; interactions cholinergic and dopaminergic neurotransmission.

EXTENT OF USE AMONG SCHIZOPHRENICS

Cannabis is still the most widely used illicit drug throughout the world, and in spite of the legal control measure in force, it continues to be readily available in the industrialized societies of the West. Recent surveys of use in the general population of North America yield lifetime prevalence rates of 33% in the USA (pop. aged 12+, 1988) and 23.2% in Canada (pop. aged 15 years +, 1989). The prevalence of active users (the year of the survey) is highest in the age bracket 18 to 25 years: 35.4% in the USA and 20.7% in Canada[1,2].

With the advent of modern psychiatric treatment approaches, most chronic psychotic patients no longer stay within institutionnal care but spend much of their lives in the community. Consequently, in recent years they have been more exposed to - and have had a greater opportunity to engage in - society's prevailing drug habits. This is certainly the case with ambulatory schizophrenics, who have been found to make use of cannabis in frequencies as high or higher than the population at large. The evidence from several North American surveys indicates that cannabis and alcohol are the psychoactive substances most frequently misused by schizophrenics, at least those registered with treatment programs in larger urban centres. Meuser et al.[3], for instance, administered structured interviews (SADS/SCID) to a sample such patients, and found that 42 % of them had abused cannabis some time in their lives, while a sizeable 22% could be defined as current abusers at the time of the inquiry. The age range sample was 18-56; younger schizophrenics have been found to present current abuse figures considerably higher ; particularly when the information is provide not by the subjects themselves, but by well informed case-managers[4].

Different explanations have been given to the apparent proclivity schizophrenics show towards the non-medical use of psychoactive substances ; some of them, of course, apply also to drug users in general, regardless of their psychiatric status. One such interpretation: drug use as an instrument of socialization, is clearly not exclusive to schizophrenics and may be reason many people engage in such habits. However, given the extremely limited social interaction abilities of many of the chronic patients, the act of seeking company to use and share drugs may well be one of the few collective activities they remain capable of. Drug use with peers must consequently be seen as a main pattern of social relation among the homeless schizophrenics, and in the low rental housing areas of the city, where these impoverished, alienated and rather withdrawn individuals tend to congregate.

Dixon et al.[5] studies patterns of cannabis use among schizophrenics and found that a significant number did so "usually" or "always" in the company of other people. On the other hand, one third of users asserted that they were "always" alone when smoking the drug. Thus, it seems that cannabism is a practice which appeals to both the socially withdrawn and the less isolated schizophrenic. The widespread availability of the drug, and its relatively low cost are features which may facilitate access even to the more exclusive patients; those who are unable or unwilling to join in cannabis smoking gatherings. By contrast, nearly all the schizophrenics who used cocaine reported doing it "with others". This less accessible habit, which was found to involve a smaller percentage of patients, is certainly more likely among schizophrenics who remain socially engaged; those who muster the higher levels of interactional skills and the psychic energy which are required to pursue it.

Another frequently mentioned reason for substance abuse among schizophrenics is the need that many of these patients experience to "self-medicate". That is, to alter the unpleasant subjective states caused by the illness itself, or by the undesirable effects of pharmacotherapy. Anhedonia, apthy, inner tension, anxiety, restlessness, insomnia, hallucinatory activity, are among the symptoms - mostly "negative" ones - or side effects schizophrenics are thought to seek to relieve through the use of drugs[6,7]. The actual effects of cannabis on such discomfort do not appear to have been clearly elucidated as yet.

Nonetheless, the Dixon et al.[5] survey did demonstrate that schizophrenics consistently agree in their description of what this drug does for them. Cannabis - the patient asserted - decreased their anxiety, lifted their "depression", rendered them "calmer", and gave them more "energy". On the other hand, those experiencing suspiciousness and lack of trust were adversely affected, as the drug served to enhance those symptoms. Such a combination of positive and negative effects suggests that - with respect to cannabis at least - the self-medication hypothesis is unlikely to explain the drug-seeking behaviour of all types of schizophrenics; or even that of a single individual throughout all stages of the illness. Of course, it is also possible that many schizophrenics take the drug in spite of the occurrence of some unwanted effects: for they also experience the beneficial ones, and are prepared to endure the former in order to enjoy the latter.

The survey data available at present does not adequately cover the schizophrenics' longitudinal history of cannabis use. It is not possible, for instance, to ascertain whether these patients go through changes in their attitude towards drug use over the course of their illness. However, in a sample we studies in Montreal, it was observed that the majority of patients who did start using cannabis in their youth, had already ceased the habit by the time they reached the age of 30 years (see Fig. 1). This is of course a tendency observed in the general population as well, but the rate of quitting among the schizophrenics is considerably higher (age corrected ratio of past to current users in the group of schizophrenics 7.2:1 vs 2.7:1 in the general population). Indeed, the comparative profile displayed in Fig. 1 indicates that while a significantly larger proportion of the schizophrenics were using the drug at a younger age, the rate of active users past the age of 30 years is quite similar to that observed in the population at large.

If the observations made on the Montreal sample were to apply to schizophrenics at large, then some general explanation must be found. The finding could be tentatively interpreted as follows: a) The schizophrenics, more than the general population, are forced to abandon the habit because of proverty and the loss of social connections. b) Cannabis use makes some unpleasant symptoms worse, (suspiciousness, paranoid feelings) and consequently exert a negative reinforcement effect. c) The progression of the illness causes the schizophrenic to become less and less responsive to the drug and leads to the extinction of the desire for it (no reward, no positive reinforcement).

EFFECTS ON THE PHENOMENOLOGY OF SCHIZOPHRENIA

Cannabis use has been linked with a number of occurences which, by and large, lead to conclude that this drug has an adverse effect on schizophrenia. Both the manifestations and the course of the illness appear to be negatively affected by such a practice.

Precipitating Factor

Data gathered through both clinical and epidemiological studies consistently suggest the probability that cannabis could act as a trigger, prompting the overt expression of schizophrenia in individuals who suffer the basic pathophysiological anomalies. In some such individuals - those observations imply - the disease may have remained limited to a "subclinical" derangement, with only minor behavioural manifestations, were it not for the enhancing action of the cannabinoids.

Figure 1:
RATES OF CANNABIS USE AMONG SCHIZOPHRENICS & IN THE GENERAL POPULATION (CANADA)[1];PER GROUP OF AGE

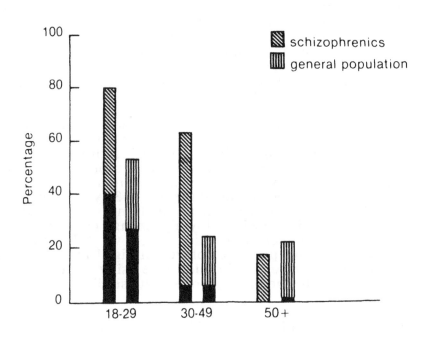

Columns represent percentages of ever users; black portion
indicates percentage of current users

1- adapted from Gallup Omnibus Study, June 1980

The evidence supporting this assumption derives from several separate and independent studies:

- a prospective epidemiological survey of a large cohort of men in Sweden (see paper by P. ALLEBECK in this volume), which demonstrated that individuals who are frequent users by age 20 years, present 6 times the risk of developing schizophrenia in the course of the subsequent 15 years, when compared with those who do not use cannabis at all.

- various studies which have consistently elicited a younger age of onset of schizophrenia in patients who are cannabis users when compared with those who are not[8,9].

- comparative clinical research which found that cannabis using schizophrenics tend to start the illness more suddenly - in the form of a clearcut psychotic outbreak - without the premorbid personality changes frequently observed in non-user counterparts[10,11].

Worsening of symptoms

A number of clinical reports have identified cannabis as a drug capable of enhancing symptoms and increasing the level of behavioural disturbance in schizophrenic patients. Cannabis use has also been blamed in instances of symptom recurrence in individuals whose illness had improved or even remitted. Rottamburg et al.[12], for example, observed higher levels of psychomotor agitation, and BERHARDSON & GUNNE[10] noted unusually aggressive and destrutive behaviour. Other authors[13,14] have published case reports with examples of relapse following renewed exposure to the drug.

In two separate surveys of large clinical samples[4,5] it was found that the patients themselves are aware of the negative effects of cannabis on paranoid symptoms. A majority of the respondents who suffer from paranoid schizophrenia report that cannabis makes them feel worse.

Our own study in Montreal, which involved a sample of 137 schizophrenics in ambulatory care[15], established that the subjects who were using cannabis during the six month observation period, did display a higher level of delusional and hallucinatory activity than those who were not. Not surprisingly, cannabis using schizophrenics were also the ones who necessitated more active care, as indicated by the higher average number of visits they made to the hospital during that period. Table I presents the results of a multiple regression analysis; it demonstrates that cannabis use, more than other relevant factors, explains the variance in symptom severity within the group studied. These data concern a subsample of younger schizophrenics only (age 30 years or less, n = 52), for it was that age group which proved to be most exposed to drug use.

The observations recorded to the present appear to indicate that cannabis effects the "positive" symptoms of schizophrenia rather specifically. The worsening of "deficit" problems such as lethargy autism, anhedonia and social withdrawal either does not occur or cannot be noted as readily. Current knowledge about the mechanisms whereby cannabis exerts its action in the brain is still limited. The existence of a cannabinoid receptor has recently reported[16]; it appears that a synthetic analog (CP55940) does in fact bind quite specifically in certain brain structures. This analog, however, was developed from a sectionof the cannabinoid molecule only; it is not certain at all that Tetrahydrocannabinol itself would behave in a similar way. Even if it did, there is no evidence that this particular receptor binding function is linked with the psychoactive effects of the drug, for the molecular component which led to the synthesis of CP55940 appears to have analgesic properties only. The adverse effects of cannabis on the symptoms of schizophrenia are more likely to be due to an interference in central neurotransmission[17].

TABLE I

Severity of clinical condition :
some factors accounting for the variance within
a group of schizophrenics aged 30 years or less (n = 52)

Factors of variance	Delusional thinking	Hallucinatory activity	Dose medication	Frequency visits
Age	NS	NS	NS	NS
Gender	NS	.04	NS	.04
Duration of treatment	NS	NS	NS	NS
Use other substances	NS	NS	NS	NS
Use of cannabis	.03	.02	NS	.05

Figures express the level of significance (PR>F) of the variance due to each factor on the list, over and above that which can be explained by the other four.
Adapted from NEGRETE, et al. (1986)

Present thinking on the pathophysiology of schizophrenia explains the development of "positive" symptoms on the basis of a state of excessive dopaminergic activity[17,18]. "Negative" symptoms, on the other hand, are believed to result from a dopaminergic deficit in the prefrontal cortex[18], or from cholinergic hyperactivity[19]. A balance between cholinergic and dopaminergic functions is of central importance in the pathophysiology of schizophrenia. These two neurotransmitter systems are known to exert a moderating effect on one another: a lowering of dopaminergic activity causes the cholinergic one to increase, and vice-versa. Positive symptoms could then be caused by a surge in Dopamine function; either spontaneous, or provoked by the effect of an exogenous dopamine agonist. They could also result from a decrease in central cholinergic activity. Both these mechanisms are possible in the case of cannabis, for this drug has been found to inhibit Acetylcoline turnover, increase Dopamine content in the corpus striatum and enhance brain reward responses, which are usually associated with an increased Dopamine afflux in the Limbic system[1,20]. Although the dopaminergic hyperactivity hypothesis provides a coherent pathophysiology model for both schizophrenia and some features of the cannabis intoxication[21], this is unlikely to be the only underlying mechanism. Indeed, several additional neurotransmitter alterations have been identified in these two conditions; it would appear that the Noradrenaline, Serotonine and Gammaaminobutyric acid systems are also affected [7,22], but their participation in the clinical expression of the problem has not to the present been clearly understood.

Interference with therapeutic response

Schizophrenics who abuse drugs tend to respond poorly to treatment; not only do they show an increased inability to benefit from psychosocial rehabilitation programs, but it would appear also that pharmacotherapy is less effective with them. Both the initial and the long term responses to therapy are adversely affected. In a well controlled study of young male schizophrenics, Bowers et al.[9] elicited a significant difference in symptom control during the first 10 days of drug therapy. Compared to the controls, patients with a history of drug abuse (cannabis and/or cocaine) did more poorly. The difference could not be explained on the bases of pharmacokinetic factors, for both groups presented similar plasma concentrations of the neuroleptic used to treat them. Instead, it was observed that drug abusing schizophrenics enter treatment presenting lower blood concentrations of the Dopamine metabolite Homovanillic acid, a finding that has been associated with poorer therapeutic response in other studies as well[23]. A drug abuse background is also associated with slower recovery on the long term. TURNER & TSUANG[24] reported a study on the outcome of schizophrenic, and noted that self-reported levels of drug use correlate positively with length of stay in hospital and with duration of pharmacotherapy. The heaviest abusers, these authors observed, were also the ones who required prescriptions for the longest period of time.

REFERENCES

1. **U.S. Department of Health and Human Services** - *Drug Abuse and Drug Abuse Research.* Third Triennial Report to Congress from the Secretary. DHHS Publication (ADM) 91-1704, Rockville, MD. 1991.

2. Health and Welfare Canada, *National Alcohol and Other Drugs Survey*: Highlights Report, Ministry of Supplies and Services, Ottawa, 1990.

3. **MUESER K.T., YARNOLD R.R., LEVINSON D.F. et al.** *Prevalence of substance abuse in schizophrenia: Demographic and clinical correlates*, schizophrenia Bulletin, 16, 31, 1990.

4. **TEST M.A., WALLISCH L.S., ALLNESS D.J. et al.** *Substance use in young adults with schizophrenic disorders.* Schizophrenia Bulletin, 15, 465, 1989.

5. **DIXON L., HAAS G., WEIDENP. et al.** - *Acute effects of drugs abuse in schizophrenic patients: clinical observations and patients' self-reports.* Schizophrenia Bulletin, 16, 69, 1990.

6. **SCHNEIER F.R, SIRIS S.G.** - *A review of substance use and abuse in schizophrenia* J. of nervous and mental diseasse, 175, 641, 1987.

7. **SIRIS S.G., kANE J.M., FRECKER K. et al.** - *Histories of substance abuse in patients with post-psychotic depressions.* Comprehensive psychiatry, 29, 550, 1988.

8. **TSUANG M.T., SIMPSON J.C., KRONFOL Z.** - *Subtypes of drug abuse with psychosis*, Archives of general psychiatry, 39, 141, 1982.

9. **BOWERS M.B., MAURE C.M., NELSON J.C. et al.** - *Psychotogenic drug use and neuroleptic response.* Schizophrenia Bulletin, 16, 81, 1990.

10. **BERNHARDSON G., GUNNE L.M.** - *Forty six cases of psychosis in cannabis abusers.* International journal of addiction, 7, 9, 1972.

11. **ANDREASSON S., ALLEBECK P., RYDBERG U.** - *Schizophrenia in users and non-users of cannabis.* Acta psychiatrica scandinavica, 79, 505, 1989.

12. **ROTTANBURG D., ROBINS A.H., BEN-ARIE O. et al.** - *Cannabis associated psychosis with hypomanic features*, Lancet, ii, 1364, 1982.

13. **TREFFERT D.A.** - *Marijuana use in schizophrenia: a clear hazar*, American Journal of Psychiatry, 135, 1213, 1978.

14. **DAVIDSON K., WILSON C.H.** - *Psychosis associated with cannabis smoking.* British Journal of Addictions, 67, 225, 1972.

15. **NEGRETE J.C., KNAPP P.W., DOUGLAS D.E. et al.** - *Cannabis affects the severity of schizophrenic symptoms: results of a clinical survey.* Psychological

Medecine, 16, 515, 1986.

16. JOHNSON M., DEVANE W., HOWLETT A. et al. - *Structural studies leading to the discovery of a cannabinoid binding site.* In Problems of drug dependence 1989, Harris L.S. (Ed.) NIDA Research Monograph Series n° 90, DHHS Publication, Rockville MD. 1990, p.p. 129-135.

17. ASHTON H. - *Actions of cannabis: Do they shed light on schizophrenia ? Biological aspects of schizophrenia and addictions,* Hemmings G. (Ed.) John Wiley & Sons, London 1982, p.p. 225-241.

18. DAVIS K.L., KAHN R.S., KO G. et al. - *Dopamine in schizophrenia: a review & reconceptualization,* American Journal of Psychiatry,, 148, 1974, 1991.

19. TANDON R., GREDEN J.F. - *Cholinergic hyperactivity and negative schizophrenic symptoms.* Archives of General Psychiatry, 46, 745, 1989.

20. GARDNER E.L. - *Marijuana's interaction with brain reward systems: update 1990.* Paper presented at "Marijuana's 90, an International Conference of Cannabis and Cannabinoids" - Kolympari, Chania, Greece, July 1990.

21. LIEBERMAN J.A., KINON B.J., LOEBEL A.D. - *Dopaminergic mechanisms in idiopatic and drug-induced psychosis,* Schizophrenia Bulletin, 16, 97, 1990.

22. VAN KAMMEN D.P. - *The biochemical basis of relapse and drug responses in schizophrenia: Review & hypothesis,* Psychological Medicine, 21, 881, 1991.

23. DAVIDSON M., KHAN R., KNOTT P., et al. - *Effects of neuroleptic treatment on symptoms of schizophrenia and plasma homovanillic acid concentrations,* Archives of General Psychiatry, 48, 910, 1991.

24. TURNER W.M., TSUANG M.T. - *Impact of substance abuse on the course and outcome of schizophrenia.* Schizophrenia Bulletin, 16, 87, 1990.

SCHIZOPHRENIA AND CANNABIS: CAUSE-EFFECT RELATIONSHIP?

Peter Allebeck.

Associate Professor, Department of Community Medicine, Karolinska Institute at Huddinge University Hospital, Sweden.

ABSTRACT

Whether or not cannabis consumption is associated with an increased risk of developing schizophrenia has since long been a matter of controversy. In this paper findings from a longitudinal study of 50,645 Swedish conscripts on the association between cannabis and schizophrenia are presented. The accumulated evidence for a causal association is analyzed using the criteria proposed by A.B. Hill. Baseline data from a questionnaire survey of all conscripts in 1969-70 were available on use of cannabis, alcohol and other drugs as well as social background factors. The cohort was followed in the national register for psychiatric care, where data on hospital admissions for schizophrenia were retrieved. The case histories were analyzed in detail for a subgroup of patients hospitalized in Stockholm County.

The relative risk of developing schizophrenia was 6.0 (95% confidence interval 4.0-8.9) among high consumers of cannabis (use on more than 50 occasions) compared to non-users. The association persisted after control for use of other drugs and social background factors. Scrutiny of medical records confirmed the diagnosis of schizophrenia in all cases and showed that use of cannabis preceded the development of schizophrenia, and not vice versa.

In conclusion, the accumulated evidence is in favor of a causal association between cannabis and schizophrenia. We believe that cannabis might trigger schizophrenia in vulnerable individuals. The interplay between genetic, personality and toxic factors in the causation of schizophrenia should be further elucidated.

KEY WORDS

Cannabis, drug abuse, schizophrenia, psychosis, personality.

It is well known that cannabis consumption is associated with a variety of psychopathological disturbances. The nature and characteristics of different supposedly cannabis associated psychopathological disorders are indeed complex. These disorders have not always been easy to define and classify according to standard psychiatric nosolgy. Thus, the classifications proposed by Negrete (1) have been valuable in providing a map of orientation in this complex field. Furthermore, the specific role of cannabis in the causation of psychiatric disturbances is difficult to assess: There is the problem of other factors that may confound the association, and there is the problem of the causal direction -whether the psychiatric disorder is an effect of cannabis abuse or rather an antecedent to substance abuse.

We have been particularly interested in trying to find out whether or not cannabis consumption might play an etiological role in the causation of schizophrenia. There is evidence from several studies that cannabis may be a risk factor for the development of schizophrenia. As pointed out by Thornicroft (2) there are, however, considerable methodological problems in the investigation of this association. The availability of some population based health registers in Sweden enabled us to address some of these methodological problems.

I will briefly summarize two studies which, in our opinion, give support to the hypothesis according to which cannabis may contribute to the development of schizophrenia.

LONGITUDINAL STUDY OF MILITARY CONSCRIPTS

Data was available on a cohort of 50 465 conscripts, examined in 1969-70 at age 18-19 years. All conscripts filled in a questionnaire on use of alcohol, tobacco and drugs, as well as family background, school and work conditions and other psychological and social factors. They were all seen by a psychologist for a structured interview, and those who presented mental disturbances were also examined by a psychiatrist. The cohort was followed in the national register of psychiatric care through 1983, in order to identify cases of schizophrenia in this cohort (3).

Table 1 shows a dose response relation between cannabis consumption at conscription and admission for schizophrenia during the follow-up. Given the association between cannabis consumption and different psychological and social factors, we allowed for these factors in a log linear model? As shown in table 2, there was still a significantly increased risk of schizophrenia after control for possible confounders, although at a lower level. In order to assess more in detail the cases identified in the registers, we examined 8 cases of schizophrenia for whom we had a documentation of exposure to cannabis. In all these cases, we confirmed that a regular cannabis consumption was present before the onset of schizophrenia. Furthermore, all cases fulfilled the DSM III criteria for schizophrenia. Thus, there was no evidence that these were cases of toxic psychosis or other psychopathology related to toxic effects of ongoing cannabis abuse. We also compared the cases of schizophrenia in which there was a history of cannabis abuse with 13 other cases of schizophrenia within the cohort without any record of previous cannabis consumption. We noted that the cannabis associated schizophrenics in general had a more rapid onset and more positive symptoms than other cases of schizophrenia, although all cases identified did fulfill the DSM-III criteria for schizophrenia (4).

Our findings have been questioned on grounds that partly have to do with the DSM-III criteria themselves, i.e. that our cases might not conform to a more rigorous definition of schizophrenia, we agree with the remark that it is possible that cannabis might contribute to the development of certain types of schizophrenia. It is generally admitted that the diagnostic term schizophrenia comprises a heterogeneous group of disorders. What is not known is to what extent different types of schizophrenia correspond to different etiological factors (6). This problem is an important area for further research.

In order to study the nature and the course of schizophrenia in persons with previous cannabis abuse and to address the problem of the causal direction, we performed another longitudinal study, based on a larger number of hospitalized cases.

LONGITUDINAL STUDY OF PATIENTS TREATED IN STOCKHOLM COUNTY.

Cases were selected from the Stockholm County in patient register. The register covers all persons admitted in the County of Stockholm, comprising about 1.5 million inhabitants. We selected all cases treated during the period 1971-83 with :
1) a diagnosis of cannabis abuse (ICD code 304,50 according to the Swedish version of ICD-8), and
2) a diagnosis of psychosis (ICD codes 292-299), independently of the time period between the two diagnoses, and independently of which diagnosis preceded the other. Thus, a diagnosis of psychosis could either precede or follow a diagnosis of cannabis abuse.

We identified 229 cases in this selection. Medical records of all cases were retrieved and scrutinized. We assessed the nature and the duration of the substance abuse, and we assessed a diagnosis of psychosis by the Research Diagnostic Criteria (7) as well as the DSM-III-R criteria.

The examination of medical records revealed that 112 cases were schizophrenics. These cases were defined using a strict application of DSM-III-R, in the sense that we required a documentation of at least six months' disease history without any use of drugs. 117 of the cases were other types of psychoses, of which 16% were classified as schizophreniform disorders, 19% as unspecific psychotic disorders and a few cases of manic and other disorders.

Table 3 shows the distribution of time lapse between onset of regular cannabis consumption and first psychotic symptoms. Of the 112 cases of schizophrenia, 12 appeared before cannabis consumption, and 12 in the same year. In the remaining cases, i.e. 69% of all cases, regular cannabis consumption preceded the onset of psychosis by at least one year.

Table 4 shows the main symptoms among the schizophrenics as well as the non-schizophrenics. The schizophrenics in general had open, positive symptoms, and often with rapid onset. The majority of schizophrenics had records of auditory, hallucinations and commenting voice. Many cases also had olfactory, gustatory and tactile hallucinations.

We are currently pursuing the analysis of these cases. We admit that with this design we could not assess the relative risk of schizophrenia among cannabis users compared to non-users. This would require either a classical cohort study, with base line assessment of cannabis abuse and follow-up of a well defined population, or a case control study, in which one of the main problems would be to assess cannabis exposure in retrospect. Nevertheless, our study has the advantage of being performed on a population based register comprising all cases of treated for cannabis abuse <u>and</u> psychosis in a defined population and in a defined period of time. The study showed on one hand that the majority of cases of psychoses among cannabis abusers indeed were cases of schizophrenia according to standard criteria and, on the other hand, that regular cannabis consumption in general preceded the onset of psychosis by several years.

In conclusion, we think that these studies provide evidence in favour of the hypothesis that cannabis consumption might play an etiological role in the development of schizophrenia. Cannabis is certainly not a sufficient cause of schizophrenia. But our results are compatible with the theory of vulnerability (8), according to which some individuals by a genetic influence or other predispositions, are vulnerable to schizophrenia, but that another factor, a trigger, is needed to start the schizophrenic process. In this sense, it is quite possible that cannabis might be one of several risk factors for schizophrenia.

REFERENCES

1. Negrete J.C., Psychiatric effects of cannabis use. In: <u>Cannabis and health hazards</u>. O'Brien Fehr K. and Kalant H., eds., Addiction Research Foundation, Toronto 1983, 577.

2. Thornicroft G. Cannabis and psychosis: Is there evidence for an association? <u>Br. J. Psychiatry</u>, 157, 25, 1990.

3. Andréasson S., Allebeck P., Engstrom A., Rydberg U. Cannabis and schizophrenia: A longitudinal study of Swedish conscripts. <u>Lancet</u>, 2, 1483, 1987.

4. Andréasson S., Allebeck P., Rydberg U. Schizophrenia in users and non users of cannabis: A longitudinal study in Stockholm County. <u>Acta Psychiatr. Scand.</u>, 79, 505, 1989.

5. Defer D. Intervention. In: <u>Physiopathology of Illicit Drugs</u>., Nahas G.G. and Latour C., eds., Pergamon Press, 1991, 13.

6. Andréasson N.C. The diagnosis of schizophrenia. <u>Schizophrenia Bulletin</u>, 13, 9, 1987.

7. Spitzer R.L. et al., <u>Research diagnostic criteria (RDC) for a selected group of functional disorders</u>, Washington, American Psychiatric Association, 19875.

8. Zubin J., Spring B., Vulnerability: A new view of schizophrenia. <u>J. Abnorm. Psychol.</u>, 86, 103, 1977.

CANNABIS AND SCHIZOPHRENIA.
HOW CAUSAL A RELATIONSHIP?

Bernard Defer

Centre Hospitalier Paul Guiraud, Villejuif, France

ABSTRACT

Cannabis, because of its destructuring effect on the state of consciousness, is susceptible to favor upheavals into authentic schizophrenia in personalities who presented a psychotic falt. However, French and Moroccan psychiatrists remain unconvinced that cannabis plays a specific role in the etiology of chronic psychotic schizophrenic process.

KEY WORDS

Cannabis, psychosis, schizophrenia.

S. Andreasson, P. Allebeck and their collaborators carried out a study of considerable dimension and great relevance by following 50.000 young swedes during 14 years.

They conclude that cannabis users present schizophrenic evolutions more frequently, especially if the intake of cannabis is high.

Methodologically speaking, they strove to isolate various parameters and strove not to confuse cause and effects, knowing full well that vulnerable adolescent personalities will easily resort to cannabis when feeling the anxiety of psychotic wavering.

The question of the relationship between cannabis and schizophrenia has been asked many a time ever since clinicians have been observing cannabis induced psychopathological patterns.

In the 19th Century, along with descriptions of acute intoxication and despite their rather elaborate character, J. Moreau de Tours proceeded to create a remarkable clinical study of cannabis effects, experimenting on himself.

Later, during their travels or practice in countries where cannabis was of traditional usage, european psychiatrists took interest in the toxic effects of cannabis and evoked the possibility of schizophrenic evolutions because of the dissociative or discordant aspects of certain clinical patterns.

As far as we are concerned from 1958 to 1965 in Morocco, which was a real laboratory due to the chronic sometimes massive use of cannabis and often exclusive of other drugs, and where we performed experimental intoxications, we analysed many cases of acute cannabis intoxication in Moroccan and sometimes European users, 600 cases of acute psychosis and many cases of chronic intoxication.

Acute cannabis intoxication in its minor form is determined by an isolated intake of more or less important quantity and is expressed by a jubilating and blissful euphoria sought out by users finding thereby satisfaction and carries with it hypomanic elements.

With a more important dose, and more sensitive subjects, the destructuration of consciousness is deeper and creates a fleeting dreamlike state along with disturbances in time/space sense, hyperesthesia, synesthesia, illusions, fantastic hallucinations, depersonalization and irruption of the imaginary giving to the subject the impression of participating in some kind of fascinating performance but towards which he believes to have kept a deceptive detachment.

There are also manic forms, or rarely anxio-depressive, confusion inducing, stupor inducing and lastly discordant forms, and the subject, more often than not remembers the toxic experience.

If the use of cannabis is habitual and prolonged an important dose is susceptible to provoke an acute psychotic pattern the evolution of which is more long lasting.

We may then observe the most numerous dreamlike forms, enriched with experiences of delirium and hallucinations, and in order of frequency : manic forms, excited and confused forms, discordant, confused, anxio-depressed, stupor-like and confused dream-like forms, and so it is important to remember the instability and polymorphism of these manifestations.

More than half of these acute episodes that last a few weeks, the experience of which remains usually amnesic, contain dissociative and discordant elements confirmed by objective tests.

These characteristics have been observed by many authors some of which had offered the notion of dissociative state, schizophrenia or precocious cannabis induced dementia.

In fact, the destructuralisation of consciousness remains relatively superficial and transitory, without real lasting autistic subversion of the personality.

Beyond that, massive and prolonged cannabis intoxication is susceptible to induce evolutions of a morbid deficient type emphasized by acute and subacute episodes, infiltrated by post dream-like residues and matched with emotional dulling, existential disinvestment, withdrawal if not intellectual dementia, the discordant aspects of which take on a pseudo-paranoid aspect.

These observations have not allowed us to retain the notion of cannabis induced schizophrenia, in the sense the French school defines the psychotic schizophrenic procedure.

In their work, our Swedish colleagues mention that they have based their diagnosis of schizophrenia in a restrictive manner and by reference to the DSM3-R criteria.

These seem to be less rigorous than those demanded by the French tradition, particularly concerning the evolution. The cannabis consumption of the young Swedes, inadequately specified, is most likely inferior to that of the moroccan users, among whom authentic schizophrenic psychosis is no more frequent than in other populations according to our observations and those of moroccan psychiatrists (Prof D. Moussaoui, Dr M. Bouzekraoui).

It is important to note that the swedish researchers have recognized specific clinical aspects in the schizophrenic patterns they have observed in cannabis users: these aspects confirm past conclusions and one may still doubt these observations correspond to real chronic dissociative process.

We cannot deny that due to its destructuring effect on the state of consciousness, cannabis is susceptible to favor upheavals wherefrom begins the entrance into authentic schizophrenia in personalities who previously presented a psychotic fault and tried compensating their anxiety by the euphoric effects of cannabis.

In our eyes, this important study, rich with information, does not enable us to attribute to cannabis a causal and specific role in the etiology of chronic psychotic schizophrenic processes.

CANNABIS-ASSOCIATED DEATHS IN MEDICO-LEGAL POSTMORTEM STUDIES. PRELIMINARY REPORT.

Jovan Rajs[1], Anna Fugelstad[2] and John Jonsson[3]

[1]Department of Forensic Medicine, Karolinska Institute, Stockholm, [2]Department of Psychiatry, St Göran's Hospital, Stockholm and [3]National Laboratory of Forensic Chemistry, Linköping, Sweden.

ABSTRACT

The aim of this study was to find out whether or not cannabis can be associated with the cause or manner of death in users. During a four year period (1987-1990), 13 417 medico-legal autopsies were made, 26% of which after violent deaths. The autopsies were supplemented with toxicological analyses with regard to presence of alcohol, medicinal and narcotic drugs as well as of tetrahydrocannabinol (THC), or its compounds as indicated by past history and police records. In 24 deaths, THC and/or cannabinoids were found, as a single drug in 8 cases, and along with alcohol, alcohol and medicinal drugs, and only medicinal drugs in 10, 5, and 1 cases, respectively. These 24 deaths among cannabis users were compared to random samples of deaths during the same period where no THC, but alcohol, amphetamine or opiates, alone or in combination, was found. Of the 24 cannabis-associated deaths, only one was due to a non-violent cause (alcohol-associated liver-cirrhosis). Ten deaths resulted from suicides, 8 from accidents and 5 were homicides. Of the 10 suicides, 4 were committed by jumping from a high place. Two suicides were preceded by the killing of other people. In the alcohol-, amphetamine-, and opiate-associated groups, suicides occurred at 4, 2 and 3 instances, respectively; no one was committed by jumping from a high place or preceded by homicide. Deaths due to accidents or homicide were approximately as frequent in the alcohol and amphetamine abuser groups as in the cannabis group, while the opiate group was dominated by fatal collapse upon intravenous injection of heroin (20 out of a total of 24 cases). All 5 cannabis-associated homicides occurred in connection with non-premeditated impulsive outbursts. In 3 of these, the perpetrator was also a cannabis user.

KEY WORDS

Cannabis; Alcohol; Amphetamine; Heroin; Cause and manner of death.

INTRODUCTION

At the Department of Forensic Medicine in Stockholm various
studies on drug-related mortality are carried out. One important object is to create a register of deaths related to
illicit drug use in the Stockholm area. This register includes
users of heavy drugs chiefly amphetamine and heroin.

However, in recent years a number of cases were found with
tetrahydrocannabinol but no other illicit drugs, in blood or
cannabinoids in urine.

At closer investigation, it appeared that manners of death
among the cannabis users were unexpectedly violent compared to
the ways of death among other drug addicts. The proportion of
suicides, 10 out of 24, was particularly high. This finding was
remarkable, since cannabis use is not generally associated with
violence.

The literature contains several descriptions of delusional
states, paranoid symptoms and personality disturbances in
cannabis users[1]. These abnormal mental states are of
comparatively short duration, usually lasting for just a few
hours. Even first-time users may be affected.

These so called cannabis psychoses are not very frequent and
are often induced by long periods of intensive cannabis use. It
is uncertain whether or not these psychoses can be separated
from shizophrenia or other known paranoid states[3]. The entity
distinguishing cannabis psychosis from other mental disorders
is the notably subsiding the symptoms already after some days
or a few weeks. In most cases, the patient recovers totally but
remains at risk for relapse in connection with repeated
cannabis use.

In case studies of cannabis psychoses, it is not uncommon to
see descriptions of violent and destructive outbursts, suicide
attempts and assaults on other persons, often with major
violence.

In a Swedish follow-up study of conscripts with a history of
cannabis use, an excess mortality was noted 15 years after
conscription. The excess mortality from suicides was
particularly high, and the share of suicides increased in
proportion to the magnitude of the cannabis abuse[4].

PATIENTS AND METHODS

The study was based on material collected at the Government
Institute for Forensic Medicine in Stockholm during a four year
period (1987 to 1990). According to Swedish law, a medico-legal
investigation must be made on deceased persons who have died
outside a hospital as a result of external violence, poisoning
or otherwise suddenly, without previously known fatal disease,
or under unclear or suspicious circumstances.

During these four years, a total of 13 417 medico-legal autopsies were made on males and females of all ages (about 26% after violent deaths). A total of 1.9 million people resided in the service area in 1987; of these, about 1.6 million lived in Stockholm county and 668,810 in the city of Stockholm. Information about deceased persons was obtained from police reports and was, when possible, supplemented with information from clinical journals, families and friends as well as from social workers.

Complete autopsies were carried out, with few exceptions, when the blood tests analyses concerning HIV-infection were positive[3]. The autopsies were supplemented with histological investigations. Toxicological analyses were made, with the purpose of disclosing the presence of alcohol, barbiturates, tranquilizers, opiates, central stimulants and cannabis, as suggested by autopsy findings, past history, police records, or circumstances of death. Toxicological analyses were also carried out when insufficient information about the circumstances of death was at hand, for instance in cases of violent death, or when the postmortem findings were inconclusive, and also when information about unusual behaviour or symptoms was available.

Analytical methods

Ethanol was determined in blood and urine by head-space gas chromatography according to Machata et al 1975 [6], using two glass columns (2 mm x 1,5 m) packed with 0.2 per cent CW Carbopack C and 5 per cent CW Carbopack B (Supelco). Spectrophotometry was used to measure the carbon monoxide concentration in the blood[7]. Drugs in the blood were extracted with butylacetate at pH 7 and 9, and the content determined by gas chromatography[8]. Tetrahydrocannabinol (THC) was extracted from the blood with pentane and the concentration determined by mass spectrometry, essentially according to Christophersen (1986)[9]. Analyses with regard to presence of cannabinoids in urine were introduced as a routine method at this laboratory in 1987, and blood analysis for THC in 1988. In 1991, a total of 477 analyses were made (mostly in persons below the age of 40-50 years). In 113 instances, cannabinoids were found in the urine; in 63 of these also THC in the blood. Mass spectrometry was also used for the determination of amphetamine (as its trifluoroacetyl-derivative) and morphine[10]. An immunological assay, EMIT, (SYVA) was used for the screening of narcotics in urine.

Of all the investigated medico-legal autopsy cases, the following were included in the present series:

I: Instances with findings of THC in the blood and/or cannabinoids in the urine, either solely or in combination with alcohol and/or medicinal drugs, comprising a total of 24 deaths. These cases, referred to as <u>cannabis users</u>, were compared to those in the following control groups:

II: Instances with or without a history of chronic alcoholism,
       with findings of alcohol in the blood - solely or in
       combination with medicinal drugs - but with no THC,
       cannabinoids, centralstimulants or opiates in any post-
       mortem sample, in all 231 cases (by random selection
       reduced to 24 cases). This group will be referred to as
       alcohol users in the following.

III: Instances with findings of amphetamine in the blood,
       solely or in combination with alcohol and/or medicinal
       drugs, but exclusive of cocaine, opiates or, THC with
       it's compounds in any postmortem sample, in all 27 deaths
       (by random selection reduced to 24 cases). This group
       will be referred to as amphetaminists, hereinforth.

IV: Instances with findings of opiates in the blood, solely
       or in combination with alcohol and/or medicinal drugs, but
       exclusive of alcohol, central stimulants, or THC with
       it's compounds in any postmortem sample, in all 85 cases
       (randomly reduced to 24 cases). This group will be refer-
       red to as heroinists in the following.

Since all the 24 cannabis users were HIV-negative, HIV-positive
cases in the control groups were removed before random selection.

During the entire study period, the number of deaths with
findings of cocaine and LSD was less than 24; due to the
limited number of cases, no control groups consisting of
cocaine and LSD users could be established. Cases in which more
than one narcotic drug was detected, or a narcotic drug was
detected in addition to THC, were not included in this
preliminary study.

The autopsies were made by different forensic pathologists, and
there were no strict rules for indications for toxicological
examinations. Consequently, in some instances, the autopsist
was satisfied with information about cannabis use and did not
demand toxicological confirmation of the drug. Such cases were
not included in this study. For the same reason, the manner of
death, i.e. whether it was of accidental, homicidal, suicidal
or undetermined origin, was reascertained by the authors
according to the criteria described before[11].

RESULTS

During the four-year period of 1987-1990, cannabis was found to
be the only narcotic drug in postmortem samples in 24 cases; in
8 of these, cannabis was the only finding, while alcohol,
alcohol and medicinal drugs, or only medicinal drugs was
demonstrated together with cannabis in 10, 5 and 1 instance,
respectively. The number of deaths among these 24 cannabis
users during the study period is shown in Figure 1. There were
23 males and 1 female, 20-43 years of age (mean age 29.6
years). About two thirds (15) of the persons in this series
were, at the time of death, under the influence of alcohol.

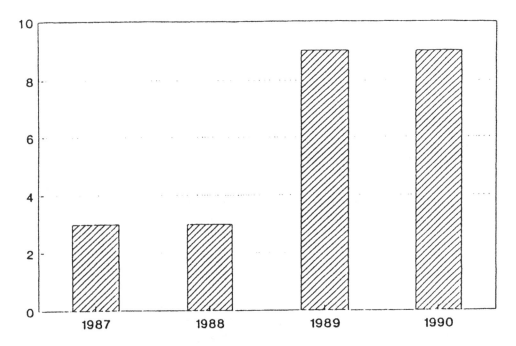

Fig. 1. Numbers of cannabis user deaths in Stockholm
1987 - 1990.

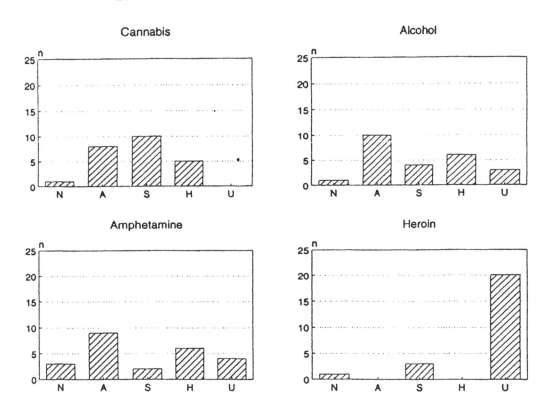

Fig. 2. Manners of death among 24 cannabis users in comparison to the same number randomly selected alcohol
users', amphetaminists' and heroinists' deaths.
N = Natural deaths. A = Accidental deaths. S = Suicides. H = Homicides. U = Undetermined deaths.

Among the 24 cannabis users, only one died from a non-violent
cause, while the remaining 23 (96%) died as a result of
violence, compared to a total of 26% of the entire number of
13 417 investigated postmortems during the same period (Table
I). No stigmata which could be indicative of cannabis use were
noted at the postmortems. Non of the cannabis users had blood
tests that were positive for HIV.

The manners of death among 24 cannabis users were compared with
those in the same number of users of alcohol, amphetamine and
heroin (Figure 2); a similar distribution for alcohol users and
amphetaminists was noted. There were obvious differences
between cannabis users and heroinists, with suicides dominating
for cannabis users, while deaths of undetermined origin,
following collapse in connection with intravenous drug
administration, predominated among heroinists.

Non-violent deaths

One single cannabis user out of the total of 24 died of a non-
violent cause, i.e. from alcohol associated liver cirrhosis
with a consecutive hemorrhage from esophageal varicosities.
Corresponding figures for alcohol users, amphetaminists and
heroinists were 1, 3, and 1, respectively. The alcohol user
died of alcoholic cardiomyopathy, 2 amphetaminists died of
chronic, probably amphetamine-related, cardiomyopathy, and one
44-year old amphetaminist of a ruptured aortic aneurysm, while
the heroinist, who also abused alcohol, died in an epileptic
seizure. Thus, all the six "natural" deaths in this series,
consisting of 96 alcohol-, cannabis- and heavy drugs-influenced
deaths, resulted from organ lesions caused by chronic abuse of
alcohol and/or amphetamine.

Accidental deaths

Eight of the 24 cannabis users died in accidents, 7 males and
one female, ranging in age from 22 to 43 years (mean 32.5
years). Two users died of alcohol intoxication. At the time of
death, the concentrations of THC in the blood were only 0.8 and
2.0 ng/g. One male with 0.5 ng THC/g blood and 2.7 g ethanol/l
blood died from drowning after suddenly loosing his balance and
falling into the water. Five cannabis users died in traffic
accidents, 3 as drivers of motor vehicles, and 2 as passengers
with cannabis-influenced drivers. One of the deceased car
drivers had 30 ng THC/g blood (and no other drug), while the
remaining 4 traffic accident victims had 1-4 ng THC/g blood in
combination with alcohol (0.3, 0.9, 2.8 and 3.2 g/l), in two
cases also with low concentrations of chlormezanon and parace-
tamol. In one of the automobile accidents, high speed (3 times
exceeding the speed limit through central Stockholm) preceded
collision with another car; in an another, the cannabis-
influenced driver passed another car, at twice the speed limit,
and just continued to drive on the left side of the road until
he collided with a car coming from the opposite direction. In
the two others, the drivers were not able to keep their cars on
the road in a minor curve; and in yet another, the car was
simply driven into a ditch.

TABLE I.

CAUSE AND MANNER OF DEATH AMONG 25 CANNABIS-ASSOCIATED DEATHS
DURING THE 4-YEAR PERIOD 1987 - 1990.

Cause and manner of death	Total n=24
Non-violent deaths, total	1
Complications of alcohol abuse (303)	*1*
Violent deaths, total	23
Accidental deaths, total	8
Traffic accidents (E 819)	*5*
Alcohol intoxication (E 859)	*2*
Drowning (E 910)	*1*
Suicide, total	10
Tablet intoxication (E 950)	*3*
Carbon monoxide intoxication (E 951, E 952)	*2*
Gunshot suicide (E 955)	*1*
Jumping from high place (E 957)	*4*
Homicide, total	5
Stabbing (E 966)	*5*

In Table II, accidental deaths among cannabis users have been compared with those in control groups. There were approximately as many deaths caused by traffic accidents among alcohol users as in the amphetaminist group. Deaths following a drug-free interval after an accidental overdose of amphetamine, with consecutive myocardial lesions or cerebral hemorrhage, were common among amphetaminists (4 out of 10 accidental deaths), while carbon monoxide intoxication in connection with fire, caused by smoking in bed, occurred only in the alcohol-associated control group. No heroinist death has as so far been classified as accidental (see below).

Suicides

Ten males, 22-37 years of age (mean age 29 years) and no female out of the 24 cannabis users committed suicide - the dominating manner of death in this group; comparable figures were 4, 2 and 3 suicides among alcohol users, amphetaminists and heroinists, respectively. Method of suicide, related to the respective drug, is shown in Table III.

It is seen that suicides among cannabis users were more frequent than in all three control groups taken together. Jumping from a high place was resorted to 4 of the cannabis-associated suiciders, but not by any of the controls. In 3 of these 4 cases, information about mental depression and suicidal intention was obtained from the police report or from the medical history; in 1 case, no such information was at hand. In 5 of the remaining 6 suicides, there was verbal information about suicidal intent, or a suicidal note. However, the suicidal act came suddenly and unexpectedly in most of the cases. Another notable difference between cannabis user suicides and controls was homicidal actions preceding suicide among cannabis users. In one case, a cannabis user suspected that one of his four children had been subject to incestuous behaviour by a family member. This prompted him to kill his wife, all four children and himself by means of household gas. In another case, the desperate self-murderer shot an unknown person, trying to prevent him from committing suicide and, immediately thereafter, shot himself. Suicide was never preceded by homicide in control groups.

Known motives for suicide were the following:
(a) Dissatisfaction with the medical treatment; (b) being deserted by his girlfriend; (c) being drafted for military service. In one case, a young man in the midst of discussing mortal fear with friends, suddenly left the room and jumped from the balcony.

Among the 10 cannabis user suicides, THC was found in the blood or cannabinoids in the urine in 5 cases, in amounts between 1 and 3 ng/g blood in 5 cases, together with ethanol in 5 cases, and with medicinal drugs in 5 cases.

TABLE II

ACCIDENTAL DEATHS IN RELATION TO MAIN DRUG USED.

| Main drug | Intoxication | | | Drowning | Fall | Traffic accidents |
	Alcohol	Amphetamine	Carbon monoxide			
Cannabis n=8	2	0	0	1	0	5
Alcohol n=10	1	0	2	0	2	5
Amphetamine n=9	0	4	0	0	1	4
Heroin n=0	0	0	0	0	0	0
Total	3	4	2	1	3	14

TABLE III

SUICIDES. TYPE OF METHOD IN RELATION TO MAIN DRUG USED.

| Main drug | Methods of suicide | | | | | |
	Intoxication	Overruning by train	Jumping from high place	Drowning	Hanging	Gunshooting
Cannabis n=10	5	o	4	0	0	1
Alcohol n=4	0	1	0	1	1	1
Amphetamine n=2	2	0	0	0	0	0
Heroin n=3	1	1	0	0	1	0
Total n=19	8	2	4	1	2	2

Homicides

Five out of 24 cannabis users were murdered, 4 males and 1
female, ranging in age from 22 to 37 years (mean age 26.6
years). Corresponding figures were 6, 6 and 0 cases in groups
of alcohol users, amphetaminists and heroinists respectively. A
knife was the most frequent murder weapon in all groups, a
firearm was used only in one instance in the alcohol- and
amphetaminists groups, while lethal blunt injuries were noted
in 2 alcohol user homicides.

Cannabinoids were detected, but only qualitatively, in the
urine of 5 homicide victims.
Two cannabis users were murdered by their respective fiancé, 1
was beaten and stabbed to death on the street by 3 for him
unknown cannabis user, 1 was stabbed to death by a drinking
companion and 1 murder was not solved. Two murderers who were
the respective fiancés of their victims, were cannabis users as
well. The settings were similar: Mutual emotional relationship
between the future victim and the perpetrator, a sudden
stabbing to death preceeded by a short and trivial quarrel. In
one of these cases, the murderer was shocked and regretful; in
the other, the murderer was, upon arrival of the police,
confused and disoriented. Three out of 5 cannabis users were
killed by other cannabis users. Four killings occurred in
connection with non-premediated, impulsive outbursts.

Fights between alcohol-influenced persons, using fists, blunt
weapons or knives was the characteristic pattern for the
alcohol-user control group. Murders of amphetaminists – usually
perpetrated by another amphetaminist – were the most violent in
this series, manifested by infliction of numerous injuries,
suggestive of rage and ruthlessness in the perpetrator. No
homicides of heroinists were observed in this series.

Undetermined deaths

There was sufficient information about all cannabis user
deaths, and no case was referred to the group of undeter-mined
deaths. In the alcohol user and amphetaminist groups, there
were 3 and 4 undetermined deaths, respectively, the
differential diagnosis being death by accident or suicide.
Totally, 20 of the 24 heroinist deaths occurred following
collapse upon intravenous injection of heroin. Although in no
case homicide was reasonably suspected, or confirmed, and in
none of the cases could a suicidal origin be ascertained, these
20 deaths were all regarded to be of undertermined origin,
(E980), despite the notion that the majority probably were
accidental.

DISCUSSION

The methods for detection and quantitative analysis of THC in
the body fluids have been improved in recent years. However,
the methods are still rather complicated and expensive and

subsequent screening for cannabis has not been routinely made in cases of sudden violent death.

It appears that screening for illicit drugs is more common at the Department of Forensic Medicine at Karolinska Institute in Stockholm than in a number of other forensic departments. The autopsy frequency in the Stockholm department is also rather high. A recent study [4] of hospitalized drug addicts in Stockholm showed that 90% of the deceased addicts that were autopsied were examinationed post mortally at the department of forensic medicine. This may be the reason why there are so few other studies on cannabis-related mortality.

During the course of the study, information has appeared that indicates a connection between cannabis use and some of the most spectacular murder cases in other parts of remaining Sweden. In these cases, the perpretators were cannabis users. However, our knowledge is incomplete regarding the magnitude of the abuse and the actual relation between ingestion of the drug and the crimes in question.

Use of cannabis may have a stronger association to violent crimes than what is presently acknowledged. One possible explanation of these violent and impulsive misdeeds, previously described, could be the confusional states or cannabis psychosis in users giving rise to rapid fluctuations in mood, panic attacks and paranoid delusions. This seems to make the cannabis-related deaths different from deaths related to influence of alcohol, amphetamine and heroin. Another difference is the high incidence of suicides among cannabis users, though occuring unexpectedly, these acts were in most instances preceeded by a history of mental depression and explicit suicidal intent.

The present study represents a pilot project in a limited research area. However, hopefully, it will lead to extended studies in other parts of Sweden and elsewhere.

We also hope that this study may contribute to a new practise among the police and in the judicial system, including a comprehensive screening for presence of THC in both perpetrators and victims of violent crimes and acts of violence where cannabis use cannot be excluded.

REFERENCES

1. Pålsson, Å., Thulin, S. O., Tunving, K., Cannabis psychoses in south Sweden, <u>Acta Psychiatrica Scandinavica</u>, 66, 311, 1982.

2. Thornicroft, G., Cannabis and Psychosis. Is there Epidemiological Evidence for an Association?, <u>British Journal of Psychiatry</u>, 157, 25, 1990.

3. Andréasson, S., Allebeck, P., Engström, A., Rydberg, U., Cannabis and Schizophrenia. A longitudinal Study of Swedish Conscripts, <u>The Lancet</u>, 26, 1483, 1987.

4. Andréasson, S., Allebeck, P., Cannabis and mortality among young men: A longitudinal study of Swedish Conscripts, <u>Scand J Soc Med</u>, 18, 9, 1990.

5. Rajs, J., Karlsson, T., Eklund, B., Bergendahl, K., HIV-related deaths outside medical institutions in Stockholm, <u>Forensic Sci Int</u>, 41, 269, 1989.

6. Machata, G., The advantage of automated blood alcohol determination by head space analysis, <u>Z Rechtsmedizin</u>, 75, 229, 1975.

7. Maehly, A. C., Analyse von Kohlenoxyovergiftungen, <u>Deut Z ges gerichl Med</u>, 52, 369, 1962.

8. Eklund, A., Jonsson, J., Schubert, J., A procedure for simultaneous screening and quantification of basic drugs in liver, utilizing capillary gas chromatography and nitrogen sensitive detection, <u>J Anal Toxicol</u>, 7, 24, 1983.

9. Christophersen, A. S., Tetrahydrocannabinol stability in whole blood: plastic versus glass containers, <u>J Anal Toxicol</u>, 10, 129, 1986.

10. Schubert, J., Schubert, J., Gas chromatographic-mass spectrometric determination of morphine, codeine and 6-monoacetylmorphine in blood extracted by solid phase, <u>J Chromatogr</u>, 490, 444, 1989.

11. Rajs, J., Fugelstad, A., Suicide related to human immunodeficiency virus infection in Stockholm, <u>Acta Psychiatr Scand</u>, 85, 234, 1992.

12. Annell, A., Fugelstad, A., Ågren, G., Mortality in relation to HIV-infection and type of drug abuse among drug addicts in Stockholm 1981-1988, <u>Submitted</u>, 1992.

5. Effects on immune function and carcinogenesis

DELTA-9-TETRAHYDROCANNABINOL SUPPRESSES MACROPHAGE EXTRINSIC ANTI-HERPESVIRUS ACTIVITY *

G. A. Cabral and R. Vasquez

Department of Microbiology and Immunology,
Medical College of Virginia/VCU
Richmond, Virginia, 23298, U.S.A.

ABSTRACT

The effect of Delta-9-tetrahydrocannabinol (THC), the major psychoactive component of marijuana, on macrophage intrinsic and extrinsic anti-herpesvirus activities was examined. THC had no effect on the capacity of the macrophage-like cells RAW264.7, J774A.1, and P388D$_1$ to take up virus. In addition, replication of virus within macrophages did not occur regardless of drug treatment indicating that THC had no effect on macrophage intrinsic antiviral activity. In contrast, the cannabinoid exerted a dose-dependent inhibition of macrophage extrinsic anti-viral activity. This activity describes that macrophage function by which these cells suppress virus replication within xenogeneic cells in an interferon-independent manner. The inhibitory effect of THC on extrinsic antiviral activity was greatest on RAW264.7 and J774A.1 cells followed by P388D$_1$ cells. These macrophage-like cells regained their extrinsic antiviral activity in a time-related fashion following removal of the drug.

These results indicate that THC inhibits macrophage extrinsic anti-herpesvirus activity but has no effect on intrinsic antiviral activity. However, the suppressive effect of THC on extrinsic antiviral activity is reversible upon removal of the drug.

* Reprinted with permission from Proceedings Experimental Biology and Medicine, 1992, 192:205-263.

KEYWORDS

Marijuana; delta-9-tetrahydrocannabinol; THC; macrophages; extrinsic antiviral activity; herpesviruses

INTRODUCTION

Delta-9-tetrahydrocannabinol (THC), the major psychoactive component of marijuana, has been shown to elicit a variety of immunosuppressive effects in vivo and in vitro. These include dysfunction in lymphocyte response to mitogens and particulate antigens (1,2), decrease in T-cell rosette formation (3,4), inhibition of natural killer cell activity (5,6), suppression of leukocyte migration (7), and perturbation of macrophage morphology, function, and motility (8-10). THC, in addition, has been shown to decrease host resistance to bacterial and virus infections. Morahan et al.(11) demonstrated that mice treated with various cannabinoids experienced a dose-related decrease in resistance to both Listeria monocytogenes and herpes simplex virus type 2 (HSV2). We have demonstrated that exposure of mice and guinea pigs to THC exacerbates primary HSV2 genital infection (12).

Macrophages play a critical role in host immunity, especially at the primary site of infection. These cells are actively involved in resistance to intracellular infection (13), act as accessory cells in the immune response in their capacity as antigen-presenting cells and as producers of immunomodulatory signals (14,15), and function as cytotoxic effectors against tumor cells (16) and virus-infected cells (17). Macrophages also exert intrinsic and extrinsic antiviral activities. Intrinsic activity is the process by which macrophages ingest virus, degrade virus, and thereby are nonpermissive for productive virus infection (18-21). This degradation of virus by macrophages facilitates clearance of virus from the local site of infection. Extrinsic antiviral activity is the process by which macrophages elicit an interferon-independent, cell contact-dependent reduction of virus growth in infected, virus-susceptible cells (22). Both intrinsic and extrinsic antiviral activities play an important role in the control of localized virus infection and in limiting the spread of virus within the infected host.

Drug-induced alteration of macrophage antiviral activities could have a major effect on host resistance to venereally-transmitted pathogens, particularly at the primary site of exposure. Thus, the objective of this study was to define the effect of THC on macrophage intrinsic and extrinsic antiviral activities. Results of this investigation demonstrate that THC has no effect on intrinsic anti-HSV2 activity but suppresses macrophage extrinsic antiviral activity to HSV2.

MATERIALS AND METHODS

Cell Cultures

The macrophage-like cell lines Raw 264.7 (TIB71), J774A.1 (TIB67), and P388D₁ (TIB63) were obtained from the American Type Culture Collection (ATCC; Rockville, MD). Macrophages were grown in complete RPMI 1640 medium (RPMI 1640 supplemented with 10% fetal bovine serum (FBS); 1% L-glutamine; 1% nonessential amino acids; 1% antibiotic/antimycotic (penicillin 10,000 U/ml; Amphotericin B 25 ug/ml, and Streptomycin 10,000 ug/ml); 1.5% Hepes; 1.5% sodium bicarbonate, and 1% MEM vitamins). Green Monkey Kidney (Vero) cells were grown in RPMI 1640 supplemented as described above. Cells were grown and maintained at 37C in a humidified 5% CO_2 atmosphere.

Virus

Herpes simplex virus type 2 (HSV2; strain HSV-FMC-P18O) was isolated from cervical tissue of a gynecological patient with carcinoma in situ at the Medical College of Virginia/VCU. The virus was plaque-purified and was typed as HSV2 by partial restriction endonuclease digestion as described previously (23).

Virus stocks were prepared by infecting confluent monolayers of Vero cells at a multiplicity of infection (MOI) of 0.01 and harvesting the virus when 90% of the cells manifested typical HSV-induced cytopathic effect (CPE). Virus titer was determined by plaque assay (24) using Vero cells grown in 24 well tissue culture plates (Corning, Inc. Corning, NY.). A virus stock with a titer of 1 X 10^8 plaque-forming units per ml (PFU/ml) was employed in these studies.

Delta-9-tetrahydrocannabinol Preparation.

Delta-9-tetrahydrocannabinol (THC) (FW = 316) was obtained from the National Institute on Drug Abuse (NIDA), Rockville, MD. THC was prepared from a stock solution of 100 mg/ml in 95% ethanol. The stock solution was appropriately diluted in complete RPMI 1640 medium such that 10 μl was added per ml of medium to yield final concentrations of 10^{-5}M, 10^{-6}M, or 10^{-7}M THC and 0.1% ethanol. The vehicle consisted of l0 μl of 9.5% ethanol per ml of RPMI 1640 medium to yield a final concentration in medium of 0.1% ethanol. Placebo cultures consisted of RPMI 1640 medium inoculated with 10 μl/ml of the same medium.

In vitro Exposure of Macrophages to Delta-9-tetrahydrocannabinol.

For assessment of the effect of THC on intrinsic antiviral activity, confluent monolayers of macrophages in 24-well cluster plates (5 X 10^5 cells/well) were pretreated with THC (10^{-5}M, 10^{-6}M, or 10^{-7}M), vehicle (0.1% ethanol), or placebo (medium only) for 48 hr. Following washing with sterile phosphate-buffered saline (PBS), the macrophage-like cells were exposed to HSV2 and were evaluated for intrinsic antiviral activity. For assessment of the effect of THC on macrophage extrinsic antiviral activity, macrophages in 75 cm^2 culture flasks were exposed to THC (10^{-5}M, 10^{-6}M, or 10^{-7}M), vehicle, or placebo for 48 hr. Alternatively, macrophages were exposed to the drug or vehicle for 48 hr and then were maintained in THC-free medium for 48 or 120 hr. Cells, then, were scraped into the medium with a rubber policeman, assessed for viability by trypan blue exclusion, enumerated, and added to virus-infected Vero cell monolayers for assessment of extrinsic antiviral activity.

Assay of Intrinsic Antiviral Activity.

Following exposure to THC, vehicle, or placebo, macrophage-like cells were washed with sterile PBS and a 0.1 ml inoculum calculated to contain 50 PFU of HSV2 was added to each confluent macrophage monolayer in 24 well tissue culture plates. Replicate cultures containing Vero cell monolayers were inoculated with an equal volume of HSV2 in order to quantitate the number of input infectious virus on the basis of plaques elicited. Macrophages were adsorbed with HSV2 for 1, 2, 3, 4, 24, and 48 hr. At the end of each of the designated time periods, the supernatant from each culture was removed and the corresponding cell monolayer was lysed by three rapid freeze-thaw cycles. The corresponding supernatant and cell lysate of each macrophage culture were quantitated for infectious virus by plaque assay.

Assay of Extrinsic Antiviral Activity.

Confluent Vero cell monolayers (5 X 10^5 cells) in 24-well cluster plates were inoculated with a 0.1 ml volume of HSV2 such that each of the first 2 columns of wells received 500 PFU. The remaining 4 columns of wells received 250, 125, 60, or 30 PFU, respectively. Following incubation for 1.5 hr at 37C in a 5% CO_2 environment to allow for virus adsorption, the Vero cell monolayers were washed three times with sterile PBS (37°C) to remove unabsorbed virus . Suspensions (1 ml) of placebo-, vehicle-, or THC-treated macrophages, then, were added to the Vero cell monolayers such that the 4 rows of wells contained macrophage:Vero cell co-cultures at effector cell:target cell (E:T) ratios of 5:1, 3:1, and 1:1. Duplicate plates contained wells with virus-infected (500 PFU to 30 PFU) Vero cells or with uninfected Vero cells in the absence of macrophages. Additionally, uninfected Vero cells were co-cultured with macrophages at E:T ratios of 5:1 to 1:1 to monitor for potential macrophage-mediated cytotoxicity against the Vero cells. Trypan exclusion staining demonstrated that the macrophage lines did not effect cytolysis of uninfected Vero cells. Following the addition of macrophages, the culture plates were centrifuged (800 rpm, 15 min) to allow for effector cell:target cell contact and 1 ml of 2% methylcellulose in complete RPMI 1640 medium was added to each well. After three days of incubation at 37C in a 5% CO_2 environment, the plates were fixed with 4% formaldehyde in PBS, stained with 0.5% crystal violet, and enumerated for virus plaques. Each series of experiments was performed six times. The number of plaques elicited in HSV2-infected Vero cell monolayers in the absence of macrophages was designated as 100%. Extrinsic antiviral activity was measured by calculating the percentage plaques elicited in macrophage:HSV2-inoculated Vero cell co-cultures when compared with similarly HSV2-inoculated Vero cell monolayers maintained in the absence of macrophages. The lower the percentage plaques noted in co-cultures, the greater the extrinsic macrophage antiviral activity.

Statistics.

Virus titration data were expressed as log_{10}. Means ± S.E.M. were obtained from data from the individual cultures. Student's two tailed t-test was employed to determine statistical significance of the data (25).

RESULTS

Effect of Delta-9-tetrahydrocannabinol on Intrinsic Antiviral Activity.

Virus uptake experiments were performed in order to directly evaluate the effect of THC on intrinsic antiviral activity. Infectious virus in each macrophage culture supernatant and corresponding cell lysate was quantitated by plaque assay. Virus plaques produced from the cell lysates served as a measure of infectious virus taken up by macrophages at the designated time periods. In addition, these plaques served as a measure of virus "survival" within the recipient macrophages. In contrast, virus plaques elicited by the culture supernatants (1 to 4 hr postinoculation (pi)) served as a measure of input infectious virus not taken up by the macrophages. Plaques elicited by the 24 hr or 48 hr pi culture supernatants served as a measure of productive virus infection within macrophages since by that time log_{10}-fold amounts of HSV2 are released into the culture medium by virus-permissive cells. Results from a representative experiment of the effect of THC on intrinsic antiviral activity of J774A.1 macrophages are listed in Table I.

Table I. In vitro Effect of THC on J774A.1 Macrophage Intrinsic Antiviral Activity

Time(HR)	Placebo[2]	VH10^{-5} M	THC10^{-6}M	THC10^{-7}M	THC
Treatment[1]					
1	28(14)[3]	28(12)	27(18)	27(16)	26(16)
2	28(13)	24(12)	17(15)	20(14)	24(15)
3	27(13)	21(11)	15(13)	15(12)	21(14)
4	20(9)	20(11)	15(10)	13(9)	20(10)
24	14(4)	14(9)	12(10)	12(9)	13(10)
48	12(2)	10(4)	12(5)	8(5)	11(4)

[1]J774A.1 macrophages were treated with THC, vehicle, or placebo for 48 hr, washed, and inoculated with approximately 50 PFU of HSV2. The number of input PFU was corroborated by counting plaques elicited on Vero cell control monolayers following their exposure to the same inoculum. At each of the indicated times post-virus inoculation (pi) the supernatant and monolayer lysate from each culture were recovered and were assessed for infectious virus by plaque assay.

[2]Placebo = medium only.

[3]The numbers outside the parentheses denote PFU elicited by the culture supernatant while those within parentheses denote PFU produced by the cell lysate from the corresponding culture.

Plaque assay of control Vero cell monolayers yielded 45-55 PFU following inoculation with a 0.1 ml volume of virus suspension calculated to contain 50 PFU. At 1 hr pi, the sum of infectious virus in the culture supernatants and corresponding cell monolayers, regardless of treatment regimen, ranged from 40 - 45 PFU. Thus, each culture was inoculated with a comparable amount of infectious virus. Lysates of vehicle-treated J774A.1 cells yielded 12 plaques following a one hour absorption period. By 48 hr post-virus inoculation (pi), lysates of the J774A.1 cells yielded only 4 plaques. A parallel decrease in infectious virus from culture supernatants accompanied the decrease in infectious intracellular virus. By 24 hr, culture supernatants plus cell lysates yielded a total of 23 PFU. These results indicate that within the first 4 hr approximately 50% of the input virus was taken up and degraded by the macrophages. By 48 hr pi approximately 30% of the total original input infectious virus remained in the macrophage cultures. Thus, the input virus was readily taken up and degraded by the macrophages. A similar temporal reduction in infectious virus was observed for J774A.1 cells treated with THC (10^{-5} to 10^{-7}). Even after exposure to 10^{-5}M THC, macrophage intrinsic anti-HSV2 activity remained unaffected. At one hour pi, cell lysates yielded 18 PFU of HSV2. By 48 hr pi, cell lysates produced only 5 PFU. A temporal decrease in extracellular virus from the culture supernatants also was recorded (e.g. 27 PFU at 1 hr pi versus 12 PFU at 48 hr pi). Similar results were obtained when

P388D$_1$ macrophages were employed in the intrinsic antiviral studies (Data not shown). Thus, these results indicate that THC in the doses employed did not affect virus uptake by macrophages. Furthermore, the macrophages remained nonpermissive for HSV2 replication.

Effect of THC on Extrinsic Antiviral Activity.

The effect of THC on extrinsic antiviral activity of P388D$_1$, RAW264.7, and J774A.1 macrophage-like cells was assessed. Macrophages were incubated with THC (10^{-5}M, 10^{-6}M, or 10^{-7}M), vehicle, or placebo for 48 hr and, then, were co-cultured with virus-inoculated Vero cell monolayers at effector cell:target cell (E:T) ratios of 5:1, 3:1, and 1:1. Phase contrast microscopy and Nomarski optics microscopy demonstrated that THC treatment of macrophages had no effect on the attachment of macrophages to the virus-infected cells. Macrophage extrinsic antiviral activity was measured on the basis of suppression of plaque formation in macrophage:HSV2-infected Vero cell co-cultures when compared with plaque formation in similarly HSV2-infected Vero cell monolayers maintained in the absence of macrophages.

Figure 1 illustrates results of a typical experiment for assessing the effect of THC on macrophage extrinsic antiviral activity. Co-cultures of vehicle-treated P388D$_1$ macrophages and HSV2-inoculated Vero cells at E:T ratios of 5:1 yielded a minimal number of virus plaques (Figure 1A). Even at an E:T ratio of 1:1, extrinsic antiviral activity was exerted since less than 50% of the input virus elicited plaques. In contrast, decreased extrinsic antiviral activity was evident for co-cultures containing macrophages exposed to THC (Figure 1B). Co-cultures at the highest E:T ratio (e.g., 5:1) expressed a much greater number of virus plaques than the comparable co-cultures containing vehicle-treated macrophages. Co-cultures at E:T ratios of 5:1, 3:1, and 1:1 expressed a greater than two-fold increase in plaques when compared with co-cultures containing vehicle-treated macrophages.

Figure 2 illustrates quantitatively the effect of THC on macrophage extrinsic antiviral activity. Vero cell control monolayers, were inoculated with aliquots of virus suspension calculated to contain 120, 60, or 30 PFU of HSV2. The respectively-infected monolayers, maintained in the absence of macrophages were designated as producing the maximal number of virus plaques (i.e., 100%). The percentage of virus plaques elicited in co-cultures of macrophages and virus-inoculated Vero cells, then, was determined by comparison to the 100% yield of the corresponding virus-inoculated Vero cell monolayer. Infected Vero cells co-cultured with J774A.1 macrophages which were treated with placebo or vehicle yielded a minimal number of plaques when compared with HSV2-inoculated Vero cell control monolayers indicating that the J774A.1 cells exerted extrinsic antiviral activity. Co-cultures containing vehicle-treated or placebo-treated macrophages, and maintained at E:T ratios of 5:1 and 3:1, yielded less than 30% the number of plaques recorded for HSV2-infected Vero cell controls. Even at the lowest E:T ratio (1:1), macrophage extrinsic antiviral activity was exerted by placebo-treated and vehicle-treated cells since co-cultures yielded less than 58% of the plaques noted in control HSV2-infected Vero cell cultures. In contrast, co-cultures containing THC-treated J774A.1 macrophages and HSV2-infected Vero cells yielded a dose-related increase in viral plaques indicative of a dose-related decrease in macrophage extrinsic anti-HSV2 activity.

At E:T ratios of 1:1, THC eliminated nearly all of the macrophage extrinsic antiviral activity. Indeed, co-cultures containing macrophages treated with 10^{-5}M THC yielded greater than 90% the number of plaques recorded for monolayers of control HSV2-infected Vero cells.

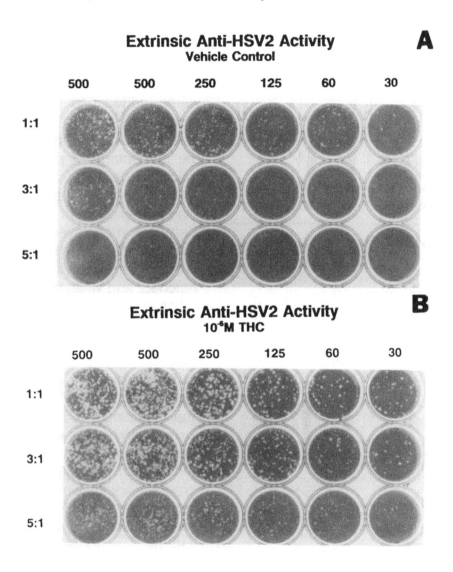

Figure 1. Effect of THC on Macrophage Extrinsic Antiviral Activity. P388D$_1$ macrophage-like cells were treated with (A) vehicle or with (B) 10^{-6}M THC for 48 hr. Macrophages, then, were added to HSV2-infected Vero cell monolayers to yield effector cell:target cell (E:T) ratios of 5:1, 3:1, and 1:1. The number over each column designates the calculated number of PFU of HSV2 added to each Vero cell monolayer in that column. The number assigned to each row designates the macrophage:Vero cell E:T ratio for each row. (A). Co-cultures Containing P388D$_1$ Macrophages Treated with Vehicle. There is a decrease in the number of virus plaques in direct correlation with increasing E:T ratios. AT an E:T ratio of 5:1 a minimal number of plaques was elicited in all of the co-cultures regardless of the input number of infectious HSV2. (B). Co-cultures Containing P388D$_1$ Macrophages Pretreated (48 hr) with 10^{-6}M THC. Note the increase in the number of plaques in all co-cultures at all E:T ratios indicative of a decrease in extrinsic antiviral activity.

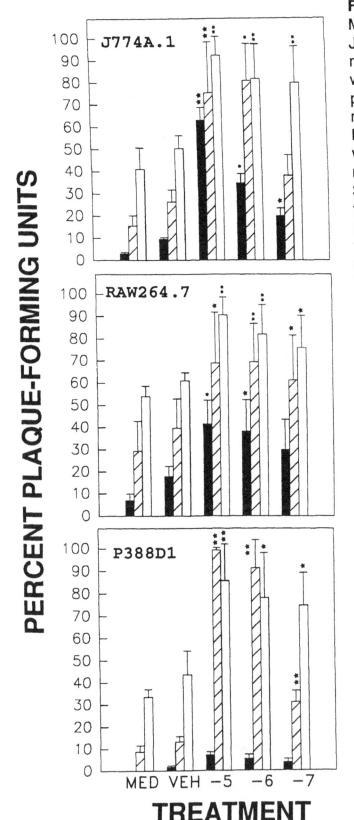

Figure 2. Effect of THC on Macrophage Extrinsic Antiviral Activity. J774A.1, RAW264.7, and P388D$_1$ macrophage-like cells were treated with THC (10^{-5}M - 10^{-7}M), vehicle, or placebo for 48 hr. Vero cell monolayers were inoculated with 60 PFU of HSV2. Macrophages, then, were added to the Vero cell monolayers to yield E:T ratios of 5:1, 3:1, and 1:1. The ordinate represents the percentage of plaques elicited in macrophage:Vero cell co-cultures when compared with the number of plaques elicited in HSV2-inoculated Vero cell monolayers not containing macrophages. Each bar designates the average of 6 experiments. The black, hatched, and white bars represent E:T ratios of 5:1, 3:1, and 1:1, respectively. The asterisks (* or **) designate significant ($p < 0.1$ or $p < 0.05$, respectively; Student's t-test) increases in the percentage of plaques formed for each E:T ratio when compared with the matching E:T ratio of co-cultures containing vehicle-treated macrophages. A drug dose-related increase in the relative percentage of plaques, indicative of a decrease in extrinsic anti-HSV2 activity, was noted for macrophages treated with THC (10^{-5})M - 10^{-7}M). MED=medium, VEH=vehicle; -5, -6, or -7 represent 10^{-5}M, 10^{-6}M, or 10^{-7}M THC, respectively.

Similar results were obtained when RAW264.7 macrophages were employed as effector cells. At E:T ratios of 5:1 and 3:1, a minimal number of plaques was elicited in co-cultures containing macrophages treated with vehicle or placebo (i.e., less than 40%). In contrast, a drug dose-related increase in plaque formation was noted at all E:T ratios for co-cultures containing macrophages treated with 10^{-5}M through 10^{-7}M THC. At E:T ratios of 3:1 and 1:1, co-cultures containing macrophages treated with 10^{-5}M or 10^{-6}M THC expressed greater than 70% the number of plaques observed in control HSV2-infected Vero cell cultures indicative of a loss of most extrinsic macrophage antiviral activity. This decrease was also noted for macrophages which were treated with 10^{-7}M drug and then were added to HSV2-infected Vero cells at an E:T ratio of 1:1.

THC, also, was shown to diminish the extrinsic antiviral activity of the $P388D_1$ macrophage-like cells. Co-cultures of placebo-treated or vehicle-treated $P388D_1$ macrophages and HSV2-infected Vero cells at E:T ratios of 5:1 and 3:1 produced less than 20% the number of PFU produced by HSV2-infected Vero cell monolayers. Even at an E:T ratio of 1:1, less than 50% the number of plaques recorded for virus-infected Vero cell cultures was noted for co-cultures containing $P388D_1$ cells treated with placebo or vehicle. THC exerted a dose-related increase in the percentage of virus plaques formed in co-cultures with E:T ratios of 3:1 and 1:1 when compared with HSV2-infected Vero cell monolayers maintained in the absence of macrophages indicative of a decrease in macrophage extrinsic anti-HSV2 activity. Macrophages treated with 10^{-5}M or 10^{-6}M THC, and co-cultured with HSV2-infected Vero cells at E:T ratios of 3:1 or 1:1, expressed greater than 70% the number of PFU expressed in HSV2-infected Vero cell cultures alone. However, at an E:T ratio of 5:1 THC exerted no major effect on macrophage extrinsic antiviral activity. For these latter co-cultures, less than 10% the number of plaques noted for Vero cell monolayers was recorded.

Temporal Effect of THC Exposure on Macrophage Extrinsic Antiviral Activity.

To assess whether macrophages treated in vitro with THC could recover their extrinsic antiviral activity following removal of the drug, an additional set of experiments was performed in which macrophages were maintained in THC-free medium for 48 hr or 120 hr following the 48 hr drug exposure period. The macrophages, then, were added to HSV2-infected Vero cell monolayers and were assessed for extrinsic antiviral activity. A time-dependent recovery of extrinsic anti-HSV2 activity by the macrophage-like cells was recorded. Results of the effect of drug removal on the recovery of extrinsic antiviral activity of J774A.1 macrophages at an E:T ratio of 3:1 are shown in Figure 3. As expected, co-cultures containing macrophages treated with vehicle elicited a minimal number of plaques at all time periods. These observations indicate that the J774A.1 cells exerted extrinsic antiviral activity throughout the experimental period. In agreement with previous experiments, co-cultures containing macrophages treated with THC for 48 hr exhibited a dose-related increase in the percentage of plaques formed indicative of a dose-related decrease in extrinsic macrophage antiviral activity. Co-cultures containing J774A.1 macrophages, regardless of the dose of drug exposure, maintained in drug-free medium for 48 hr yielded less than 40% the number of plaques which was recorded for HSV2-infected Vero cell monolayers cultured in the absence of macrophages. Co-cultures containing macrophages maintained in drug-free medium for 120 hr exhibited less than

15% the number of plaques recorded for virus-infected Vero cell monolayers. These observations indicate a nearly total recovery of extrinsic antiviral activity by the THC-pretreated macrophages at all E:T ratios.

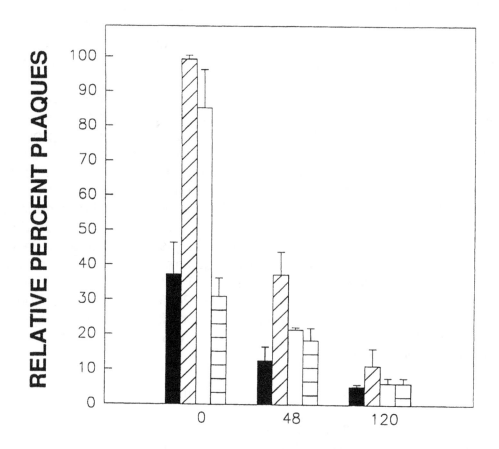

HRS POST-THC EXPOSURE

Figure 3. Recovery of Extrinsic Anti-HSV2 Activity by Murine J774A.1 Macrophage-like Cells. Macrophages were treated with THC (10^{-5}M - 10^{-7}M), vehicle, or placebo for 48 hr. The macrophages, then, were incubated in THC-free medium for either 48 hr or 120 hr. Approximately 60 PFU of HSV2 were added to each target Vero cell monolayer. The ordinate designates the percentage of plaques elicited by macrophage:Vero cell co-cultures (E:T ratio = 3:1) when compared with plaques elicited in Vero cell monolayers not containing macrophages and inoculated with 60 PFU. A time-related decrease in plaque formation, indicative of a recovery of extrinsic anti-HSV2 activity, was observed for macrophages pretreated with 10^{-5}M - 10^{-7}M THC. The bars are representative of the following: solid = vehicle, lines rising to the right = 10^{-5}M THC, open = 10^{-6}M THC, horizontal lines = 10^{-7}M THC.

DISCUSSION

Macrophages exert two important antiviral functions, intrinsic antiviral activity and extrinsic antiviral activity, in addition to their ability to bring about effector cell:target cell contact-dependent lysis of virus-infected cells (26-29). In intrinsic antiviral activity, macrophages engulf and degrade virus and thereby remain nonpermissive for virus replication (18-21,30). In extrinsic antiviral activity, macrophages inhibit virus replication within infected target cells in a cell contact-dependent, interferon-independent manner (22,31).

Macrophage cell lines were employed in this study since they are free of contaminating other cell types and can be cultivated to relatively high numbers. These cell lines differ in degree of macrophage maturity and functional properties. The J774A.1, P388D_1, and RAW264.7 cells represent relatively "mature" macrophage-like cells and exhibit many of the macrophage-virus interactions attributable to primary macrophages. For example, J774A.1 cells characteristically produce lysozyme and superoxide radicals which are important components of the intrinsic antiviral process (32). P388D_1 cells express F_c receptors and support growth of influenza A NWS virus (33) and West Nile virus (34) consequent to virus-antibody complex interaction with these receptors. Both J774A.1 and P388D_1 cells have been shown to produce acid-stable interferon when inoculated with Newcastle Disease virus (35).

In the present investigation, THC was shown to have no effect on macrophage intrinsic antiviral activity. An overall reduction in infectious virus in both culture supernatants and cell lysates was recorded for macrophages pretreated with doses of THC as high as 10^{-5}M. In addition, there was no significant difference in virus uptake by macrophages treated with THC as compared with macrophages treated with vehicle or placebo. These results indicate that THC causes neither a quantitative nor a temporal effect on HSV2 uptake by macrophages. Furthermore, the J774A.1 cells did not support productive virus infection regardless of the drug treatment regimen. Thus, THC exposure did not alter or inhibit those cellular compartments which effect macrophage nonpermissiveness for virus replication. Sarmiento et al. (30) have shown that macrophage intrinsic resistance to HSV1 infection is due to restriction of viral macromolecular synthesis. HSV replication was found to be inhibited in macrophages at multiple points in the virus growth cycle before the onset of virus DNA synthesis. The results of this study indicate that THC does not target those macrophage factors which are involved in restriction of HSV macromolecular synthesis. Similar results were obtained when P388D_1 cells were employed in the HSV2 uptake experiments indicating that the failure of THC to alter intrinsic macrophage antiviral activity was not limited to the J774A.1 cells.

In contrast, THC affected the capacity of macrophages to exert extrinsic antiviral activity. The mechanism by which macrophages exert this activity remains unresolved although some aspects have been defined. Morahan et al. (31) indicated that the extrinsic antiviral mechanism does not involve direct virus inactivation by extrinsic factors. Neither macrophage supernatants nor lysates alone inactivated virus infection suggesting an effector cell:target cell contact-dependent mechanism. Furthermore, the extrinsic antiviral activity was unlike interferon-mediated inhibition since it was exerted against virus-infected xenogeneic cells (36). Furthermore, Vero cells, which were used as targets in these experiments, have been shown not to produce interferon (37). Thus, the macrophage extrinsic antiviral activity was not species-specific in that the macrophages serving as effectors were of murine origin while the virus-infected target cells (i.e., Vero cells) were of primate origin. Other secretory products of macrophages, such as interleukin-1 (IL-1) (38-41) and tumor necrosis factor (TNF) (42-45), have been shown to possess anti-viral properties. However, the action of these molecules is neither dependent on effector cell:target cell conjugation nor on cell contact as demonstrated for extrinsic antiviral activity (36). Thus, it seems unlikely that interferon (IFN), Il-1, or TNF accounts for the majority, if any, of the extrinsic antiviral activities attributed to macrophages. Nevertheless, other secretory products such as neutral proteases and prostaglandins could be responsible for non-specific antiviral effects of macrophages on virus or virus-infected cells.

THC did not prevent any of the macrophage cell types from attaching to the virus-infected xenogeneic Vero cells. These results are in agreement with our previous studies which demonstrated that THC, injected intraperitoneally into mice which were administered Propionibacterium acnes (P. acnes), did not prevent peritoneal macrophages from attaching to virus-infected cells or to tumor cells (46). Similarly, macrophages treated in vitro with 10^{-5}M to 10^{-7}M THC remained unaffected in their ability to attach to target cells (46). Thus, these observations indicate that THC affects macrophage extrinsic antiviral activity at a step other than effector cell:target cell attachment. The mechanism by which THC effects this inhibition, however, remains unresolved. THC has been shown to affect macromolecular synthesis in a variety of cellular systems (46-50). We have shown that the cannabinoid inhibits differential expression of macrophage proteins in response to external stimuli, such as bacterial lipopolysaccharide (LPS) and P. acnes (10). These observations suggest that THC may alter extrinsic antiviral activity by suppressing macromolecular synthesis and/or expression by macrophages. Alternatively, THC may bring about a morphological disruption of cellular membranes (51,52) or of cytoskeletal elements (53). Perturbations of cellular membranes could alter cellular compartments of protein synthesis and/or post-translational events and thereby suppress the synthesis of these macrophage factors which block virus replication at the level of expression of the early to delayed-early virus genes. It has been proposed that an intact cytoskeletal system is requisite for natural killer cell and for cytotoxic T-lymphocyte cell contact-dependent killing of target cells. Following conjugation of the effector cell to the target cell, a rapid reorientation of the microtubular organizing center and Golgi apparatus complex occurs within the target cell. This orientation is effected toward the target cell and results in the direction of secretory vesicles containing cytolytic components along the cytoskeletal tracts to the target cell (54,55). Thus, THC, in addition to suppressing protein synthesis, could bring about a disruption of cytoskeletal elements within the effector cell with the consequent failure in transport of effector molecules to the virus infected cell. We have shown that THC administered in vitro elicits membrane perturbation of cell surface and cytoplasmic membranes in rat neuroblastoma cells (53).

The THC-induced inhibition of macrophage extrinsic antiviral activity was shown to be reversible. J774A.1 and P388D$_1$ macrophages treated with THC for 48 hr, and then incubated in drug-free medium for 48 or 120 hr, regained extrinsic antiviral activity in a time-dependent fashion. These observations are indicative of an ongoing "repair" process by which macrophages gradually regain their functional competence against virally-infected cells following removal of THC from the environment. The results showing the reversible effects of THC on macrophage extrinsic antiviral activity may be attributed to macrophage regeneration and "sloughing off" of putatively drug-damaged cell membranes. A similar process has been shown to occur in rat neuroblastoma cells (53). Ejection of membranous structures was most evident when cells were treated in vitro with 10^{-5}M and 10^{-6}M THC. The rat B103 neuroblastoma cells were shown to express surface blebs following exposure to drug. Scanning electron microscopy revealed extracellular globular membranous bodies surrounding the drug-exposed cells. In addition, transmission electron microscopy demonstrated cytoplasmic membranous inclusions seemingly in the process of extrusion from the drug-treated cells. Similar extruded globular structures and surface blebs were observed by Nomarski optics microscopy for macrophages treated in vitro with THC, especially at higher drug doses (i.e., 10^{-5}M). While these effects elicited by 10^{-5}M THC could represent the outcome of exposure to a relatively high drug concentration, they nevertheless may depict the upper limit of a continuum of THC-induced effects which bring about repair of drug-damaged macrophage membranes. The elimination of damaged membranes following removal of the drug from the extracellular environment may allow for cells to regenerate cellular membranes with a consequent temporal-dependent reestablishment of extrinsic antiviral activity.

The THC doses which were employed in this in vitro study are achievable in humans. Agurell et al. (56) have shown that serum THC peaks at concentrations as high as 500 ng/ml within 30 min of smoking marijuana. This peak period is followed by an equilibrium phase during which concentrations of THC may range from 0.5 ng/ml to 0.25 ng/ml in one to two days. Wall et al. (57) measured baseline levels of THC of approximately 1 ng/ml in regular marijuana users. The rapid early decline of THC in plasma is a consequence of extensive metabolism of the drug (58) and of uptake by tissues due to its highly lipophilic nature (59). For example, relatively high concentrations of THC could be anticipated in lung alveolar macrophages as a consequence of direct exposure to marijuana smoke.

BIBLIOGRAPHY

1. Nahas GG, Sucia-Foca N, Armand SP and Morishima A. Inhibition of cellular mediated immunity in marihuana smokers. Science, 1974, 183:419-420.
2. Klein TW, Newton CA, Widen R and Friedman H. The effect of delta-9-tetrahydrocannabinol and 11-hydroxy-delta-9-tetrahydrocannabinol on T-lymphocyte and B-lymphocyte mitogen responses. J Immunopharm, 1985, 7(4):451-466.
3. Gupta G, Grieco M and Cushman P. Impairment of rosette-forming T-lymphocytes in chronic marihuana smokers. N Eng J Med, 1974, 291:874-876.
4. Cushman P and Khurana R. Marijuana and T-lymphocyte rosettes. Clin Pharmacol Ther, 1976, 19:310-317.
5. Specter SC, Klein TW, Newton C, Mondragon M, Widen R and Friedman H. Marijuana effects on immunity: suppression of human natural killer cell activity by delta-9-tetrahydrocannabinol. Intl Soc Immunopharm, 1986, 8(7):741-745.

6. Klein TW, Newton C and Friedman H. Inhibition of natural killer cell function by marijuana components. J Tox Environ Health, 1987, 20:321-332.

7. Schwartzfarb L, Needle N and Chavez-Chase M. Dose related inhibition of leukocyte migration by marihuana and Delta-9-tetrahydrocannabinol. J Clin Pharmacol, 1974, 14:35-41.

8. Mann PEG, Cohen AB, Finley TN and Ladman AJ. Alveolar macrophages. Structural and functional differences between non-smokers and smokers of marijuana and tobacco. Lab Invest, 1971, 25:111-120.

9. Lopez-Cepero M, Friedman M, Klein T and Friedman J. Tetrahydrocannabinol-induced suppression of macrophage spreading and phagocytic activity in vitro. J Leukocyte Biol, 1986, 39:679-686.

10. Cabral GA and Mishkin EM. Delta-9-tetrahydrocannabinol inhibits macrophage protein expression in response to bacterial immunomodulators. J Toxicol Environ Health, 1989, 26:175-182.

11. Morahan PS, Klykken PC, Smith SH, Harris LS and Munson AE. Effects of Cannabinoids on host resistance to Listeria monocytogenes and herpes simplex virus. Infect Immun, 1979, 23:670-674.

12. Cabral GA, Mishkin EM, Marciano-Cabral F, Coleman P, Harris L and Munson, AE. Effect of Delta-9-tetrahydrocannabinol on herpes simplex virus type 2 vaginal infection in the guinea pig. Proc Soc Exp Bio Med, 1986, 182:181-186.

13. Mackaness GB. The immunological basis of acquired cellular resistance. J Exp Med, 1964, 120:105-112.

14. Unanue ER. The regulatory role of macrophages in antigenic stimulation. Part two: Symbiotic relationship between lymphocytes and macrophages. Adv Immunol, 1981, 31:1-136.

15. Howie S and MacBride WH. Cellular interactions and thymus-dependent antibody responses. Immunology Today, 1982, 3:273-278.

16. Evans R and Alexander P. Mechanisms of extracellular killing of nucleated mammalian cells. In Nelson DS, Ed. Immunobiology of the Macrophage. New York, Academic Press, 1976, p535.

17. Probert M, Stott EJ and Thomas LH. Interactions between calf alveolar macrophages and parainfluenza-3 virus. Infect Immun, 1977, 15:576-585.

18. Johnson RT. The pathogenesis of herpesvirus encephalitis. II. A cellular basis for the development of resistance with age. J Exp Med, 1964, 120:359-374.

19. Stevens JG and Cook ML. Restriction of herpes simplex virus by macrophages. An analysis of the cell-virus interaction. J Exp Med, 1971, 133:19-38.

20. Selgrade MK and Osborn JE. Role of macrophages in resistance to murine cytomegalovirus. Infect Immun, 1974, 10:1383-1390.

21. Mims GA and Gould J. The role of macrophages in mice infected with murine cytomegalovirus. J Gen Virol, 1978, 41:143-153.

22. Morahan PS, Glasgow LA, Crane JL and Kern ER. Comparison of antiviral and antitumor activity of activated macrophages. Cell Immunol, 1977, 18:404-415.

23. Hamelin C, Chagnon A and Farwell M. Identification of atypical strains of herpes simplex virus using a simplified DNA finger printing method. Microbiol Immunol, 1984, 28:723-729.

24. Rapp F. Variants of herpes simplex virus: Isolation, characterization, and factors influencing plaque formation. J Bacteriol, 1963, 86:985-991.

25. Snedecor GW and Cochran WG. In: Statistical Methods (sixth edition). Ames, Iowa, Iowa State University Press, 1967, Chapter 4.

26. Stanwick TL, Campbell DE and Nahmias AJ. Spontaneous cytotoxicity mediated by human monocyte-macrophages against human fibroblasts infected with herpes simplex virus- augmentation by interferon. Cell Immunol, 1980, 53:413-416.

27. Koff WC, Showalter SD, Seniff DA and Hampar B. Lysis of herpesvirus-infected cells by macrophages activated with free or liposome-encapsulated lymphokine produced by a murine T cell hybridoma. Inf Immun, 1983, 42(3):1067-1072.

28. Leblanc PA. Macrophage activation for cytolysis of virally infected target cells. J Leuk Biol, 1989, 45:345-352.

29. Leblanc PA, Heath LS and Um H. Activated macrophages use different cytolytic mechanisms to lyse a virally infected or a tumor target. J Leuk Biol, 1990, 48:1-6.

30. Sarmiento M. Intrinsic resistance to viral infection. Mouse macrophage restriction of herpes simplex virus replication. J Immunol, 1988, 141:2740-2748.

31. Morahan PS, Morse SS and McGeorge MB. Macrophage extrinsic antiviral activity during herpes simplex virus infection. J Gen Virol, 1980, 46:291-300.

32. Damiani G, Kiyotaki C, Soeller W, Sasada M, Peisach J and Bloom BR. Macrophage variants in oxygen metabolism. J Exp Med, 1980, 152: 808-822.

33. Ochiai H, Kurokawa M, Hayashi K and Niwayama S. Antibody-mediated growth of influenza A NWS virus in macrophagelike cell line P388D1. J Virol, 1988, 62: 20-26.

34. Cardosa MJ, Gordon S, Hirsch S, Springer TA and Porterfield JS. Interaction of West Nile virus with primary murine macrophages: role of cell activation and receptors for antibody and complement. J Virol, 1986, 57: 952-959.

35. Salo RJ, Bleam DK, Greer VL and Ortega AP. Interferon production in murine macrophage-like cell lines. J Leuk Biol, 1985, 37: 395-406.

36. Morse SS and Morahan PS. Activated macrophages mediate interferon-independent inhibition of herpes simplex virus. Cell Immunol, 1981, 58:72-84.

37. Desmyter J, Melnick JL and Rawls WE. Defectiveness of interferon production and of rubella virus interference in a line of African green monkey kidney cells (Vero). J Virol, 1968, 2:955-961.

38. Van Damme J, De Ley M, Van Snick J, Dinarello CA and Billiau A. The role of interferon- beta 1 and the 26-kDa protein (interferon-beta 2) as mediators of the antiviral effect of interleukin 1 and tumor necrosis factor. J Immunol, 1987, 139(6):1867-1872.

39. Keller F, Schmitt C and Kerin A. Interaction of mouse hepatitis virus 3 with Kupffer cells explanted from susceptible and resistant mouse strains. Antiviral activity, interleukin-1 synthesis. Microbiol Immunol, 1988, 1(2):87-95.

40. Van der Meer JW, Rubin RH, Pasternack M, Medearis DN, Lynch P and Dinarello CA. The in vivo and in vitro effects of interleukin-1 and tumor necrosis factor on murine cytomegalovirus infection. Biotherapy, 1989, 1(3):227-231.

41. Ruggiero V, Antonelli G, Gentile M, Conciatori G and Dianzani F. Comparative study on the antiviral activity of tumor necrosis factor (TNF)-alpha, lymphotoxin/TNF-beta, and IL-1 in WISH cells. Immunol Lett, 1989, 21(2):165-169.

42. Mestan J, Digel W, Mittnacht S, Hillen H, Blohm D, Moller A, Jacobsen H and Kirchner H. Antiviral effects of recombinant tumour necrosis factor in vitro. Nature, 1986, 323 (6091):816-819.

43. Koff WC and Fann AV. Human tumor necrosis factor alpha kills herpes virus-infected normal cells. Lymphokine Res, 1986, 5:215-221.

44. Paya CV, Kenmotsie N, Schoon RA and Leibson PJ. Tumor necrosis factor and lymphotoxin secretion by human natural killer cells to antiviral cytotoxicity. J Immunol, 1988, 141(6):1989-1995.

45. Duerksen-Hughes P, Wold WS and Gooding LR. Adenovirus E1A renders infected cells sensitive to cytolysis by tumor necrosis factor. J Immunol, 1989, 143 (12):4193-4200.

46. Cabral GA and Vasquez R. Effects of marihuana on macrophage function. In: Friedman H, Specter S, Klein TW, Eds. Advances in Experimental Medicine and Biology. Vol. 288. Drugs of Abuse, Immunity, and Immunodeficiency. Plenum Press, New York, 1991, pp 93-105.

47. Blevins RD and Regan JD. Delta-9-tetrahydrocannabinol: Effect on macromolecular synthesis in human and other mammalian cells. Arch Toxicol, 1976, 35:127-135.

48. Nahas GG, Morishima A and Desoize B. Effect of Cannabinoids on macromolecular synthesis and replication of cultured lymphocytes. Fed Proc, 1977, 36:1748-1752

49. Mon JJ, Jansing RL, Doggett S, Stein JL and Stein GS. Influence of Delta-9-tetrahydrocannabinol on cell proliferation and macromolecular synthesis in human cells. Biochem Pharmacol, 1978, 21:1759-1765.

50. Desoize B, Leger C and Nahas GG. Plasma membrane inhibition of macromolecular precursor transport by THC. Biochem Pharmacol, 1979, 28:1113-1118.

51. Poddar MK, Mittra G and Ghosh JJ. Delta-9-tetrahydrocannabinol-induced changes in brain ribosomes. Toxicol Appl Pharmacol, 1978, 46:737-757.

52. Meyers WA and Heath RG. Cannabis sativa: ultrastructural changes in organelles and neurons in brain septal region of monkeys. J Neurosci Res, 1979,4:9-17.

53. Cabral GA, McNerney PJ and Mishkin EM. Interaction of delta-9-tetrahydrocannabinol with rat B103 neuroblastoma cells. Archiv Toxicol, 1987, 60:438-449.

54. Kupfer A and Dennert G. Reorientation of the microtubule-organizing center and the Golgi apparatus in cloned cytotoxic lymphocytes triggered by binding to lysable target cells. J Immunol, 1984, 133(5):2762-2766.

55. Kupfer A, Dennert G and Singer SJ. The reorientation of the Golgi apparatus and the microtubule-organizing center in the cytotoxic effector cell is a prerequisite in the lysis of bound target cells. J Mol Cell Immunol, 1985, 2:37-49.

56. Agurell S, Lindgre JE, Ohlsson A, Gillespie HK and Hollister LE. Recent studies on the pharmacokinetics of Δ^1-tetrahydrocannabinol in man. In: Agurell DJ, Dewey WL, Willette RE, Eds. The Cannabinoids: Chemical, Pharmacologic, and Therapeutic Aspects. Orlando, FLA, Academic Press. 1984.

57. Wall ME, Brine D, Bursey JT and Rosenthall D. Detection and quantitation of tetrahydrocannabinol in physiological fluids. In: Vinson JA, Ed. Cannabinoid Analysis in Physiological Fluids. ACS Symposium Series 98, American Chemical Society, Washington, DC, 1979, pp 39-58.

58. Mechoulam R, McCallum NK and Burstein S. Recent advances in the chemistry and biochemistry of cannabis. Chem Rev, 1976, 76: 75-112.

59. Wing DR, Leuschner JTA, Brent GA, Harvey DJ and Paton WDM. Quantification of <u>in</u> <u>vivo</u> membrane associated Delta-1-tetrahydrocannabinol and its effect on membrane fluidity. In: Harvey DJ, Ed. Proceedings of the 9th International congress of Pharmacology 3rd satellite symposium on cannabis. Oxford, IRL Press, 1985, pp411-418.

ACKNOWLEDGEMENTS

The authors thank Dr. Louis Harris and Dr. Billy Martin of the Department of Pharmacology and Toxicology at the Medical College of Virginia/Virginia Commonwealth University for kindly providing Delta-9-tetrahydrocannabinol. This research was supported by Grant R01 DA05832 from the National Institute on Drug Abuse.

A CASE-CONTROL STUDY OF ACUTE NON-LYMPHOBLASTIC LEUKEMIA - EVIDENCE FOR AN ASSOCIATION WITH MARIHUANA EXPOSURE[1]

Jonathan Buckley

Department of Preventive Medicine, University of Southern California, Los Angeles.

ABSTRACT

A case control study of potential in utero and post-natal exposures associated with acute non-lymphocytic leukemia (ANLL) was conducted by the Children Cancer Study Group. Data were analysed for 204 case control pairs. Maternal use of mind altering drugs prior to and during pregnancy was found to be associated with an eleven fold increased risk (p = 0.003) of ANLL in offspring when compared to offspring of controls. Ten of the eleven mind-altering drug exposures were either marijuana exclusively (nine) or included marijuana (one). Mothers of ANLL cases were ten times more likely to have used marijuana just preceding or during the pregnancy when compared to control mothers (p = 0.005). Marijuana exposed cases of ANLL differed significantly from non-marijuana exposed cases with respect to age at diagnosis and morphologic subtype. The results of this previously reported study (Robison et al, 1989) suggest the possibility that maternal marijuana use during pregnancy may play an etiologic role in childhood ANLL. Studies to follow up these findings are currently underway.

KEY WORDS

Marijuana; Childhood; Acute non-lymphocytic leukemia.

[1] *Reprint from "I. Internationales Symposium gegen Drogen in der Schweiz", 19-20 November 1990, Kongresshaus Zürich, VPM Zürich 1991, p.563.*

The Children's Cancer Study Group recently completed a study which showed an unexpected association between the use of marihuana by a pregnant woman, and the likehood that her child would develop acute nonlymphocytic leukemia (ANLL). I plan to describe to you today the design of the study, the most interesting results, and discuss problems that arise in the interpretation of the findings.

The Children's Cancer Study Group, or CCSG, is a group of collaborating hospitals and universities in the United States and Canada that conducts research on the causes and treatments of cancer in children. Approximately 100 institutions contribute to the studies, and approximately 50% of all cases of cancer in children in the U.S. are treated at one of these hospitals. (Figure 1).

In 1985, CCSG began a study of the role of environmental exposures to children with ANLL, including exposures to the parents of these children before their birth. The principal investigator for this study was Dr Leslie Robison of the University of Minnesota. Leukemia, which is a malignancy of the white cells in the blood, has many subtypes, but the most important distinction is between the lymphocytic leukemias and the nonlymphocytic (also called the myelocytic) leukemias. A second distinction is between rapid onset, or acute, leukemia, and a slowly progressing or chronic form of leukemia.

In adults, the most common types are acute nonlymphocytic and chronic leukemias. In contrast, children are very rarely seen with chronic leukemia, and the acute variety is most often a lymphocytic leukemia. (Figure 2).

Thus of over 2000 new cases of acute leukemia in children in the U.S. each year, only 300-400 of these are of the nonlymphocytic variety. The causes of leukemias in man are not well understood (particulary for children), but the research done so far in adults has suggested that environmental factors can be important. In particular, exposure to radiation, pesticides and chemical solvents have been implicated. For this reason, we decided to study these factors in children with ANLL.

The only practical means of investigating the causes of cancers that occur in less than one in 100 000 children per year is to use a case-control design. This involves collection of information about a group of cases with a disease - in this instance ANLL- and collection of the same information on a comparable group of normal children. The ways in which these two groups differ, with respect to family medical history, socio-demographic characteristics or reports of unusual environmental exposures will point to possible causative factors for the disease.

In our study, all information was obtained by telephone interview, which meant that the case (and control) was only eligible if the family had a telephone. To make sure that the control was selected from the

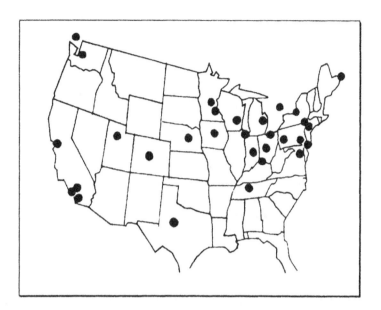

Figure 1: Children's Cancer Study Group. Major member institutions.

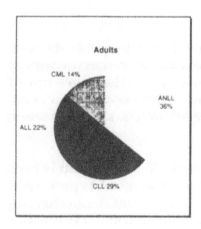
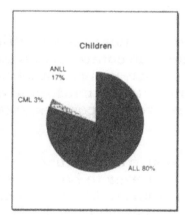

Figure 2: Distribution of leukemia in adults and children.
CML = Chronic myeloic leukemia, ALL = Acute lymphocytic leukemia,
CLL = Chronic lymphocytic leukemia, ANL = Acute nonlymphocytic leukemia.

neighborhood of each case using a random telephone procedure, and the control was required to be the same race and sex as the case and close to the case in age.

A total of 204 cases and 204 controls were available for analysis. Interview data were obtained from both parents. Although the interview concentrated on radiation, pesticides and solvents, a broad range of other topics were covered in the hour-long interviews. These included demographic details, complications of the pregnancy and delivery, medications used, parental occupation, medical history, household exposures, and diet. We could not find any link to radiation exposures. But we did show increased frequency of exposure of the fathers of cases (compared to fathers of controls) to pesticides and solvents at work. In addition, fathers of cases were more likely to have been exposed to petroleum and plastics, and lead.(Figure 3).

The case mothers were more likely to have been exposed to pesticides, metal dusts and paints (Figure 4).
These results are summarized in Table I.

Cases were also more likely to have been exposed to pesticide use in the home, either before birth (through the mother's exposure) or as a child. (Figure 5) (Table 2).

Results such as these can be difficult to interpret, since differences between a case and control group merely be chance observations of no biological relevance. In this instance, however, there are data from other studies to implicate pesticides, and our data show an increased risk for both home and work exposures, suggesting that these associations are genuine.

Another reason to believe that the pesticide association is real is that the risks of ANLL appear to be the largest for a subtype of ANLL, acute monocytic leukemia (odds ratio, OR=13.6 p=0.007), and arising in young children (OR=11.4 for children under age 5, p=0.003). (Table 3).

The association with maternal marihuana use was entirely unexpected. Part of the interview included questions on medications and non-prescription drugs used during the pregnancy. This included a single question on "recreational drug use, such as marihuana and cocaine". Eleven mothers of the cases reported such use : in 10 instances the drug was marihuana. In contrast only 1 control mother used marihuana. Although this difference seems quite large, it needs to be examined carefully before accepting it at face value.

The first possibility is that it is a chance finding, especially since we had no prior data to suggest this association. this remains a possibility, but a strong argument against this is that the increased risk was not seen

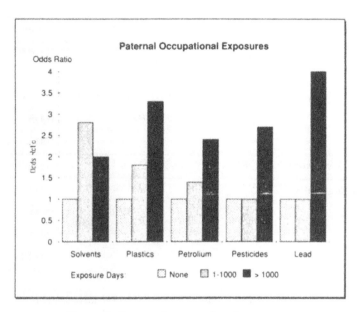

Figure 3: Paternal occupational exposures.

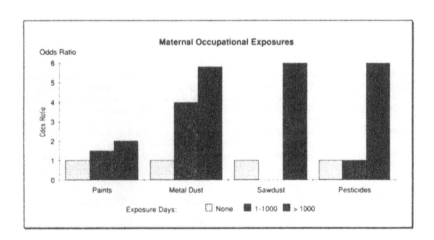

Figure 4: Maternal occupational exposures.

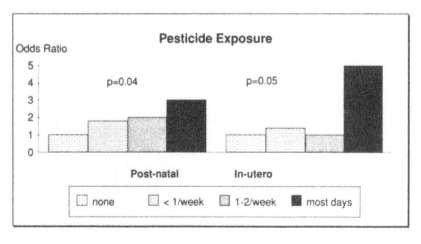

Figure 5: Pesticide exposure.

Paternal Exposures:	OR	95% CI	p-value
Solvents	2.0	1.2-3.8	0.003
Plastics	3.3	1.0-10.3	0.02
Petrolium products	2.4	1.3-4.1	0.002
Pesticides	2.7	1.0-7.0	0.06
Lead	6/0		0.03
Maternal Exposures:	OR	95% CI	p-value
Paint and pigments	2.2	0.9-5.4	0.05
Metal dusts	6.0	0.7-49.8	0.02
Sawdust	5/0	–	0.03
Pesticides	7/0	–	0.008

Table 1: Parental occupational exposures (>1000 day) associated with ANLL risk. For odds ratios (OR) that are infinite, the numbers of exposed case and controls are shown (eg. 6/0).

Exposure	Frequency	Case	Control	OR
Pesticide* (mother)	none	134	148	1.0
p=0.05	<1/week	50	40	1.4
	1-2/week	12	15	0.9
	most days	8	0	–
Pesticide (child)	none	128	148	1.0
p=0.04	<1/week	46	33	1.8
	1-2/week	13	9	2.0
	most days	8	3	3.5

Table 2: Odds ratios for significant household exposures.
(In-utero exposure of the child)*

	Duration	Case	Control	OR
All	0	142	156	1.0
	1-1000	13	15	1.2
	> 1000	23	7	3.8
				p = 0.004
Age < 5	0	65	76	1.0
	1-1000	6	5	1.4
	> 1000	11	1	11.4
				p = 0.003
N4/N5	0	42	50	1.0
	1-1000	5	2	6.6
	> 1000	6	1	13.6
				p = 0.007

Table 3: Acute myeloblastic leukemia;
paternal occupational pesticide exposure.

Maternal marihuana use:	Yes	No	
Av. age of case	19 months	93 months	p = 0.007
Monocytic/myelomonocytic	70%	31%	p = 0.02

Table 4: Comparison of case age and ANLL subtype for children with
ANLL, based on maternal marihuana use during pregnancy.

	Percentage:		
	Cases	Controls	OR
ANLL study	5.4	0.5	11.0
Other childhood cancers	1.6	1.8	0.9

Table 5: Comparison of the odds ratio for use of mind altering drugs
in the ANLL study, and in other similar studies of childhood cancers.

equally for all types of <ANLL, and with ANLL occurring early in life
(Table 4).

The second explanation is that case mothers were simply more open
and honest than the control women. This is a form of recall bias, and for
something as sensitive as illegal drug use, could be significant factor.
After all, the case mothers are much motivated to help than are control
mothers. Furthermore, the figure of 1 control mother in 204 using
marihuana is much lower than we would expect, based on other U.S.
surveys of marihuana use. One way to test this is to compare the case
mothers to the mothers of children with other cancers - that is, a different
control group. recall bias should not be a factor here, since both the cases
and "controls" are equally motivated. Fortunately, we had data from
several other CCSG studies of childhood cancers, in which exactly the
same question had been asked.

It can been seen from the data in Table 5 that the mothers of the ANLL
cases used marihuana significantly more frequently than mothers of
children with other cancers. The marihuana-exposed cases showed a
high frequency of cytogenetic abnormalities, specifically those involving
chromosomes 8 and 11.

We conclude that, although the association of marihuana exposure in
utero and subsequent development of ANLL has not been firmly
established, the evidence is strong enough to justify further study. In our
current case-control study of ANLL, we will include many more children
(approximately 630), and will include a much more detailed section in the
questionnaire on all aspects of illicit drug use. We are working closely
with behavioral psychologists who have experience in surveys of drug use,
and expect that the revised interview will substantially reduce the extent of
under-reporting.

If marihuana does have anything like the effect on ANLL risk that our
study suggests, it can be calculated that a substantial fraction of the cases
of ANLL in the U.S. could be attributed to this exposure.

REFERENCES

Abel E.L. (1983) Fetal, neonatal and adult effects of prenatal exposure to
    marijuana. In : *Marijuana, Tobacco, alcohol and reproduction*, Boca
    Raton, Florida, CRC Press, 31-33.
Aksoy M., Erdem S. (1978) Follow-up study on the mortality and the
    development of leukemia in 44 pancytopenic patients with chronic
    exposure to benzene. *Blood*, 52:285-292.
Bader J.L., Miller R.W. (1978) Neurofibromatosis and childhood leukemia.
    J. Pediatr. 92:925-929.
Bogdanoff B., Rorke L.B., Yanoff M., Warren W.S. (1972) Brain and eye

abnormalities. Possible sequelae to prenatal use of multiple drugs including LSD. *Am.J.Dis.Child* 123:145-148.

Brandt L., Nilsson P.G., Mitelman F. (1978) Occupational exposure to petroleum products in men and acute non-lymphocytic leukemia. *Br.Med.J.* 1:153.

Buckley J.D., Robison L.L., Swotinsky R., Garabrant D.H. *et col.* (1989) Occupational exposure of parents of children with acute non-lymphocytic leukemia: A report from the Children's Cancer Study Group. *Cancer Res.* 49:4030-4037.

Bunin G.R., Kramer S., merrero O., Meadows A.T. (1987) Gestational risk factors for Wilms' tumor: Results of a case/control study. *Cancer Res.* 47:2972-2977.

Carakushansky G., Neu R.L., Gardner L.I. (1969) Lysergide and cannabis as possible teratogens in man. *Lancet* 1:150-151.

German J. (1983) Patterns of neoplasia associated with the chromosome-breakage syndromes. In *Chromosome Mutation and Neoplasia*, German J. Ed., New York, Alan R. Liss, pp.97-134.

Golomb H.M., Alimena G., Rowley J.D., Vardiman J.W., Testa J.R., Sovik C. (1978) Correlation of ocupation and karyotype in adults with acute non-lymphocytic leukemia. *Blood* 52:1229-1237.

Hatch E.E., Bracken M.B. (1986) Effect of marijuana use in pregnancy on fetal growth. *Am.J.Epidemiol.* 124:983-986.

Hecht F., Beals R.K., Lees M.H., Jolly H., Roberts P. (1968) Lysergic acid diethylamide and cannabis as possible teratogens in man. *Lancet* 2:1087.

Hingson R., Zuckerman B., Amaro H. (1986) Maternal marijuana use and neonatal outcome: Uncertainty posed by self-reports. *Am.J. Public Health* 76:667-669.

Ishimaru T., Otake M., Ichimura M. (1979) Dose-response relationship of neutrons and gamma rays to leukemia incidence among atomic bomb survivors in Hiroshima and Nagasaki by type of leukemia.(1950-1971°. *Radiat. Res.* 77:377-394.

Kramer S., Meadows A.T., Jarrett P., Evans A.E. (1983) Incidence of childhood cancer: Experience of a decade in a population-based registry. *J.Natl.Cancer Inst.* 70:49-55.

Kramer S., Ward E., Meadows A.T., Malone K. (1987) Medical and drug risk factors associated with neuroblastoma. A case-control study. *J. Natl.Cancer Inst.* 78:797-804.

Linn S., Schoenbaum S.C., Monson R.R., Rosner R., Stubbelfield P.C., Rian K.J. (1983) The association of marijuana use with outcome of pregnancy. *Am.J.Public Health* 73:1161-1164.

Reimer R.R., Hoover R., Fraumeni J.F., Young R.C. (1977) Acute leukemia after alkylating-agent therapy of ovarian cancer. *N.Engl.J. Med.* 297:177-180.

Robison L.L., Nesbit M.E., Sather H.N., Level C. *et col.* (1984a) Down syndrome and childhood acute leukemia: A ten year retrospective survey from the Children's Cancer Study Group. *J. Pediat.* 105:639-644.

Robison L.L., Daigle A.E. (1984b) Control selection using random digit dialing for cases of childhood cancer. *Am.J.Epidemiol.* 120:164-166.

Robison L.L., Buckley J.D., Diagle A.E., Wells R., *et col.* (1989) Maternal drug use and risk of non-lymphoblastic leukemia among offspring. *Cancer* 63:1904-1911.

Silverberg E., Lubera J. (1986) Cancer statistics. *CA-A journal for Clinicians* 36:9-25.

Vigliani E.C., Morni A. (1976) Benzene and leukemia. *Environ.Res.* 11:122-127.

MARIJUANA AND UPPER AERODIGESTIVE TRACT MALIGNANCY IN YOUNG PATIENTS

PAUL J. DONALD, M.D., F.R.C.S.(C)

Department of Otolaryngology-Head & Neck Surgery
University of California, Davis School of Medicine
2500 Stockton Boulevard, Sacramento, California 95817

ABSTRACT

Twenty cases of advanced head and neck cancer in young patients, who were regular marijuana users, are presented. Numerous carcinogens, as well as respiratory irritants, are found in marijuana smoke. The active euphoria-producing agent, $\triangle$-9 tetrahydrocanabinol produces altered T-cell function in some in vitro systems and has been implicated in altered DNA, RNA, and protein synthesis and consequent chromosomal aberrations. Key perturbations may be in altered arginine metabolism resulted in disturbed nuclear histone synthesis and perhaps the unmasking of oncogene suppression.

KEY WORDS

Marihuana; Cancer : Head, neck, tongue; Young adults; DNA.

Malignancies of the upper aerodigestive tract are uncommon tumors, with about 55,000 to 60,000 new cases reported per year in the U.S. (Silverberg, 1984). This anatomical region includes the pharynx, larynx, oral cavity and sinonasal tract. Squamous cell cancer makes up the majority of these cases, and is characterized by slow growth and a trend toward regional metastases to adjacent lymph nodes. This is in contrast to many other tumor types throughout the body such as malignancies of the blood and lymph systems, carcinoma of the lung, and melanoma that tend to be more systemic in their behavior. The predominantly regional nature of squamous cell carcinoma makes it more amenable to control by local resection and or radiation therapy. Unfortunately these tumors often present with subtle symptoms, often suggestive of benign disease processes. If the patient is given a somewhat cursory examination, they are tragically too frequently treated as such. The symptoms invariably persist and the tumor is often quite large when finally diagnosed, necessitating wide-field resection possibly combined with postoperative radiation therapy. The result of resecting these lesions, especially the large ones, may leave the patient with devastating cosmetic and functional impairments.

The average age at which patients with squamous cell carcinoma of the upper aerodigestive tract present is 64. Tumors of this type are exceedingly unusual in anyone under the age of 40. The principal etiologic factors in the genesis of these tumors is the use of tobacco and the excessive consumption of alcohol. Smokeless tobacco seems to have a carcinogenic effect equivalent to that when smoked. Whether these factors act as a primary progenitor or trigger other factors that then act in concert to produce this disease is unclear at this time. The suppression of local immune mechanisms has been demonstrated and some form of genetic predisposition may be significant.

In 1986, the author first reported what was felt to be an association between smoking marijuana and the occurrence of squamous cell carcinoma of the head and neck region (Donald, 1986). It became suddenly apparent after a 32-year old patient asked if he thought his smoking of marijuana was the cause of his tumor. I then reviewed my total experience with these lesions in patients under the age of 40 and identified seven patients. Of these, only one had not used marijuana in the past or at the time of presentation. Parenthetically, she had irradiation to the face during adolescence for the therapy of acne and one could possibly invoke this as the most likely etiological factor.

Stimulated by this apparent association, I conducted a review of the literature searching for evidence of factors that could implicate marijuana as a possible cause of the disease. Abundant evidence became available from numerous research efforts indicating marijuana as a possible carcinogen and thereby as a cause of upper aerodigestive tract malignancy in the young.

Marijuana has an abundance of primary irritants and carcinogens. △-9 Tetra-hydrocannabinol (THC) has been found to cause chemical aberrations in DNA and RNA, resulting in perturbations in chromosomes. It has also been found to cause alterations in T-cell function.

Methods and Materials

In a review of the author's 20-year experience of over 3000 head and neck cancer cases, only 22 were identified who had squamous cell carcinoma and were under the age of 40. Those with adenocarcinoma, melanoma, and sarcoma were not included in the study. The tobacco smoking, alcohol intake, and marijuana smoking activities were researched. Their course of treatment and treatment outcomes were recorded.

Results

Table I describes the patient population who were marijuana smokers. Their average age was 26.2 years. Table II describes their treatment and eventual outcome. Of the 13 patients under age 40, only two were not marijuana smokers. The first patient was 38 years of age when first diagnosed with her extensive maxillary sinus tumor. She was not a marijuana smoker and did not use tobacco or alcohol. As previously described, she was treated as an adolescent with irradiation to the face for acne. The other non-marijuana user was 32 years of age at presentation with his relatively small (2 cm in diameter) carcinoma of the lateral tongue. Six months following wedge resection of the tongue, he developed a metastasis to the neck for which he underwent radical neck dissection. Pulmonary metastasis occurred four months later; and he succumbed to his disease within the following six months. He and patient #1 were the only deaths in the series. One patient is alive with disease four years postoperatively and is being maintained on chemotherapy. All others are alive and well without disease, for a 81.8% tumor free survival rate with only three patients followed for less than two years. One patient who was free of his original tumor for 15 years has now a second primary carcinoma of his temporal bone.

Of the marijuana smokers (Table I), only 50% (four patients) were also tobacco smokers and three patients were heavy imbibers of alcohol. In this group, only one patient died, one patient is alive and well with disease, while the others remain free of cancer (Table II). The vast majority of head and neck cancers recur locally between one and two years postoperatively. We have one patient who is disease free at one year, and another at two years. One patient has a small local recurrence, but refuses further surgery because of probable loss of her remaining eye. She has been maintained on chemotherapy for two years. The remainder are alive and well between three and ten years postoperatively.

TABLE I

PT. #	MARIJUANA CONSUMPTION	TOBACCO USE	ALCOHOL
1	HEAVY DAILY USE, ALSO JOINTS IN "EMBALMING FLUID"	4 YRS	3-4 YRS
2	HEAVY DAILY USE	2-5 PKS DAILY	"HEAVY"
3	2 JOINTS/DAY "MANY YEARS"	12 PK/YRS	"SOME"
4	3-7 JOINTS/WK "MANY YEARS"	0	0
5	"HEAVY USER"	6 PK/YRS	7-10 SHOTS HARD LIQUOR DAILY
6	1-2 PIPES/DAY "MANY YEARS"	0	OCC. SOCIAL DRINKER
7	OCC. USE IN HIGH SCHOOL (2-3 YRS)	0	OCC. SOCIAL DRINKER
8	OCC. USE IN HIGH SCHOOL AND COLLEGE	0	0
9	15 YRS USE, PLUS COCAINE AND METHAMPHETAMINE	0	OCC. SOCIAL DRINKER
10	FREQUENT USER	1 PK DAILY	2 BEERS DAILY
11	FREQUENT USER	8 PK/YRS	"HEAVY INTAKE"
12	FREQUENT USER	0	0

TABLE I continued

PT. #	MARIJUANA CONSUMPTION	TOBACCO USE	ALCOHOL
13	FREQUENT USER	1-2 PK/DAY	ALCOHOLIC
14	FREQUENT USER	1 PK/DAY	ALCOHOLIC
15	OCC. USE IN HIGH SCHOOL AND COLLEGE	0	0
16	HEAVY USER	1-2 PK/DAY	ALCOHOLIC
17	FREQUENT USER	0	OCC. SOCIAL DRINKER
18	FREQUENT USER	1 PK/DAY	OCC. SOCIAL DRINKER
19	FREQUENT USER	8 YRS OFF AND ON	0

TABLE II

PT#	SEX	AGE	SITE	RX	OUTCOME
1	M	19	PYRIFORM SINUS	RAD	DEAD - 2 MOS
2	M	28	TONSIL	SURG/RAD	A&W - 10 YRS
3	M	27	LIP	SURG/RAD	A&W - 4.5 YRS
4	M	34	TONSIL	SURG/RAD	A&W - 5.5 YRS
5	M	20	TONGUE	SURG/RAD	A&W - 7 YRS
6	M	38	TONSIL	SURG/RAD/CHEMO	A&W - 5 YRS
7	F	26	TONGUE	SURG	A&W - 2 YRS
8	M	35	TONGUE BASE	SURG/RAD	A&W - 1 YR
9	F	33	ETHMOID SINUS	SURG/RAD/ CHEMO	ALIVE - 3.5 YRS WITH DISEASE
10	M	24	TONGUE	SURG	A&W - 4 MOS
11	M	22	TONGUE	SURG	A&W - 15 YRS*
12	M	30	TONGUE	SURG	A&W - 3 MOS
13	F	39	LARYNX	SURG	A&W - 2 MOS
14	M	41	NECK/ SOFT PALATE	SURG/RAD	A&W - 4 MOS
15	F	35	TONSIL	SURG	A&W - 8 MOS
16	M	40	TONGUE	RAD/CHEMO	AWD
17	M	23	NECK	SURG	A&W - 5 MOS
18	M	32	TONGUE	SURG	A&W - 2 YRS
19	M	39	UPPER ALVEOLAR RIDGE	SURG	A&W - 6 MOS

* NEW TUMOR OF TEMPORAL BONE AT AGE 37

Because of the extensive nature of the tumors in these patients, only one individual was treated by surgery alone. One other had radiation therapy because he was considered inoperable and incurable; and he died two months following presentation. The remainder were treated by a combination of radical surgery followed by a full course of radiation therapy. An illustration of the radical nature of this type of surgery is exemplified by the patient in Figure 1. This is basically the type of surgery employed in patients 2, 4, 5, and 6. To follow are some case reports demonstrating the mode of presentation and outcome of some of these patients.

Case Reports

Case 1. A 19-year old Vietnam war veteran was seen in January 1971 with a six-week history of a rapidly enlarging, painless mass in the neck. An incisional biopsy (by another surgical service) led to a diagnosis of squamous cell carcinoma. On referral to the ENT service at the Veterans Hospital and the University of Iowa Department of Otorhinolaryngology and Maxillofacial Surgery, examination revealed a rock-hard, slightly tender, 10 x 6 cm mass in the right side of the neck, which was fixed to the underlying structures, including the paraspinous muscles. Indirect examination revealed a large, fungating mass in the pyriform fossa. A biopsy revealed squamous cell carcinoma. Further inquiry revealed that the patient had smoked cigarettes since his early teens and began smoking marijuana while in the service. He as a heavy consumer of the drug, and recently had taken to smoking "joints" dipped in embalming fluid. Despite vigorous treatment with radiation therapy, he died within two months.

Case 3. A 27-year old white man was first seen with an enlarging lesion in the lower lip in January 1984. In June 1984, a wedge resection incompletely removed a squamous cell carcinoma that was a little more than 2 cm in diameter. In September 1984, the patient was seen again with a 2 X 2.5 cm hard nodule, extending from the vermilion border of the lower lip into the skin of the chin. The tumor was not fixed to the mandible (Fig. 1). Bilateral cervical adenopathy was palpable. The patient had smoked one package of cigarettes per day for 12 years, and had smoked two "joints" of marijuana per day for "many years." In September 1984, an extended resection of the lip and a bilateral conservation neck dissection were done. The margins surrounding the tumor in the lip and chin were clear, and the enlarged lymph nodes in both sides of the neck were uninvolved with cancer. The patient is alive and well without disease, five years after surgery.(Fig. 2).

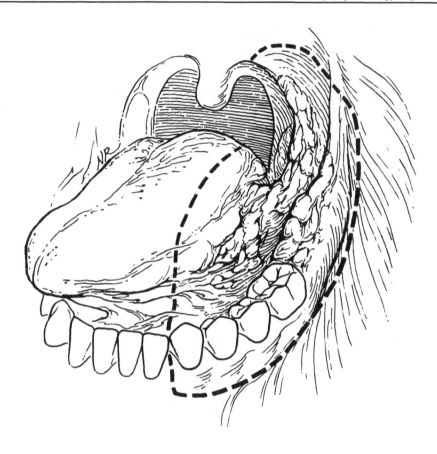

Figure 1   Elderly man with large carcinoma of the tongue, floor of
mouth, and oropharynx whose tumor extent and subsequent surgery
closely resembles patients #2, 4, 5, and 6. Proposed resection outlined.

(From Donald PJ, Chole RA: Superior based trapezius flap.
Laryngoscope, 94(7):970 & 971, 1984.)

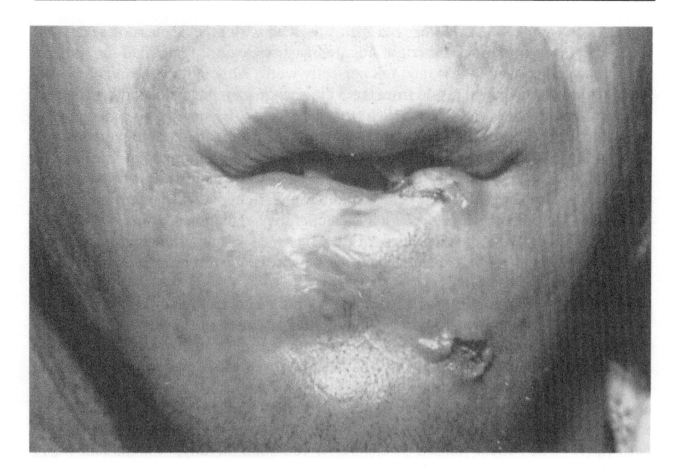

Figure 2   Case 3, showing large recurrent squamous cell carcinoma of the lower lip.

(From: Donald, P.J. (1986). Marijuana Smoking - Possible Cause of Head and Neck Carcinoma in young patients. <u>Otolaryngology-Head & Neck Surgery 94(4)</u>:518.)

Case 4. In May 1984, a 34-year old man was seen with a neck mass of seven to eight months duration in the right jugulodigastric region. He had no other symptoms and no weight loss, and felt perfectly well. Although the head and neck examination was negative (aside from the 3 cm neck mass), panendoscopy revealed a 4.5 cm endophytic tumor of the right tonsillar fossa. Biopsy led to a diagnosis of squamous cell carcinoma. The patient had never drunk alcoholic beverages or smoked tobacco, but he had regularly smoked three to seven "joints" of marijuana per week for many years. In early June 1984, he underwent a composite resection, with a lingual split of the mandible and right radical neck dissection. A postoperative fistula developed, and postoperative radiation therapy had to be discontinued at 2700 rads because of the onset of osteoradionecrosis of the mandible. This eventually cleared, and the patient is at present free of tumor.

Case 6. This 38-year old man first came to an outside facility in October 1983 with a rapidly growing mass in the left jugulodigastric area of the neck. The mass was excised in January 1984, and a "cyst with a cancer in it" was found. A panendoscopic examination (done two days later) revealed no evidence of primary neoplasm. The tonsils were not removed at this time and did not appear to be abnormal. The patient underwent radiation therapy for two weeks, but a CAT scan revealed further adenopathy and a radical neck dissection was immediately performed. The patient refused postoperative radiation therapy. The patient was referred to the Department of Otolaryngology/Head and Neck Surgery at the University of California, Davis in November of 1984 with a left neck mass of six weeks duration. He had a history of dysphagia and odynophagia for 11 weeks, as well as hoarseness and stridor. He had experienced left otalgia since the radical neck dissection of February 1984. There had been a 14 lb weight loss. Physical examination revealed a massive ulcerating neoplasm in the left lateral pharyngeal wall, extending from below the torus tubarius almost to the vallecula. Two masses were discovered in the left neck. One, measuring 5 cm in diameter, fixed to the skin with possible carotid fixation deeply, and a smaller, 1.5 cm in diameter, mobile mass just below it (Fig. 3). The patient had never smoked tobacco and was only an occasional social drinker. However, he had smoked one to two pipes of marijuana daily for many years. Panendoscopic examination revealed tumor extension from the superior pole of the tonsil to the ipsilateral aryepiglottic fold. Biopsies led to a diagnosis of squamous cell carcinoma. An exploratory operation was performed in November 1984. The tumor was found to be fixed to a long stretch of the internal and common carotid arteries. A wide debulking of the neck tumor was done, the pharyngeal neoplasm was untouched, and the skin involvement was resected. The neck wound was closed with a pectoralis major flap. The

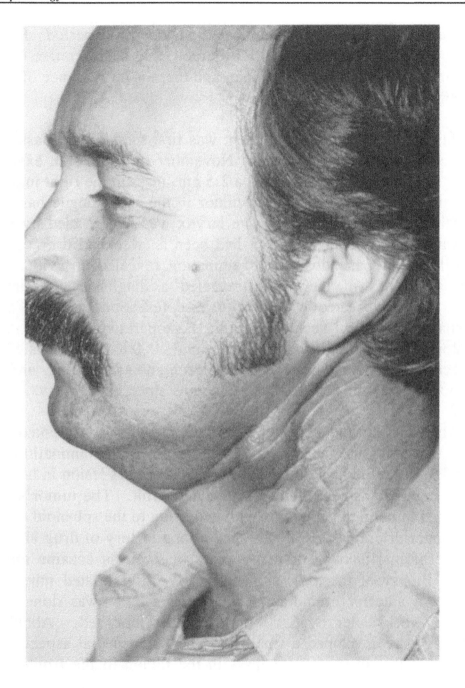

Figure 3

Case 6, showing a large mass of the neck, fixed to the overlying skin and the underlying carotid artery.

(From: Donald, P.J. (1986). Marijuana Smoking - Possible Cause of Head and Neck Carcinoma in young patients. <u>Otolaryngology-Head & Neck Surgery 94(4)</u>:519.)

patient was sent home after an uncomplicated postoperative period. He was given a combination of cis-platinum and 5-fluorouracil, and a full course of irradiation therapy. The response has been complete, and there is no evidence of gross tumor at the present time.

Case 8. This 35-year old medical student was first seen with a mass in the right neck and a complaint of sore throat in November of 1988. On examination he had a slight "hot potato" voice. He had a 2.5 cm. mass in the right jugulodigastric area and a moderately large ulcerating tumor in the right tongue base measuring approximately 3.5 cm in diameter. The larynx was not invaded by tumor. He was a nonsmoker and nondrinker, but had been a "weekend user" of marijuana while in college. A panendoscopic examination revealed no evidence of other neoplasms. Biopsies of the lesion revealed a diagnosis of squamous cell carcinoma. He was treated with tongue base resection, removal of the pre-epiglottic space and radical neck dissection. Postoperative radiation therapy was given at a dosage of 6500 rads to the primary and 4000 to both necks. The patient made an excellent recovery, has excellent speech and swallowing, and is back in medical school. He is disease free at 18 months since his surgery.

Case 9. This 33-year old checker at a food store first presented with nasal obstruction, epistaxis, and facial pain in 1986. Physical examination revealed a mass in her left nasal fossa. Radiographs revealed a mass lesion in her right nasal fossa, upper maxillary sinus, ethmoid bloc, and orbit. The tumor's geographic center appeared to be in the ethmoids with extension to the sphenoid sinus and the floor of the anterior cranial fossa. She had a long history of drug abuse, being a frequent user of marijuana over many years, as well as cocaine and methamphetamines. Biopsy of the mass showed poorly differentiated mucoepidermoid carcinoma. An anterior craniofacial resection   was done. This was followed four weeks later by 6500 rads to the tumor bed. About one year postoperatively, she developed a small mass near the medial aspect of the right eyebrow. A CAT scan revealed a mass in the region of the frontal sinus with extension intracranially into the right orbit and under the forehead skin. Further resection would require exenteration of the remaining eye and repeat craniotomy. She has been maintained on chemotherapy and to this date has declined further surgery.

Discussion

These data are presented because of the surprising coincidence of marijuana smoking and head and neck cancer in this group of young patients. With the exception of nasopharyngeal carcinoma in young Chinese, in my experience with more than 3000 head and neck cancers, there have been no other carcinomas seen in persons of this age group other than this cohort of patients. The average age in this group when the tumor first appeared was 26.2 years, with a minimum of 19 and a maximum of 38 years. This is in contrast, for example, to the average age of onset of pyriform fossa carcinoma, in our experience, (Donald, 1980)) of 64.6 years, which is generally representative of squamous cell carcinoma patients.

The frequent association of head and neck cancer with tobacco and heavy alcohol consumption is well known. In these patients, some had never smoked tobacco and were either occasional drinkers or nondrinkers. Some studies suggest that smoking marijuana and tobacco may have a synergistic effect, one that is greater than the effects of each individually (American Thoracic Society, 1975). Recent experimental work has shown that greater amounts of primary irritants to the tracheobronchial tree are found in marijuana smoke than in tobacco smoke (American Thoracic Society, 1975; Fehr, 1972; Fehr, 1983). These include ammonia, hydrocyanic acid, acrolein, and benzene. Fehr and Kalant (1983) found that the smoke condensate - or tar - from marijuana smoke had a 70% higher concentration of naphthalene than tobacco smoke. It also contained an especially abundant amount of benzopyrene, a definitive carcinogen. Other carcinogenic polynuclear aromatic hydrocarbons, such as benzanthracene, are 50% higher in marijuana (American Thoracic Society News, 1975). Nitrosamines are found in equal concentrations in both kinds of smoke. A further distressing fact is that these carcinogens may be concentrated by the paraphernalia used to enhance the euphoric effect of marijuana by increasing the actual amount of drug in the smoke. Herbicides such as Paraquat may exist as contaminants and may be carcinogenic.

Numerous experiments have been carried out on tissue cultures (and in vivo) in an attempt to ascertain the effect of marijuana smoke and its active euphoria-producing hydrocarbon $\triangle$-9-tetra-hydrocannabinol ($\triangle$-9-THC) on cellular DNA, RNA, and protein. Experiments by Leuchtenberger et al (1973 a & b) have shown that the tobacco and marijuana smoke markedly alters cellular morphology, mitosis, and DNA synthesis in fresh lung explant. They have also demonstrated atypical proliferations of cells, as well as alterations in DNA content and chromosomal complement. Other studies, quoted in the review by Fehr and Kalant

(1972), showed reduced DNA and RNA synthesis in cell cultures exposed to cannabis smoke.   This results in the production of chromosome breaks, the synthesis of hypoploid cells (cells containing a reduced number of chromosomes), and cells with micronuclei (Leuchtenberger, 1973).

A fascinating theory concerning the mechanism of action of $\triangle$-9-THC is its effect in depletion of cellular arginine.   This results in a deficiency in nuclear histone synthesis.   Nuclear histones are the central core material around which the double-helical strands of DNA are wound.   Changes in nuclear histones produce characteristic ultrastructural alterations.   Many of the enzymes responsible for DNA and RNA metabolism have arginine at their specific site of action.   With the arginine depleted because of the influence of $\triangle$-9-THC, these enzymes are partially or completely inactivated (Fehr, 1983).   This is especially manifested in spermatids and leukocytes.   Chromosomal abnormalities, growth retardation, disturbed gametogenesis and function, as well as Cushmanimmunosuppression, can all be explained on the basis of arginine deficiency.   The search for the possible depletion or masking of an oncogene suppressor that may thus result in oncogene expression is in progress.

There is abundant definitive evidence in animals that marijuana smoke and $\triangle$-9-THC decrease both humoral and cell-mediated immunity.   However, the evidence in human beings is somewhat conflicting.   Petersen et al (1975) found lowered T-cell counts in chronic marijuana smokers compared to non-smokers, as well as less effective phytohemagglutinin (PHA) stimulation of lymphocytes.   Additionally, their polymorphonuclear leukocytes were less able to phagocytose yeast cells, compared to nonsmoking controls.   However, no alterations in immunoglobulins were noted.   In a study by Morishima and Nahas (1975), marijuana smokers demonstrated impairment of orderly T-cell replication, and this was duplicated in vitro by the addition of $\triangle$-9-THC to lymphocyte cell cultures of nonsmokers.   A more recent report by Nahas (1989) describe both cellular and humoral immune suppression in mice who were exposed to marijuana smoke or received $\triangle$-9-THC parenterally.   Perturbations of monocyte and macrophage function were found as well.   Zimmerman (1989) et al found a dose dependent depression of both the inductive and productive phases of humoral immunity in mice exposed to enantiomeric cannabinoid.   Cushman et al (1975, 1976) showed a marked reduction in T-cell function when compared to either tobacco smokers or nonsmokers.   Although the exact clinical significance of rosette formation is unclear, there is an association of suppression of this phenomenon and the altered immunity often found in cancer (Hong, 1975).   Specter (1989) found a dose related suppression of natural killer cells of human peripheral blood mononuclear cells when exposed to △-9-THC in concentrations of 7.5 gm/ml or greater. Waltz et al (1977) found that △-9-THC and cannabidiol, when introduced to the blood of healthy non-drug taking volunteers, caused suppression of interleukin-1, interleukin-2 and cytokine secretion. Analysis of a group of our patients who had stopped smoking marijuana but had contracted head and neck cancer showed no T-cell abnormalities.

On the other hand, Rachelefsky et al (1977) found no effect on the DNA synthesis of resting lymphocytes or on PHA stimulation when △-9-THC was added. This was supported by the study of White et al (1975) who found no difference between the in vitro responses to PHA or pokeweed mitogen on lymphocytes from 12 healthy marijuana smokers compared to 12 age-matched controls. However, in this study the smokers used the weed for only 30 days.

The disturbing fact is that, although the effect of THC on the immune system is controversial at the present time, this may simply be a reflection of its concentration. The concentrations of this euphoria-producing chemical are much higher in the products of hybrids of cannabis plants that have been more recently produced. It has been estimated that some present day marijuana has a 600% higher concentration of THC than the product seen in the 1960's. Concentrations of THC seen in some of the sinsemilla hybrids reaches as high as 7%, equally high to double that level that is seen in hashish.

Only scant evidence exists suggesting the direct carcinogenic effect of marijuana smoke or tar. Marijuana smoke has been found to produce squamous metaplasia of ciliated respiratory epithelium in dog trachea when the animals were exposed to four marijuana cigarettes daily for 30 months (White, 1975). This was a far greater effect than when the dogs were exposed to tobacco smoke. Rosenkrantz and Fleischman (1979) demonstrated precancerous changes in the bronchial epithelium of rodents that were exposed to marijuana for months on a smoking machine. Magus and Harris (1971) demonstrated premalignant alterations in the cells of the sebaceous glands of rodent skin when painted with marijuana "tar." Hoffman et al (1975) actually produced tumors with a similar experiment. Montour et al (1981) showed a statistically increased incidence in the production of irradiation-induced tumors in rats that were injected with marijuana extract. In human subjects, Henderson et al (1972) showed atypical cells (in bronchial biopsy specimens) in chronic hashish smokers.

Information directly implicating marijuana in the genesis of human malignancies was virtually nonexistent prior to 1986. Since that time a few reports have emerged suggesting the association. Colon (1980) discovered 129 oral papillomata in 105 young male inmates in three federal prisons over a ten year period. Their ages varied between 20 to 36 years, with 71 patients being in their twenties. Al patients admitted to heavy use of marijuana of at least once per day for two years or more. No histologic evidence of malignancy was found. Endicott (1989) describes a series of 50 youthful patients with cancers of the head and neck collected from the South Florida area. All of these individuals were users of marijuana.

Ferguson (1989) and Taylor (1988) have reported on the possible relationship of marijuana to cancer. Ferguson (1989) described a young man of 27 who died of widespread metastatic carcinoma of the lung who was a heavy marijuana and hashish smoker. Taylor (1988) reported on ten cases of carcinoma in youthful marijuana smokers, two of whom had primary carcinoma of the lung and the remaining eight with tumors in the upper aerodigestive tract. All the patients were less than 40 years old. Three of the ten patients went on to die of their disease. A further study by Neglia (1990) showed that pregnant women who smoked marijuana had a 10.6 times greater chance of giving birth to a child that would eventually develop acute nonlymphocytic leukemia.

A remarkable feature of the group of patients in our series is the rather high survival rate they enjoy. Of the marijuana patients who had resectable cancers, only two have had a local recurrence and the remainder are alive and well without disease. Only two patients had small T1 (2 cm or less in diameter) lesions. Most of the remainder had T3 & T4 lesions, many of them with lymph nodal metastasis. The survival rate for these large tumors is usually around 25-35%. It is interesting to speculate what may be the reason for this high success rate other than the factor of age. However, an important caveat at this point is that five-year tumor free survival is the bench test by which we gauge results, and some of the patients presented here have not gone five years. On the other hand, local recurrence of tumor usually appears at one to two years in 80-90% of cases.

Disquieting information concerning the effects of marijuana smoke on cellular genetics and the immune system raises the level of suspicion, suggesting the implication of this drug in the genesis of cancer. Certainly the incidence of marijuana smoking in this series of 20 patients with head and neck squamous cell carcinoma intensifies the suspicion.

REFERENCES

American Thoracic Society. (1975). Position paper: Marijuana and the lungs. <u>American Thoracic Society News</u>, Spring, p 7.

Colon, P.G. (1980). Oral papilloma in marijuana users. Quintessence International - Report #1839, pp 75-80.

Cushman, P., Grieco, M., Gupta, S. (1975). Reduction in T-lymphocytes forming active rosettes in chronic marijuana smokers. <u>Int J Clin Pharmacol Biopharm 12</u>:217-220.

Cushman, P., Khurana, R. (1976). Marijuana and T-lymphocytes forming active rosettes. Clin Pharmacol Ther 19:310-317.

Donald, P.J., Hages, R.H., Dhaliwal, R. (1980). Combined treatment for pyriform sinus carcinoma using postoperative irradiation. <u>Otolaryngol Head Neck Surg 88</u>:738-744.

Donald, P.J. (1986) Marijuana smoking - possible cause of head and neck carcinoma in young patients. <u>Otolaryngology-Head and Neck Surgery 94(4)</u>:517-521.

Endicott, J. (1989). Personal communication.

Fehr, K.O., Kalant, H. (1972). Analysis of cannabis smoke obtained under different combustion conditions. <u>Can J Physiol Pharmacol 50</u>:761-767.

Fehr, K.O., Kalant, H. (1983). Cannabis health hazards: Proceedings of an ARF/WHL Scientific meeting on Adverse Health and Behavioral Consequences of Cannabis Use. Toronto. Addiction Research Foundation, pp 501-563.

Ferguson, R.P., Hasson, J., Walker, S. (1989). Metastatic lung cancer in a young marijuana smoker. <u>JAMA 261(1)</u>:41-42.

Henderson, R.L., Tennant, F.S., Guerry, R. (1972). Respiratory manifestations of hashish smoking. <u>Arch Otolaryngol95</u>:248-251.

Hoffman, D., Brunnemann, K.D., Gori, G.B., Wynder, E.L. (1975). On the carcinogenicity of marijuana smoke. <u>Recent Adv Phytochem 9</u>:63-81.

Hong, R., Horowitz, S. (1975). Thymosin therapy creates a "Hassal?" N Eng J Med 292:104-106.

Leuchtenberger, C., Leuchtenberger, R., Schneider, A. (1973). Effects of marijuana and tobacco smoke on human lung physiology. <u>Nature 241</u>: 137-139.

Leuchtenberger, C., Leuchtenberger, R., Ritter, U. (1973). Effects of marijuana and tobacco smoke on D.N.A. and chromosomal compliment in human lung explants. <u>Nature 242</u>:403-404.

Magus, R.D., Harris, L.S. (1971). Carcinogenic potential of marijuana smoke condensate. Fed Proc (Abstract) 30:179.

Mitsuyama, S.S., Jarvik, L.F., Fu, T.K., Yen, F.S. (1976). Chromosome studies before and after supervised marijuana smoking. In Braude MC and Szara S (Eds.): Pharmacology of Marijuana, Vol. 2. New York, Raven Press, pp 723-729.

Montour, J.L., Dutz, W., Harris, L.S. (1981) Modification of radiation carcinogenesis by marijuana. Cancer 47:1279-1285.

Nahas, G.G., Desoize, B., Armand, J.P., Hsu, J., Morishima, A. (1975). Natural canabinoids: Apparent depression of nucleic acids and protein synthesis and cultures of human lymphocytes. In Szara S and Braude M (Eds.): Pharmacology of Marijuana. New York, Raven Press, pp 177-186.

Nahas, G.G. (1989). Psychoactive drugs and immune response. Presentation at "Drugs of Abuse, Immunity, and Immunodeficiency." Clearwater, FLA, December 13-15.

Neglia, J. (1990) Maternal use of marijuana and incidence of leukemia. Presentation at Physiopathologie des Stupefiants, Paris, France, May 31.

Palmer, J. (1985). Personal communication.

Petersen, B.H., Lemberger, L., Graham, J., Dalton, B. (1975). Alterations in the cellular-mediated immune responsiveness of chronic marijuana smokers. Psychopharmacol Commun 1:67-74.

Rachelefsky, G.S., Opelz, G. (1977). Normal lymphocyte function in the presence of △-9-THC. Clin Pharmacol Ther 21:44-46.

Rosenkrantz, H., Fleischman, R.W. (1979. Effects of cannabis on lungs. In Nahas, G.G., Paton, W.D.M. (Eds.): Advances in the Biosciences, Vols. 22 and 23. New York, Pergamon Press, pp 279-299.

Roy, P.E., Manan-Lapoint, F., Huy, N.D., et al (1976.) Chronic inhalation of marijuana and tobacco in dogs: Pulmonary pathology. Res Commun Chem Pathol Pharmacol 14:305-317.

Silverberg, E. (1984). Cancer Statistics. Cancer 34:9-2.

Specter, S. (1989). Effects of marijuana on natural killer cell activity in man. Presented at "Drugs of Abuse, Immunity, and Immunodeficiency. Clearwater, FLA, December 13-15.

Taylor, F.M. (1988). Marijuana as a potential respiratory tract carcinogen: A retrospective analysis of a Community hospital population. Southern Med J 81:1213-1216.

Waltz, B., Scuderi, P., Watson, R.R. (1989). Influence of marijuana components (THC and CBD) on human mononuclear cells cytokine secretion in vitro. Presented at "Drugs of Abuse, Immunity, and Immunodeficiency. Clearwater, FLA, December 13-15.

White, S.C., Brin, S.C., Janicki, B.W. (1975). Mitogen-induced blastogenic responses of lymphocytes from marijuana smokers. Science 88:71-73.

Zimmerman, A.M., Titischov, N., Zimmerman, S., Mechoulam, R. (1989). Enantiomeric cannabinoid effects on the immune system of mice. Presented at "Drugs of Abuse, Immunity and Immunodeficiency. Clearwater, FLA, December 13-15.

6. Effects on reproductive function

EFFECTS OF CANNABIS ON REPRODUCTION

Herbert Tuchmann-Duplessis

Laboratoire d'Embryologie, Faculté de Médecine des Sts Pères, Paris, France.
Membre de l'Académie Nationale de Médecine, Paris.

ABSTRACT

Investigations on animals and the clinical data demonstrate that cannabis use while not producing somatic teratogenesis has noxious effects on reproduction. Spermatogenesis and ovogenesis and prenatal development are impaired, damaging effects of cannabis on reproduction observed in experimental animals are confirmed by clinical observations. Infants born from marihuana smoking mothers are of shorter height, weight less and have smaller head circumference at birth than a control group. The influence of cannabis on reproduction is determined by its action on the hypothalamus. In the absence of the release of the hypothalamic hormone LHRH, the hypophysis does not release the gonadotropic hormones FSH-LH and therefore the gonadal hormones don't get the necessary stimulation. The noxious effects of cannabis on reproduction, which are well demonstrated, clearly indicate that cannabis should be considered as a dangerous compound for humans.

KEY WORDS

Cannabis; gametogenesis; embryogenesis; animal and human fetal development; post-natal behavioral effects.

INTRODUCTION

In the 1970s, Rosenkrantz and others reported that marihuana products were toxic to fetal development in all species studied : fish, birds, rodents, hamsters, rabbits, dogs and monkeys offspring also displayed retarded development and behavioral anomalies. Others experimental and clinical studies demonstrated that cannabis does not produce somatic teratogenesis but does negatively affect all phases of reproductive function.

I. EXPERIMENTAL STUDIES

Cannabis (hashish, marijuana) has negative developmental effects on gametes and offspring if experimental exposure to it occurs during any of the three following periods:
        1 - gametogenesis (oögenesis, spermatogenesis).
        2 - embryogenesis (organogenesis, foetal development)
        3 - post-natal (via maternal milk, "exterior gestation").

Gametogenesis (oögenesis, spermatogenesis)

Animal studies by Henrich *et al*(1) (1983) showed that THC-treated female mice had a higher incidence of abnormal fertilized ova than did untreated control animals, suggesting effects on the maturing oöcyte.
In 1982, Dalterio (2) reported reduced fertility, increased pre- and post-natal foetal death and reduced litter size among the offspring of male mice treated with high doses of cannabis. She concluded that cannabinoïds can be mutagenic and teratogenic in mice. Heavy marihuana smoking is associated with decreased spermatogenesis and increase in abnormal non ovoid form of sperm (3). Consequences of these anomalies on fertility and offspring of cannabis smokers remain to be ascertained.

Embryogenesis (organogenesis, foetal development)

Since THC crosses the placental barrier, its metabolites concentrate in the foetus, which eliminates them more slowly than do either the placenta or the mother.

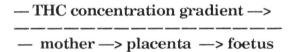

                        — THC concentration gradient —>
                 ———————————————————
                 — mother —> placenta —> foetus

THC is found in yolk sac, amniotic fluid and foetal brain, as well as in other developing tissues and organs. There also appears to be a negative effect of cannabis on maternal support (placental circulation and function) of the developing embryo and foetus in primates.

Experimental studies have reported developmental effects and lower birthweight in various species (4). Although no convincing evidence of teratogenic effect has been presented, there exists a common observation of dose-related increased incidence of foetotoxicity.

Post-natal period (via maternal milk)

Cannabis derivatives in maternal milk fed to newborn and young animals ("exterior gestation") constitutes post-natal administration of the drug, with effects similar to those reported for exposure during embryogenesis. Such cannabis administration affects the growth and development programmed to occur during and beyond the corresponding period. These effects include retarded growth and delayed development, behavioral disturbances (hyperactivity, increased irritability, hyperresponsiveness to visual, auditory and social stimuli, sleep disturbances and learning deficit.

II. OBSERVATIONS IN MAN

The effects of cannabis on foetal development in humans were reported on by Dr Steven Parker of the Developmental and Behavioral Pediatrics Division of Boston City Hospital. Dr Parker described the results of an epidemiological study (5) in which it was found that there was an average 79 gram lower birthweight among newborn infants whose marihuana-smoking mothers participated in the study. Urine tests (EMIT, followed by HPLC in positive EMIT tests) were carried out and the results compared with the mothers' self-reporting statements regarding drug use.

TABLE 1 : MARIJUANA USE DURING PREGNANCY

	URINE ASSAY		
	Positive	Negative	Total
Self-report positive	149	129	278
Self-report negative	53	895	948
TOTAL	202	1024	1226

Table 1 shows that of the 1,226 women who participated in the study, 149 (12.2%) who reported and 53 (4.3%) who did not report using marihuana during pregnancy (total 202; 16.5%) had HPLC-confirmed positive urine tests for cannabis during pregnancy or in the perinatal period.

TABLE 2 : FREQUENCY OF REPORTED MARIJUANA USE DURING PREGNANCY BY POSITIVE OR NEGATIVE URINE ASSAY FOR MARIJUANA

FREQUENCY	POSITIVE ASSAY n = 149	NEGATIVE ASSAY n = 129
3 or more times per week	42 %	19 %
1 - 2 times per week	30 %	26 %
1 - 3 times per month	19 %	24 %
Less than 1 time per month	9 %	31 %

Chi square 29.4; d.f 3; p< 0.001

Table 2 shows that if urine assays had not be carried out, 16% of the pregnant marihuana users who did not report using it would not have been identified, which would have resulted in their being misclassified as non-users, thereby reducing the significance of the results comparing users and non-users among pregnant women.

TABLE 3 : UNIVARIATE ANALYSIS: POSITIVE URINE ASSAY AND NEONATAL GROWTH, GESTATIONAL AGE AND CONGENITAL ANOMALIES

	Non-Use	Positive Assay	Significance
Birthweight (g)	3,260	2,980	<.001
Length at birth (cm)	49.8	48.3	<.001
Head circumference (cm)	34.3	33.4	<.001
Gestational age (weeks)	39.2	38.9	N.S.
Congenital anomalies ≥ 3 minor or 1 major	9 %	9 %	N.S.
Congenital anomalies ≥ 1 minor	38 %	45 %	N.S.

Table 3 demonstrates that marihuana use is correlated with

diminished birthweight (79 g), shorter length at birth (0.5 cm) and smaller head circumference.

Parker's results lead to the conclusion that maternal use of cannabis during pregnancy is associated with impaired foetal growth. Moreover, this study points out that maternal self-reporting of marihuana use and non-use must be verified by urine assay.

The foetotoxic effect of THC observed in both experimental animals and in human newborn appears to be non-specific and may be related to the general properties of cannabis, which, in micromolar concentrations is known to decrease synthesis of macromolecules. Furthermore, other toxic components of marihuana smoke may also be responsible for the observed effects. However, since THC crosses the placental barrier and binds to its specific receptors in foetal brain and cerebellum and since cannabis is known to be a neurotrope in nanomolecular concentrations, careful neurological and behavioral evaluation of infants and young children should be carried out in order to detect possible deficits in the development and maturation of the central nervous system. One might expect the offspring of women who consume marihuana during pregnancy to demonstrate behavioral deficit, which has been observed in experimental animal studies.

Working with rodents, Borgen in 1971[6], Vardaris in 1976 [7], and Sassenrath in 1979 [8], using monkeys, reported retarded post-natal development in the offspring of marihuana- or THC-treated mothers. These symptoms were designated by the authors as manifestations of behavioral teratology.

III. MECANISM OF ACTION

The main effect of cannabis is on the central nervous system. Since reproduction is under the control of the hypothalamus and of the hypophysis, it is to be expected that cannabis impairs reproduction by a central mechanism. The hypothalamic control of reproductive function is mediated through gonadotropic releasing factor LH RH. This neuro-hormone controls the gonadotropic activity of the hypophysis; not only the secretion but also the synthesis of sex dial hormones. The secretion of the gonadal hormones is under two systems of retro-control, one by a positive retro-action and the other under a negative retro-control.

CONCLUSION

The effect of marihuana on prenatal development is discrete, in contrast to the actions of numerous teratogenous substances such as antitumeral agents, thalidomide and retinoic acid, which cause immediately-detectable morphological malformations. Marihuana affects metabolic processes, resulting in a slowing-down of intrauterine growth as well as prematurity and nervous disorders.

As for every drug absorbed during pregnancy one question arises: What are the consequences on postnatal development? Since epidemiological study reveals only discrete modifications, one could consider the use of this substance to be relatively harmless. But such a conclusion would be mistaken, for it has been shown that marihuana increases the risk for developing cancer (Robison, 1989) (9). Leukemia is 10 times more frequent in children born to mothers using marihuana than in children who remained unexposed. It is therefore obvious that we are not confronted here with a harmless substance, but with a dangerous one. The struggle against its use must be continued and intensified.

The noxious effects of THC on foetal brain development and on subsequent post-natal maturation observed in experimental preparations remain to be studied in human subjects.

REFERENCES

1. Henrich, R.T., Shinohara, O., Nogawa, T., and Moroshima, A. (1983): Effects of chronic administration of THC on early embryonic development of mice. In: The Cannabinoids:Chemical, Pharmacologic and Therapeutic Aspects edited by S. Agurell, W.L. Dewey and R.E. Willette Academic Press, New York.
2. Dalterio, S., Badr, F., Bartke, A., and Mayfield, D. (1982): Cannabinoids in male mice. Effects of fertility and spermatogenesis. Science, 216: 315-316.
3. Hembree III, W.C., Nahas G.G., Zeidenberg P. and Huang H.F.S. (1979): Changes in marihuana spermatozoa associated with high dose marihuana smoking. In: Marihuana: Biological Effects, edited by G.G. Nahas and W.D.M. Paton, Pergamon Press, New York.
4. Rosenkrantz, H. (1979): Effects of cannabis on fetal development of rodents. In: Marihuana: Biological Effects, edited by G.G. Nahas and W.D.M. Paton, pp 479-499. Pergamon Press, New York.
5. Zuckerman, B., Frank, D., Hingson, R., Amaro, H;, Levenson, S., Kayne, H., Parker, S., Vinci, R., Aboagye, K., Fried, L., Cabral, H., Timperi, R., and Bauchner, H. (1989): Effects of maternal marihuana and cocaïne use on fetal growth. New Engl. J. Med. 320, 762
6. Borgen, L.A., Davis, N.M., and Pace, H.B. (1971): Effects of synthetic delta-9-tetrahydrocannabinol on pregnancy and offspring in the rat. Toxicol. Appl. Pharmacol., 20: 480-486.
7. Vardaris, R.M., Weisz, D.J., Fazel, A., and Rawitchh, A.B. (1976): Chronic administration of delta-9-THC to pregnant rats: Studies of pup behavior and placental transfer. Pharmacol. Biochem. Behav. 4: 249.
8. Sassenrath, E.N., Chapman, L.F., and Goo, G.P. (1979): Reproduction in rhesus monkeys chronically exposed to moderate amounts of delta-9-tetrahydrocannabinol. In: Marihuana: Biological Effects, edited by G.G. Nahas and W.D.M. Paton, pp 501-512. Pergamon Press, New York.
9. Robison, L.L., Buckley, J.D., Daigle, A.E., Wells, R., Benjamin, D., Arthur, D.C., and Hammond, G.D. (1989): Maternal drug use and risk of non lymphoblastic leukemia among offspring. Cancer, 63: 1904-1911.

Section II

EPIDEMIOLOGY

REPORT FROM NORTH AMERICA

Mitchell S. Rosenthal, M.D.

Phoenix House Foundation, Inc., New York, NY 10023

ABSTRACT

The most significant recent development in North America has been a dramatic reduction of marijuana use in the United States and a similar decline in Canada. While controversy over medical use of marijuana continues, there is now less interest in decriminalization. Reduced use has been achieved, in part, by more aggressive enforcement efforts aimed at both imports and domestic production. Even more effective have been prevention initiatives, including increased drug education, greater community involvement, and a national marketing campaign aimed at "de-normalizing" all illicit drug use.

KEY WORDS

Marijuana, North America, reduced use, medical use, demand reduction, prevention, denormalization, changing attitudes.

REPORT FROM NORTH AMERICA

Since Professor Jeri covered Mexico and the Caribbean in his report on Latin America, this allows me to focus on the rest of North America. Here, the most significant development in recent years has been declining marijuana use in the United States and a similar reduction in Canada.

I believe it fair to say that physicians and other medical professionals have played a role in reducing use of marijuana. But their role has not been *pivotal*. Rising concern among health care providers has reflected a more generalized increase in concern that is focused less on the health consequences of marijuana use than on its behavioral impact and on marijuana as a "gateway" drug, the use of which tends to precede involvement with such other and more disabling substances as crack cocaine.

The overwhelming majority of physicians in the United States -- and everywhere else, I would suspect -- are reluctant to deal with *any* form of illicit drug use in their practices or to look for drug abuse in their patients. Other than AIDS or hepatitis, few symptoms or conditions are likely to trigger their suspicion of drug abuse. Only a handful of pediatricians or family practitioners, for example, routinely screen for drugs and many will not consider marijuana a likely culprit even when adolescents present with chronic bronchitis.

The U.S. medical community has not focused sharply on health hazards of marijuana since 1982, when these were detailed by the National Academy of Sciences' Institute of Medicine in its report *Marijuana and Health*.[1] Although the report recommended further study, there has been relatively limited subsequent research directly related to the health consequences of marijuana use.

There has, however, been continued controversy over the medical use of marijuana -- to treat glaucoma, to stimulate the appetite of patients with wasting diseases, or to control the nausea of patients in chemotherapy. Despite legal challenges, the Drug Enforcement Agency has kept marijuana in its most restrictive category, Schedule 1, supported by the assertion of the U.S. Food and Drug Administration that marijuana has "no currently accepted medical use in treatment in the United States."[2, 3]

The same position is taken by the U.S. Public Health Service (PHS), which has been attempting to end the program under which marijuana is provided to certain "approved" patients. The PHS argues that synthetic delta-9-THC is now available. It also would like to avoid having marijuana supplied to victims of AIDS, for it recognizes that "there is potential harm that may result from smoked marijuana for immune-suppressed people."[2]

The Public Health Service has managed to reduce the number of patients receiving marijuana through its program to 13.[3] Nevertheless, medical support for certain marijuana uses remains relatively strong. A survey of U.S. oncologists last year found that 40 percent had recommended marijuana to relieve chemotherapy-induced nausea, and 48 percent would prescribe the drug in some cases were it legal to do so.[4]

While the "medical use" controversy continues, the "decriminalization" debate, which raged fiercely during the late Seventies, no longer arouses much passion or much interest. In both the U.S. and Canada, a small but aggressive "legalization" lobby today advocates various "harm reduction" formulas and alternatives to *all* drug prohibitions. But these advocates command none of the popular support enjoyed by pot apologists more than a decade ago and are unlikely to enjoy the same success.

Between 1973 and 1979, starting with Oregon, eleven of the United States eliminated criminal penalties for the possession of modest amounts of marijuana. Since 1979, however, no new state has opted for decriminalization, and one decriminalized state, Alaska, re-instituted criminal penalties by popular vote in 1990.

DECLINING USE

Not only has the political popularity of marijuana been declining since 1979, so has use of the drug. The U.S. household survey of drug use, conducted every two or three years by the National Institute on Drug Abuse (NIDA), now finds fewer than 20 million Americans reporting marijuana use in the past year and fewer than 10 million reporting use during the past month.

Among young adults 18 to 25, the incidence of marijuana use during the preceding year fell from close to 50 percent in 1979 to below 25 percent in 1991, while use during the past month dropped even more sharply from above 35 percent to 13 percent (see Figure 1).[5, 6] Among adolescents 12 to 17, marijuana use during the past year declined from 24 percent in 1979 to barely 10 percent in 1991, while use during the past month plummeted from close to 17 percent to a little more than 4 percent (see Figure 2).[5, 6]

The annual survey of high school seniors, conducted for NIDA by the University of Michigan's Institute for Social Research, shows a similar decline, with annual prevalence of marijuana use falling from a 1979 peak of 51 percent to 27 percent in 1990 and daily use by seniors dropping from a 1978 high close to 11 percent to barely more than 2 percent in 1990 (see Figure 3).[7]

Neither the household survey nor the high school senior survey can be considered a valid measure of *all* drug use in the United States, for both omit populations likely to include substantial numbers of drug abusers. The household survey does not include the homeless, criminal offenders in prison or jail, members of the armed services, students living in dormitories, or the residents of nursing homes or drug treatment programs. The senior survey clearly does not include those adolescents who have not remained in school long enough to reach 12th grade. Nevertheless, the surveys do provide a valid

FIGURE 1
MARIJUANA USE, ADULTS 18-24

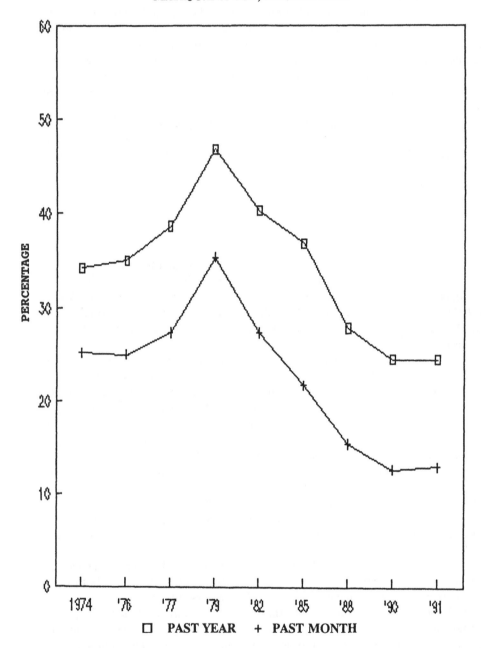

FIGURE 2
MARIJUANA USE, ADOLESCENTS 12-17

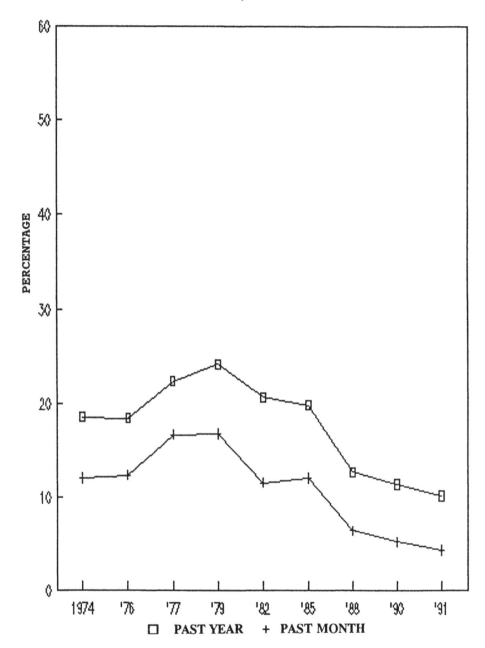

FIGURE 3
MARIJUANA USE, HIGH SCHOOL SENIORS

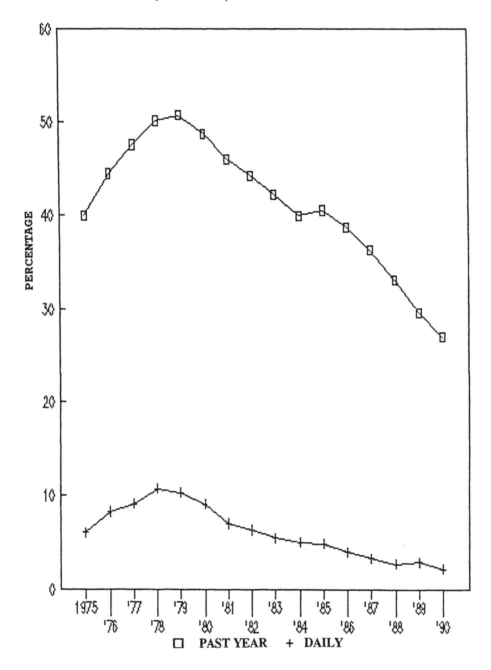

guide to trends, for they measure drug use in essentially the same populations each time.

SUPPLY SIDE EFFORTS

It is clear that marijuana use in the United States has dropped sharply over the past decade. There are a number of reasons for this, not the least of which are vigorous government efforts to reduce imports and domestic cultivation. The escalation of interdiction, aimed primarily at cutting off the illicit flow of cocaine from abroad, has made marijuana smuggling a marginal enterprise at best. Although the U.S. Customs Service was able to seize more than 3.5 million pounds back in 1979, when interdiction was far less efficient, federal agents intercepted fewer than a quarter of a million pounds in 1990.[8]

Government efforts that year focused more sharply on domestic production. In Hawaii, nearly 90 percent of the state's marijuana crop was destroyed by spray planes. Drug enforcement agents began more aggressively raiding greenhouses and tracking domestic producers through distributors of nursery supplies.[8] Peter Reuter, of the Rand Corporation, points out that arrests for the sale of marijuana were higher in 1988 than in 1979 and arrests for possession only slightly lower, in spite of a greatly reduced user base.[9]

Once readily available, marijuana is now considerably harder to find, and prices have risen sharply. In those parts of the United States where marijuana once sold for 20 to 30 dollars an ounce, it now costs close to 300 dollars. In Canada, the increase, though substantial, has been somewhat less extreme.[8]

What's being sold at these prices is a high quality product, significantly more potent than the marijuana used in the early Seventies.[10] The THC content of street-market marijuana rose gradually during the Seventies and early Eighties from less than one percent to close to six percent.[11] Reuter suggests increased potency is the result of more aggressive interdiction and seizure. One means of minimizing risks, he points out, is to increase per-unit value. Thus, optimal strategy for both smugglers and domestic producers demands a shift to higher-potency strains of cannabis.[9]

DRAMATIC DEMAND REDUCTION

While supply-side efforts provide a partial explanation for declining marijuana use, it is on the demand side that the most dramatic progress has been made. To "de-normalize" drug use, extraordinarily effective prevention initiatives have been launched in the United States during the past several years.

Significantly more drug education and prevention is now found not only in schools but also in the work place and in the community. Local community efforts, in which health care professionals play key roles, have increased in number and add a vital dimension to the "de-normalization" effort.

Demonstrating singular success has been a national marketing campaign, focused on de-normalization, by the Partnership for a Drug-Free America. A project of the advertising industry, with support from the media, the Partnership uses sophisticated advertising, with broad exposure on television and radio and in magazines and newspapers, to *un-sell* Americans on drugs. The Partnership's television commercials are among the nation's most widely recognized, and its market surveys, measuring changes in attitudes about drugs and drug use, document the significant impact of the campaign, particularly on adolescents and pre-adolescents.[9]

Changing attitudes among adolescents appears the key determinant of reduced drug use in this age group. The high school senior survey shows awareness of marijuana harm increasing each year since 1980, as use of the drug declined. Disapproval of marijuana use has increased over the same period. While barely a third of high school seniors disapproved of "any" use in 1978, more than two-thirds did in 1990 (see Figure 4).[7]

A study by the University of Michigan's Institute for Social Research, which conducts the senior survey for NIDA, examined various other factors that might have contributed to declining marijuana use by adolescents. Although lifestyle factors plainly exercise influence, the study found no lifestyle trends that would explain the drop. Nor had there been at the time of the study any decline in perceived availability of marijuana. Thus, the authors concluded that "if there had not been a distinct increase in negative attitudes about marijuana, we would not have found steadily lower levels of marijuana use in each succeeding class of high school seniors since 1979."[12]

Few adolescents try illicit drugs *other* than marijuana without prior use of marijuana, and this sequential pattern is only slightly less true for adults.[13] Thus, it might be reasonable to assume that a sharp drop in marijuana use would reduce use of more disabling substances.

Sadly, this is not the case. As marijuana use in the United States has fallen and, with it, overall use of illicit drugs, heavy and high-risk use of the most disabling substances has *increased*, and it has increased among the most vulnerable populations, the poor, the unemployed, the mentally ill, and the troubled young.

The success we have had in the prevention of marijuana use has not resolved our drug problems. Indeed, in many ways, including the role of drug abuse in crime and violence, homelessness and chronic mental illness, child neglect and abuse, and the spread of AIDS, our problems appear worse now than ever before.

Yet this does not diminish what *has* been achieved. Through intensive and aggressive efforts on both the supply and demand sides, we have been able to reduce the risk that marijuana poses to an enormous number of U.S. youngsters.

FIGURE 4
MARIJUANA USE, ATTITUDES OF HIGH SCHOOL SENIORS

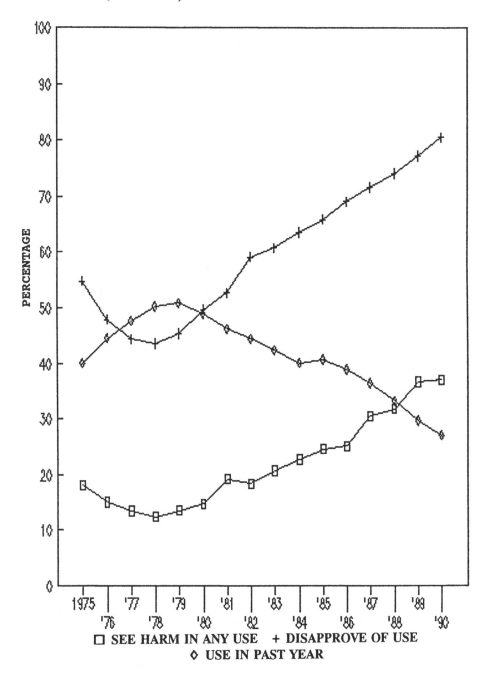

□ SEE HARM IN ANY USE   + DISAPPROVE OF USE
◊ USE IN PAST YEAR

REFERENCES

1.   *Marijuana and Health*, Report of Study by a Committee of the INSTITUTE OF MEDICINE Division of Health Sciences Policy, National Academy Press, Washington, D.C., 1982

2.   Ostrow, R. J., "Delay in Lifting Pot Ban to Aid Seriously Ill is Assailed," *Los Angeles Times*, January 31, 1992

3.   Bishop, K., "Marijuana Still a Drug, Not a Medicine," *The New York Times*, March 22, 1992

4.   Dolbin, R. and Kleiman, M. A., "Marijuana as Antiemetic Medicine: A Survey of Oncologists' Experiences and Attitudes," *Journal of Clinical Oncology*, 1991, July 9, 1314-9

5.   *National Household Survey on Drug Abuse: Main Findings 1990*, U.S. Department of Health and Human Services, Rockville, Maryland, 1991

6.   *National Household Survey on Drug Abuse: Population Estimates 1991*, U.S. Department of Health and Human Services, Rockville, Maryland, 1991

7.   Johnson, L. D., O'Malley, P. M., and Bachman, J. G., *Drug Use Among High School Seniors, College Students, and Young Adults, 1975-1990*, U.S. Department of Health and Human Services, Rockville, Maryland, 1991

8.   Treaster, J. B., "Costly and Scarce, Marijuana is a High More are Rejecting," *The New York Times*, October 29, 1991

9.   Reuter, P., "On the Consequences of Toughness," in *Searching for Alternatives*, Lazear, E. and Krauss, M., Eds., Hoover Institution Press, Palo Alto, California, 1991, pp 138-164

10.   Health Promotion Directorate, Ottawa, Canada

11.   Jones, H. C. and Lovinger, P. W., *The Marijuana Question*, Dodd, Mead & Company, Inc., New York, New York, 1985

12.   Bachman, J. G., Johnson, L. D., O'Malley, P. M., and Humphrey, R. H., "Explaining the Recent Decline in Marijuana Use: Differentiating the Effects of Perceived Risks, Disapproval, and General Lifestyle Factors," *Journal of Health and Social Behavior* 29:92-112, 1988

13.   Kandel, D. B., Yamaguchi, K., and Chen, K., "Stages of Progression in Drug Involvement from Adolescence to Adulthood: Further Evidence for the Gateway Theory," *Journal of Studies on Alcohol* (in press)

MEDICAL ASPECTS OF CANNABIS ABUSE IN LATIN AMERICA

F. Raul Jeri

Department of Neurology,
San Marcos University Medical School, Lima, Peru.

ABSTRACT

Cannabis was introduced to Latin America soon after the discovery of the New World. In the sixteen century it was incorporated to the ritual ceremonies of several Mexican and Brazilian tribes. In the eighteenth and nineteenth centuries cannabis smoking was widespread in Mexico, Costa Rica, Jamaica, Colombia and Brazil, long before it adquireed epidemic proportions in the United States and the rest of the world.

Clinically many acute and chronic psychological syndromes have been described in recreational or intensive users, ranging from acute anxiety to panic reactions, to paranoid, delirious, schizophreniform and apathic disorders. Epidemiological research done in most Latin American countries has shown that marihuana is the drug used in the majority of the population, that it precedes the consumption of other illicit drugs and that frequently its use is associated with alcohol, tobacco, cocaine paste and cocaine hydrochloride.

KEY WORDS

Cannabis, South America (Peru, Mexico, Venezuela, Costa Rica, Panama, Columbia, Brazil, Ecuador) history, epidemiology, psychosis, adverse effects.

INTRODUCTION

Cannabis is a plant that belongs to the genus cannabaceas, a family that comprises only two genera Cannabis and Humulus, the genus of hops. Many workers are of the opinion that three species of cannabis can be delimited: cannabis sativa, cannabis indica and cannabis ruderalis. Cannabis is the source of hempen fibers, an edible akens, an oil of industrial value, some popular medicines and various psychotropic preparations. The genus is thought to be native of Central Asia, especially the Turksmenkaya and Uzbezkaya ex-Soviet Republica near the border with Afghanistan. Undoubtly it is one of the oldest psychomimetics used by man and today the hallucinogen most widely disseminated around the world. It was known by the Chinese several thousand years BC and used in religious ceremonies by Assyrians, Iranian Zend Avesta and Greek civilizations[1].

The psychotropic use of cannabis have been also known for thousands of years by the inhabitants of India, the Near East, the southoriental countries of Europe and Africa. Many years later since Muslim faith -prohibited alcohol there was immediate and considerable interest in a substitute that could produce pleasant experiences without incurring in mortal sin[84]. The Arab invasions of the Minth to the twelfth centuries into North Africa, from Egypt to the north of Tunisia, Algeria and Morocco, introduced cannabis to these territories. The spaniards did not use it, they were christians and wine was not forbidden to them.

Soon after the discovery of America the Spanish Crown directly fomented the diffusion of cannabis, as the raw material for hemp rope, cord and sandals[2]. In 1554 officials of the Seville Company advised to the Council of Indies that flax and hemp seeds should be sent to the New World. Nine years later (1563), Phillip II openly ordered hemp sown in various parts of the American Empire[3]. However most attempts to establish a hemp fiber industry failed. They were tried in Peru, Chile and Mexico, but few enterprises succeeded Even in Peru hemp production was very limited and rope had to be imported from Spain[4]. Only Chile maintained and developed the capacity to export fiber[5].

One of the earliest psychotropic uses of cannabis in Latin America seems to have occurred in Brazil. West African slaves introduced cannabis smoking to that country[18]. The ritual use by some Brazilian indigenous groups indicate an early diffusion to the aboriginal civilizations[6,11,14]. In Mexico, cannabis was cultivated immediatly after the first trip of Cortez. It seems that the aztecs soon learnt about its hallucinogenic properties because by 1550 an ordinance passed prohibiting the cultivation of that plant. However crops continued to be harvested and cannabis adquired religious importance, sometimes substituting the sacred payote, which for centuries had been used by chichimecas and Toltecs[7].

Jamaica, discovered by Columbus in 1492, also received slaves brought by spaniards from West Africa, to replace the dying Arawak Indians. After 153 years of Spanish domination it was invaded by British forces and many more black slaves were imported to work in the agricultural fields. Probably these negro tribes brought cannabis smoking as part of their religious or recreational activities. When indentured labourers from India introduced to the island stronger forms of cannabis (ganga), its use was diffused to extensive areas of the population, as occurred in Egypt and Morocco. Cannabis was introduced to the Atlantic ports of Colombia by African slaves or labourers who came from Jamaica or the Bahamas[8]. Soon it apread to the Northern provinces of the country so when Jacques Joseph Moreau in 1840 used hashish to observe its mental effects, they were very well known in the major and lesser Antilles, Brazil, Mexico and Colombia. Even in Peru, a physician described in 1856 the phantasies produced by hashish[9].

By the end of the Nineteenth Century the femals cannabis plant was smoked by coolies working in the Atlantic Area of Costa Rica. At that time the word coolie was used to describe mixed individuals of Chinese, Indian or Negro descent. Fighting and disorder had been observed among these labourers and it was attributed to the smoking hemp. It must be also remembered that Jamaicans were also imported to Costa Rica as labourers to complete the Limon-San Jose Railroad. Many of them were also definite ganga smokers[8].

few decades later travel and migration facilitated the spread of cannabis use from Mexico, Jamaica, Colombia and Brazil to the Southern and Western territories of the United States of America.

CLINICAL STUDIES

Before marihuana use adquired epidemic proportions in the United States, there was considerable concern in South America about its toxic effects and the serious social problems observed in the abusers. In Mexico, it was written that many crimes were committed by persons addicted to marihuana. Use was very extensive in jails but also in persons of high social classes[18].

In Chile, the plant is known as "*pito de pango*" and in a few years it spread to the youth population. In a survey it was found that since the spring of 1968 there was massive marihuana consumption by chilean adolescents[19]. In Brazil in 1900 it was discovered that "the Brazilian negro smoked, chewed and ate a drug prepared as pills or the material was cooked". This drug was cannabis which in Brazilian language is named maconha. A paper published in 1958 by several authors[17] came to the conclusion that cannabis abuse represented a serious social problem due to the great number of affected persons, the huge quantities of drug used and the criminal activities associated with its consumption, trafficking and commerce. In 1957 only is one city (Sao Paulo) of 326 subjects investigated for maconha abuse 322 resulted positive. During the same year 1381 men, 95 women and 150 minors were imprisonned for offering, transporting or selling cannabis.

Jose LUCENA[16] studied the great diversity of hallucinations experienced during acute marihuana intoxication which he attributed to the several compounds utilized by the consumers. He also observed individuals who were chronically intoxicated by maconha and subjects who developed psychotic disorders characterized by schizophreniform, confusional, depressive, paranoid and delirious symptomatology. In the psychotic individuals he thought that there were complex interrelationships between external precipitating factors and endogenous etiological conditioners which made difficult the differentiation between schizophrenia and schizophreniform reactions in some of these subjects.

Pacheco E. Silva[18] documented schizophrenic disorders in subjects intoxicated by marihuana. Some of those had certain schizoid manifestations which developed into the full schizophrenic psychosis by use of the drug. He did not find cases with dementing evolution as had been described in Oriental literature.

In Colombia the social problems of marihuana smokers were described in the second largest city of that country[55]. As that time the illegal traffic and smoking of marihuana was Colombia's worst social problem affecting thousands of persons. He as a pharmacologist described the effects of oral and intraperitoneal administration of cannabis to dogs, rabbits, guinea pigs and ice. After a short period of stimulation he registered a state of depression which lasted several hours. Catalapsy was observed when the extract was administered intraperitoneally.

In 1962 an investigation was done of 50 former hashish addicts in Bahia (Brazil)[22]. All wee men, treated in a sanatorium and 177 were prisioners (106 men and 11 women) sentenced for various crimes. In that city abuse of hashish usually began in late childhood and during adolescence. The smoking of hashish was practiced mainly by people with an abnormal personality, strongly conditioned by constitutional factors. Familiar, social and economic factors were also important. Doubts about the existence of a mental syndrome during acute hashish poisoning were probably the result of insufficient research. These symptom were conditioned by the plant itself, by the method of smoking or by the constitution of the addict. Hashish intoxication lead to changes in a person's instincts, feelings, activity, conscience and sensory perception. The mental symptoms were mostly an expression of excitation of the cerebral cortex or of the release of the subcortical formations and, rarely in a secondary phase, of depression of the cortex. So far there were no evidence of any permanent mental changes due to hashish, at least not in the material examined. In prisoners investigated abuse of hashish was more frequently associated with theft and murder. Though the causal role of the plant in these crimes had not been proved the authors thought that it was possible that the disinhibition of desires facilitated crimes. It must be admitted that hashish intoxication can predispose to other crimes on account of the mental symptoms and this is the case with many other drugs and intoxications[22].

By the year 1969 physicians became aware of the cannabis epidemic in Peru[23]. In Chile another practitioner[19] two years later conducted an investigation of drug use in young people. He came to the conclusion that since the spring of 1968 there was massive consumption of marihuana bychilean youths (once 39.3 %, daily 1.5 %). During the same year three peruvian psychiatrists[25] described clinical features of 36 children, adolescents and youngsters who used illicit drugs, mostly for a period of 2-4 years. They were mainly referred due to behaviour disturbances, compulsive drug use, acute psychotic intoxications, anxiety attacks and suicide attempts. The main drug of abuse was marihuana which was used alone or combined with other drugs (LSD, alcohol). All of them had personality or behaviour disturbances before using drugs. The six patients who developed psychoses showed profound thought disorganization, paranoid ideation in clear consciousness which resembled schizophrenia. Four recovered promptly (2-3 weeks) but 2 remained psychotic after one year of follow up. The authors concluded that the effects of cannabis are dose relatied, depend on the material used, previous personality and previous brain pathology. In this group of patients it was documented that marihuana alone or mixed with other drugs could precipitate acute, subacute and chronic psychotic disorders.

When the World Health Organization convened a meeting in Geneva of health autorithies of 33 countries[20] about alcohol and drug dependence, Chile reported massive cannabis consumption by adolescents, apart from Chile, Mexico answered that cannabis at that time was the main drug for high school studens of middle an upper classes (regular users were 3.8 %). Venezuela responded that the most commonly used drug was marihuana, especially among teenagers and delinquants.
A few years later drug use by dependents and members of the Peruvian Armed Forces were reported[26] indicating that use of illicit substances had increased considerably in the last decade (1966-1976). This paper was based on the clinical observations of 79 drug abusers: males 86.0 %, females 13.9 %; ages 15-35, mostly 15-20 (79.7 %). In this group 94.9 % came from regular homes and 5.0 % from broken homes. The economical level was high in 20.2 %, medium in 56.9 % and low in 22.7 %. There was familial psychopathology in 71.0 % of the subjects. The main reasons for referral were school failure 35.4 %, suicide attempt 17.7 %, acute psychotic intoxication 15.1 %, job desertion 13.9 %, depressive reaction 11.3 %, repeated fugues 7.5 %. The great majority of persons abused marihuana (86.0 %). Other chemicals ingested were LSD (36.7 %), emphetamines (22.7 %), coca or cocaine paste (8.8. %) and cocaine hydrochloride (5.0 %).

Three years later buccal examination of drug abusers in Buenos Aires[27] showed a connection between oral diseases and certain types of substances, particularly amphetamines and marihuana. Male 20 year old patients had numerous caries, *obturations* and extractions. Thirty four percent suffered from gingivitis, 63 had a saburrhal tongue, 8 % presented with leukoplakia and 5 % had dental prosthesis. All these conditions were related to consumtpion of cannabis or amphetamines and poor oral hygiène.

When the syndromes produced by coca or cocaine paste were described, for the first time in the literature[86], it was noted that this substance, when dried, was mixed with tobacco or marihuana so that it may ignite.

In another small series of polydrug abusers studies in Peru, it was discovered that of 26 drug addicts 53.8 % used marihuana[28].

A group of researchers studies chronic marihuana use in Costa Rica[29]. The original sample consisted of 84 users and 156 non users, which after a year was reduced to 41 users and 41 non users carefully matched. Users smoked the substance mostly on a daily basis, for more than 10 years previous to their selection. Mean marihuana consumption for 41 users was 9.6 cigarretes per day. All subjects were male, living in the city's working class residential areas. Users learned less money, had more arrests, had been in jail or in the reformatory and had been reared bysurrogate parents. They also changed jobs more frequently, worked in part time jobs or were unemployed. Thirty nine effects were discribed by users, most were pleasant. Some developed panic reactions, others euphoria (68.2 %). Unpleasant experiences were also reported when mixing marihuana with alcohol. The great majority of users had serious adjustment problems long before they began to use marihuana. They also suffered from urinary and gastrointestinal symptoms. On examination they had lower weight and showed more cases of pterigion and atrophy of one testicle, had more surgical scars, lower levels of hemoglobin and signs of irritation of the anterior segment of the eye. They had some EEG changes during sleep but no significant univariate or multivariate differences were found on sixteen variables in a psychological test battery.

EPIDEMIOLOGICAL RESEARCH

Following guidelines suggested by the World Health Organization many epidemiological studies have been done in Central and South America to establish the magnitude of the problem, the changing patterns and the corrective measures to be applied. Thus in a survey done on 648 peruvian university students[87] it was found that 55 % had used marihuana at least once, occasionaly 37 % and frequently 15 %. Similar results were found in high school students[21]. Once more cannabis was the most used illicit substance: 16.43 % answered that they had tried at least once, 11.57 % were occasional users and 4.85 % were habitual consumers. Two other peruvian researchers[23] in a group of addicts admitted to public and private hospitals found that marihuana (27 %) was only second to sedatives and barbiturates (36 %). Comparing samples from four cities Negrete[73] found that cannabis was used much more frequently in Toronto (31.3 %) than in Medellin (24.0 %), Sao Paolo (9.2 %) or Mexico (6.4 %).

In a home survey done in Lima marihuana was consumed at least once by 3.18 %, also less frequently than sedatives[66], a figure similar (3.0 %) to what was found in high school students some years later[58]. In 1986 several investigations were done in South America. De Carvalho[60] applied an anonymous questionnaire to 2475 university students of the Sao Paolo and discovered that cannabis had been employed by 41.2 % of them at least once, 13.8 % were occasional consumers and 4.2 % were regular macaonha smokers.

Several Government officials assembled in 1986 in Quito found that cannabis was the illegal psychoactive drug most used in Venezuela, Ecuador, Bolivia and Peru[33-34] until 1983, but after that year cocaine paste surpassed marihuana in

Bolivia, Peru and the northern sector of Colombia [42-43]. Three years later the situation had not changed much, marihuana continued to be the most used psychotropic substance in Costa Rica, Chile, Colombia, Argentina, Mexico, The Bahamas, Peru and Bolivia. However street children in Mexico, Brazil, Colombia and Peru began drug experiences with inhalants followed by coca (cocaine) paste in Bolivia, Colombia and Peru, while in Mexico marihuana followed the use of inhalants, and later heroin or cocaine hydrochloride[70-77].

In 1990 several Latin American epidemiologists met in San Antonio, Texas, to compare recent investigations in their own countries. Brazil reported that high school and university students preferred inhalants, benzodiazepines, amphetamines and cannabis[40]. Some used syntetic cholinergic but very few used cocaine[66-78-82]. Brazilian street children preferred inhalants, cannabis and anticholinergics[40]. Comparing student and addicts it was found that the former smoked maconha (2.9 % in 1987 and 3.4 % in 1989) while people in hospitale were mostly depend on cannabis (31.1 %) in 1987 and 29.4 % in 1989).

In emergency Rooms in Buenos Aires[37-38] cocaine and marihuana (10 % each) citations were only secondary to alcohol (65 %).

Another epidemiologic investigation conducted in Ecuador[77] interviewing 6000 individuals found outlawed drug used at least once in the following proportions: marihuana 4 %, inhalants 2 %, native plants 1 %, cocaine 1 %.

In Mexico several surveys done in schools, factories, prisons and homes showed in the seventies predominance of marihuana everywhere[68]. Home surveys done in 1974 - 1985 demonstrated that the commonly abused substances were barbiturates and marihuana[72]. Heroin and cocaine was abused in Northwest Mexico[80]. In 1986 marihuana was the main drug of abuse in Northern mexican students. Between 1986 and 1988 national rates increased significantly for inhalants, amphetamines and marihuana[80]. Another group also found, in the same country, that the most common illegal drug used was marihuana[71].

In Peru several surveys also showed that the predominant drug of abuse was marihuana from the sixties to the eighties [21, 23, 25]. After 1980 the preferred illegal substances were coca (cocaine) paste and cocaine hydrochloride[39-86].

In Venezuela a high school survey (sample 21 842 students, 9 456 males, 12 386 females) showed that the most frequently abused substances by the 11 -13 years old were glue and solvents. Use of marihuana, cocaine and basuko (coca paste) increased with age. Marihuana 11-13 years old 0.6 %, 14-17 years 2.2 %, older than 18 years 5.6 %. In the last six months marihuana was consumed by 3.3 % in males and by 1.3 % in females [65].

In over 600 peruvian addicts to coca (cocaine) paste and cocaine hydrochloride it was found that before using those substances 305 patients (49.5 %) had smoked marihuana. When dependent on cocaine twenty six individuals preferred using paste mixed with marihuana, while 18 usually combined, in one session, cocaine hydrochloride with coca paste mixed with cannabis. Thus 44 subjects continued to smoke marihuana while addicted to coca paste or cocaine[39].

A home survey done in 1988 in Peru (5 195 houses) showed that marihuana had been tried by 4 % of the urban population of 12 to 50 years old being particularly preferred by man (10.7 %) than by women (1.3 %). Most consumers were 20-39 years old and had university education (12.6 %) superior non university education (5.3 %), high school education (4.5 %) and elementary education (1.3 %). Though marihuana was the most consumed illegal drug it seemed that it was becoming less fashionable, according to the authoress[79], when comparing the results of the 1988 survey with the 1986 investigation[69]. In 1986 it was used at least once by 8.3 % of the population, in 1988 it had descended to 5.3 %.

Recently teenagers were selected by mexican investigators[41] from a file of the National Addiction Survey and identified 1 475 subjects 14-17 years old. Most consumed marihuana (63.5 %) and inhalants (33.3 %). The use of two or more illicit drugs was reported by 24.2 % of the subjects. Of these individuals who consumed two drugs 71.4 % employed marihuana and inhalants and almost a third (28.6 %) used cocaine and marihuana. Teenagers 16-17 years old consumed drugs twice as much as those aged 14-15.

COMMENT

Though cannabis sativa probably originated wild in the level grounds of the Altaj Mountains in Central Asia, it was cultivated by the Chinese and Indian peoples and soon its use, as a ceremonial and religious incense, was extended to many oriental territories. Later it was discovered that its many preparations could be eaten, smoked and drank. Conquest migration and travel spread its use to the Middle East and to Northern and Western Africa. When America was discovered cannabis was introduced by the Spaniards to the new colonies (Mexico, Cuba, Hispaniola, Jamaica, Colombia, Peru, Chile) with the purpose of establishing the hemp industry. The Portuguese very early in the sixteenth century, brought thousand of slaves from Guinea to Brazil. These people (Malinka, Yoruba, Balante, Fulani, Mandyako, Pepel and others) brought their religious costumes and practices, including cannabis smoking which was called maconha from thence.

Cannabis need w warm dry climate and adequate moisture in the soil. Thus it had been cultivated in the Lower Himalayas, the Rif sector of Morocco and Mount Lebanon. In the new world it was extensively grown in almost all Central and Northern South American countries (the Antilles, Mexico, Colombia and Brazil).

Hundred of years before its use was extended to the west and south ern regions of the United States, cannabis was smoked in Mexico and Brazil. In the nineteenth century marihuana use was prevalent in the Atlantic coast of Honduras, Jamaica, Haiti and Santo Domingo. Finally the inhabitants of those countries, when they emigrated to Georgia, Louisiana, Alabama and California, they introduced the habit to Black Americans, musicians and performers, as is well Known.

In south America the psychotropic effects of cannabis smoking were well known by the middle of the nineteenth century[9], but it was by the clinical work of Brazilian physicians that the acute and toxic disorders produced by marihuana were reported to the rest of spanish speaking doctors of Latin America[11-17].

Before the marihuana epidemic of the sixtees acquired explosive characteristics, cannabis used increased considerably in Brazil, Central America and Mexico. Intensive users developed many types of mental disorders in those countries, as well as in Colombia and Peru.

It is very difficult to compare clinical work of different physicians, in different countries, with different patients. However some results are similar in diverse environments. For example M. Souief, Professor of Psychology at the University of Cairo, did a psychosocial study of 850 hashish users and 839 controls[84], he found among hashish users that there was a definite pattern of oscillation of temperamental traits, swinging between two opposite poles: that of social case, acquiecence and elation under the immediate effects of the drug; and that of ascendancy, seclusiviness, negativism, depression and pugnacity when the subject was deprived of it. Concerning the various tests given to the volunteers it was found that the more educated the user, the greater the impairment caused by the drug. While an illeterate peasant might perform five percent less efficiently than a non smoking control of the sam educational background, a man with equivalent of grade school education might do thirty percent less well than his smoking counterpart. It seemed therefore that the drug was more taxing on those who have most to lose.

In another cosmopolitan setting, children and adolescents who had high intelligence and special abilities, representing a group whose potential was of greatest personal and social value, rapidly showed inhibition of learning, thought fragmentation, loss of fluency of verbal expression, alienation, lack of motivation and direction, depression and confusion[31]. The tendancy to employ other drugs after the inicial encounter with marihuana was also confirmed, as well as the development of acute and chronic psychoses. In a smaller town at least five different kinds of psychoses associated with marihuana were documented[30].

In areas where heavy cannabis consumption is prevalent, in dispossed and illiterate people,toxic psychoses are frequent. In Morocco for example one clinician reported on 140 cases of toxic psychotic reactions. These mental disturbances were very short, with autonomic symptoms and changes of mood varying from euphoria to depression, sometimes with agresive outbursts and associated with vivid hallucinations[24]. More elaborate work, done in South Africa, comparing 20 psychotic men who had urinary cannabinoid levels on admission with 20 psychotic men free of cannabis, showed in the former significant more hypomania and agitation and significantly less affective flattening, auditory hallucinations and incoherence of speech than did the controls. After one week the cannabis group showed marked improvement whereas the controls remained virtually unchanged. They concluded that a high intake of cannabis may be related to a rapid resolving psychosis manifesting marked hypomanic features, though often presenting as schizophrenia like illness[32].

In our report we observed six psychotic disorders in 36 children and adolescents. Four psychotic patients recovered rapidly, while two remaided psychotic for over a year. The paranoid psychoses resembled schizophrenia[25]. Acute and chronic cannabis psychoses have been also documented by brazilian, and colombian investigators.

In Costa Rica 41 chronic users who smoked marihuana for at least ten years, about 9 cigarettes per day, showed no significant mental changes, as judged by a neuropsychological battery of 16 variables, when compared with 41 controls. They admitted that acute psychotic reactions could occur with heavy consumption but doubted that chronic use could affect the higher cortical functions[29]. The tests were applied by three psychologists. These results are contradicted by clinical and comparative studies done by many investigators[16,22,24-26,30-32,34,40]. Carter excluded users who had any physical or psychological disorder and he commented that sociocultural and idiosyncratic conditions may contribute to pathological findings in other studies.

Clinical investigators regularly take account of genetic, constitutional, environmental, physical and characterological conditions of every subject, as well as the psychoactive substances consumed. But irrespective of economical, cultural or social factors, it has been demostrated repeteadly that high doses of cannabis, ingested acutely or chronically, can produce objective psychopathological disorders[22-26,30-32,37,40].

Carter also wrote that his chronic cannabic users were not interested in other drugs[29], apart from tobacco and alcohol, but this assessment is against contradicted by the patients seen by many clinical investigators[11-26,30-32]. In fact, it is well known by any person who has experience with drug dependent patients in the last 30 years, that cannabis is frequently the first psychoactive substance followed sooner or later by LSD, heroin, mescaline, cocaine or designer drugs, whatever is the current fashion in the subculture[21,25,35,39,40,46,51,59].

Concerning basic research, 31 years ago a pharmacologist considered that Colombia's worst social problem was cannabis smoking. He administered marihuana extracts to several small mammals and found that after a period of stimulation followed a period of depression, which lasted several hours. He described catalespsy when the extract was administered intraperitonsally[55].

In Panama three researchers injected several concentrations of marihuana extracts diluted with sesame oil to 2400 hen eggs and only sesame oil to 2400 control eggs. Embryos injected with cannabis showed a 90 % developmental delay, 80 % had malformations of the spinal cord and 70 % damage to otocytes and eyes, 15 % developped as double monsters. Higher concentrations produced more malformations. They concluded that marihuana has teratogenic effects in the development of chick embryos[52]. In these experiments there was no possibility of maternal toxicity[53] nor were they limited to a few days of observation[54].

Five per cent of emergencies seen recently at three general hospitals in Buenos Aires were associated with psychoactive substances of abuse: alcoholic 65 %, cannabis 10 %, cocaine 10 %. Marihuana and cocaine were often found in combination with alcohol and with other drugs[37]. Lack of education, unemployment and marital separation were important factors for men to abuse drugs, while for woman personal crises predisposed them to over doses[38]. In Mexico the Information Reporting System (Health and Justice) found that the most common used drugs were marihuana (64-75 %) and solvent inhalants[71].

In the same country the surveillance system of addictions was applied to three high risk groups: teenagers, juvenile delinquants and people living in slums[41]. The most consumed drugs were marihuana (63,5 %) and inhalants (33,3 %) in the first group. In juvenile reformatories marihuana was the first drug of choice, ages of onset 9-19 for 51 % of the subjects. Marihuana and inhalants were also predominant in the shanty towns.

These studies, as well as those done in general and psychiatric hospitals, probably mean that home and school surveys, done by questionnaires, are underreporting the current drug use in Latin American countries. However even today use of drugs by those nations probably is less prevalent that what occurs in the United States, Canada[65,79-82] and Europe.

In the future it would be convenient that all countries should adopt a similar procedure to assess drug abuse patterns and trends at yearly intervals, considering a variety of sources such as: (1) Drug related deaths reported by medical examiners, morgues, local coroners and public health agencies; (2) Drug related emergency room episodes in general and specialized hospitals; (3) Primary substances of abuse reported by patients at admission to treatment facilities; (4) urinary drug analysis done to people arrested by the Judiciary System for delinquent activities, disordered conduct or bizarre behaviour; (5) Seizure, price, purity, prescription/distribution and arrest data from national, state or local law enforcement agencies; and (6) Other city specific data gathered from ethnographic research, surveys, criminal justice and correctional sources, public health sources, and other sources unique to local areas[85]. Reports of drug abuse in general were summarized in Brasil[18] and Peru[83]. Recent reviews of cannabis use in Brasil were published a decade ago[45], an excellent study on the health effects followed two years later[47] and a study on the epidemiological surveys done in 12 Latin American countries[44].

In conclusion, it can be said that cannabis has been used in Latin American for several centuries, that it was introduced to the United States in the thirties and that nowadays in the main drug of consumption in almost all those nations. Marihuana has produced acute and chronic psychological disorders in many groups of individuals, including psychoses, which have been documented by several investigators. Cannabis was the first illegal drug used by later polydrug, heroin or cocaine addicts. Cannabis is frequently use in association with alcohol, toabacco and other illicit psychoactive substances.

REFERENCES

1. **SCHULTES RE, HOFFMANN AT.** : *The Botany and Chemistry of Hallucinogens*, 2nd Ed., C.C. Thomas, Springfield, Ill, 1980.
2. **AYALA MJ.** : *Mobiliario Hispano Americano del siglo XVII*. Coleccion de documentos Ineditos para la Historia de Ibero America, Volumen II, Madrid, 1930.
3. **CAPPA R.** : *Estudios criticos acerca de la dominacion espanola en America*, Libreria Catolica, Madrid, 1890.
4. **COBO B.** : *Historia del Nuevo Mundo*. Soc. Bibliog. Andaluces, Sevilla, 1891.
5. **PATINO UM.** : *Plantas cultivadas y animales domesticos en America Equinoccial*. Imprenta Departemental, Cali, 1969.
6. **WAGLAY C, GALVAO E.** : *The Tenehara Indians of Brazil*. Columbia Press, New York, 1949.
7. **ARDILA F.** : *Aspectos medico-legales y medico-sociales de la marihuana*. Tesis Doctoral, Univ. de Madrid, 1965.
8. **STEWART W.** : *W. Keith and Costa Rica*, Univ. New Mex Press. Alburquerque.
9. **CORPANCHO MN.** : *Fantasia poetica producida por el hashish*. Graceta Med. Lima 1, 2, 1856.
10. **ROSSELLI H.** : *Barba Jacob y la historia de la marihuana*. Acta psiquiat. Psicol. Lat. Amer. 32, 259, 1986.
11. **VASCONCELOS S.** : *Algunas notas sobre la maconha en el Brasil*. Comis. Nac. Fisc. Narcot., 88, Rio de Janeiro, 1951.
12. **HASSELMAN J., RIBEIRO O.** : *La accion toxica de la maconha cultivada en el Brasil*. Comis. Nac. Fisc. Narcot. 326, Rio de Janeiro, 1951.
13. **PEREIRA JR.** : *Contribucion para el estudio de las plantas alucinatorias*. Comis. Nac. Fisc. Narcot. 146, Rio de Janeiro, 1951.
14. **MORENO G.** : *Aspectos del maconhismo en Sergipe*. Comis. Nac. Fisc. Narcot., 203, Rio de Janeiro, 1951.
15. **LUCENA J.** : *Algunos nuevos datos sobre los fumadores de maconha*. Comis, Nac. Fisc. Narcot., 98, Rio de Janeiro, 1951.
16. **LUCENA J.** : *Maconhismo y Alucinaciones en el Brasil*. Comis, Nac. Fisc. Narc., 116, Rio de Janeiro, 1951.
17. **SIQUIERA A., WASCKY R., NEURBERN TA.** : *Aspectos scientificos del problema de lamaconha*. Communic. Soc. Bras. Inv. Scien., 1958.
18. **PACHECO E., SILVA AC.** : *Intoxicacion cronica in America Latina*. Rev. Psiquiat, Peruan, 2, 159, 1959.
19. **RICHARD P.** : *El consumo de marihuana : un problema social en Chile*. Inst. Sociol., Univ. Catol., 1951.
20. **MOSER J.** : *Problems and programmes related to alcohol and drug dependence in 33 countries*. World Health Organization, Geneva, 1974.
21. **BOGGIANO MA.** : *Algunos aspectos del uso de la marihuana y otras drogas en los estudiantes de educacion secundaria en Lima Metropolitana*. Psiquiat. Peruan. III, Tall. Graf. Villa Nueva, Lima, 1975.
22. **PIRES DA VEIGA R., RUBINS DO PINHO A.** : *Contribucao ao estudo do maconhismo na Bahia*, Neurologia 25, 38, 1962.
23. **CARAVEDO B., ALMEIDA M.** : *Alcoholismo y toxicomanias*. Minist. Salud Lima, 1972.
24. **CHRISTOZOU C.** : *L'aspect marocain de l'intoxication cannabique d'après des études faites dans les conditions d'un hôpital psychiatrique de malades chroniques.*

Maroc Med, 44, 630, 1965.
25. **JERI FR., CARBAJAL C., SANCHEZ CC.** : *Uso de drogas por adolescentes y escolares.* Rev. Neuropsiquiat. 34, 243, 1971.
26. **JERI FR., SANCHEZ CC., Del POZO CC.** : *Consumo de drogas peligrosas por miembros y familiares de la Fuerza Armada y de la Fuerza Policial Peruana.* Rev. Sanid. Min. Int. 37, 104, 1976.
27. **DI CUGNO F.** : *Efecto del consumo de drogas sobre tejidos y organos bucales.* Bol. Ofic. Sanit. Pan. Am. 87, 300, 1979.
28. **ALMEIDA M.** : *Uso y abuso de alcohol y drogas : algunos aspectos de sus interrelaciones en la clinica.* Bol. Ofic. Sanit. Pan. Am. 87, 45, 1980.
29. **CARTER W.** : *Cannabis in Costa Rica.* A study of chronic Marihuana Use, Inst. Stud. Human Issues, Philadelphia, 1980.
30. **CARRANZA F.** : *Marihuana induced psychoses, in drug abuse in the modern world*, Pergamon Press, New York, 1981.
31. **MILMAN DH** : *Effect on children and adolescents of mind altering drugs with special reference to cannabis, in drug abuse in the modern world*, Pergamon Press, New York, 1981.
32. **ROTTANBURG D., ROBINS AH., BEN-ARIS O., TEGGIN A., ELK R.:** *Cannabis associated psychosis with hypomanic features.* Lancet 2, 1364, 1982.
33. **SAMANIEGO N.** : *Desarrollo de los programas de attencion de farmaco-dependientes en el Ecuador*, in Convenio Hipolito Unanue 1-16 Quito, 1986.
34. **OLIVARES A.** : *Informe de Colombia*, in Conv. Hipolito Unanue, Quito, 1986.
35. **MADALENO M., FLORENZANO R., ALVO D., LARRAIN G., JUSTINIANO JC., ADRISSOLA G.** : *Estructura de morbilidad de adolescentes consultantes en el sector oriente de Santiago.* Rev. Chil. Pediat. 58, 164, 1987.
36. **ALVAREZ A., SAGUIER ML, FOBLIA JL, BOFFI H., ANDRADE JH.** : *Prevalencia del consumo de drogas en varones de 18 anos de la ciudad de Buenos Aires.* Bol. Acad. Nac. Med. Buenos Aires. 67, 427, 1989.
37. **MIGUEZ HA, GRIMSON RW.** : *Consultas por abuso de sustancias paicoactivas en hospitales de Buenos Aires.* Bol. Ofic. Sanit. Pan Am. 107, 296, 1981.
38. **MIGUEZ HA.** : *Improper use of psychoactive substances : some strategies for obtaining information, Epidemiologic trend in drug abuse*, NIDA, Rockville, 1990.
39. **JERI FR., PEREZ JC.** : *Dependencia a la cocaina en el Peru. Observaciones en un grupo de 616 pacientes.* CEDRO, Lima, 1990.
40. **CARLINI EA., NAPPO SA.** : *Illicit use of psychotropic drugs in Brazilian cities, in Epidemiologic trends in drug abuse*, NIDA, Rockville, 1990.
41. **TAPIA R., KURI P., NAZAR A., CRAVIATO P., HUERTA S.** : *Epidemiologic surveillance system of addictions in Mexico, in Epidemiologic trends in drug abuse*, NIDA, Rockville, 1991.
42. **MADRIGA E.** : *Prevencion y control de sustancias psicoactivas y de bebidas alcoholicas.* Convenio hipolito unanue, Quito, 1986.
43. **CASTRO J.** : *Plan nacional de prevencion de la farmacodependencia.* Convenio Hipolito Unanue, Quito, 1986.
44. **FLORENZANO R.** : *Uso de drogas en America Latina y en el Caribe.* Bol. Vigilanc. Epidemiol. Chile 13, 266, 1986.
45. **CARLINI EA** : *Maconha (cannabis sativa) : mito a realidade, fatos e fantasias.* Med. E. Cultura 36, 1, 1981.
46. **WRAY SR., MURTHY NVA.** : *Review of the effect of cannabis on mental and physiological functions.* West Ind. Med. J. 36, 195, 1987.
47. **NEGRETE, JC** : *Efectos del canabismo sobre la salud.* Acta psiquiat. Psicol. Am. Lat. 29, 267, 1983.
48. **SILVA R., MAGALHAES MP., OLIVEIRA R., AZEVEDO RB., ALMEIDA SP., SILVA MT.** : *Avaliacao dos efeitos da maconha por usuarios de populacao estudiantil.* Cienc. E. Cultura 41, 652, 1989.
49. **BUSTAMENTE LP.** : *La informacion sobre el dano a la salud producido por las drogas y su uso. Estudio en un grupo de estudiantes de medicina y medicos internos de pregrado.* Salud. Pub. Mex. 26, 553, 1984.
50. **MAGALHAES MP., SILVA R., OLIVEIRA R., AZEVEDO R., ARAUJO MT.** :

Padroes de frequencia do uso de maconha por estudantes universitarios. Rev. Asoc. Bras. Psiquiat. 11, 35, 1989.

51. **MURILLO GA.** : *Consideraciones antropologicas acerca del uso de drogas en el adolescente chileno.* Rev. Psiquiat. Hosp. Horwitz 5, 13, 1988.

52. **SOUSA F., FOMEZ JA., GRANDI C.** : *Teratogenesis producida por la marihuana (cannabis sativa) en diversos estados de desarrollo de embrio de pollo.* Rev. Med. Panama. 7, 223, 1982.

53. **HEISHMAN SJ., HUESTES MA, HENNINFIELD JE, CONE EJ.** : *Acute and residual effects of marihuana : profiles of plasma THC levels. Physiological, subjective and performance measures.* Phar. Biochem. Behav. 37, 561, 1990.

54. **HUTCHINGS DE, DOW-EDWARDS D.** : *Animal models of opiate, cocaine and cannabis use.* Clin. Perinatol. 18, 1, 1991.

55. **CANO G.** : *La marihuana y sus peligros.* Antioquia Med. 11, 23, 1961.

56. **FLORENZANO R., MADRID V., MARTINI AM, ZALAZAR ME, MANTELLI E.** : *Prevalencia y caracteristicas del consumo de algunas sustancias quimicas en estudiantes de ensenanza media en Santiago de Chile.* Rev Med Chile 109, 1051, 1981.

57. **BELTRAN N., ARCINIEGAS J., FADUL MA., LLANOS CA., LLANOS A., LIZCANO F., VELEZ F., MARTINEZ H., PORTELA G., ANGULO S.** : *Prevalencia de Farmacodependencia en estudiantes del sexto ano de bachillerato en colegios oficiales y privados.* Salud Uninorte 4, 1, 1988.

58. **HUAMAN MJ., TUEROS M., VILLANUEVA M., FRIAS C., RAMOS M.** : *Aspectos psicosociales relacionados con el uso de drogas en los adolescentes de Lima Metropolitana,* Impren. Univ. Catol., Lima, 1984.

59. **DA SILVA AM, BRANCACIO C., MARINI D., BORGES LE, de LELLOS M., MESQUITA ME.** : *Prevalencia do uso de alcohol, cigarro e maconha nos alunos da Faculdade de Medicina de Universidade de Sao Paulo.* Arquiv. Coord. Saude Ment do Est. Sao Paulo. 45, 134, 1985.

60. **De CARVALHO FV.** : *El consumo de drogas entre estudiantes universitarios en el Estado de Sao Paulo.* Bull Estupefac. 38, 37, 1986.

61. **BUCHER RE., LANDINI M.** : *Conocimiento y uso de drogas en Brasilia*; acta Psiquiat. Am Lat. 34, 113, 1988.

62. **CASTRO ME., ROJAS ML., de la SERNA J.** : *Estudio Epidemiologico sobre si uso de drogas y problemas asociados entre la problacion estudiantil que asiste a los colegios de bachilleres.* Salud Mental 11, 35, 1988.

63. **SMART R., PATTERSON SD.** : *Comparacion del abuso de alcohol, tabaco y drogas entre estudiantes y delincuentes en las Bahamas.* Bol. Ofic. Sanit. Panam. 107, 514, 1989.

64. **FLORENZANO R.** : *Alcoholismo y abuso de otras drogas : Programas de prevencion en Santiago, Chile.* Bol. Ofic. Sanit. Pan Am 107, 577, 1989.

65. **KRAMER S.** : *Second national survey among high school students by régions in Venezuela.* In epidem. trends in drug abuse, NIDA, Rockville, 1990.

66. **CARBAJAL C., JERI FR., SANCHEZ CC., BRAVO C., VALDIVIA L.** : *Estudio epidemiologico sobre uso de drogas en Lima.* Rev. Sanid. Fuerz. Polic, 41, 1, 1980.

67. **VELASQUEZ E., TORRES Y., SANCHEZ MM., RAMIREZ H., HERNANDEZ NE., REBAGE L., BUSTAMANTE LE.** : *Estudio epidemiologico de uso de drogas en la problacion general de Medellin.* Rev. Colomb psiquiat. 14, 116, 1985.

68. **MEDINA ME, TAPIA R., SEPULVEDA J., OTERO R., SOLACHE G., LEZCANO F., VILLATORO J., LOPEZ EK., de la SERNA J., ROJAS E.** : *Encuesta Nacional de Adicciones en Mexico.* Salud Mental. 12, 7, 1989.

69. **JUTKOWITZ JM., ARELLANO R., CASTRO R., DAVIS PB., ELINSON J., JERI FR., SHAYCROFT M., TIMANA J.** : *Uso y abuso de drogas en el Peru urbano*, CEDRO, Tall. Graf. Lima, 1987.

70. **TORRES Y., MURELLE L.** : *Consumo de sustancias que producen dependencia en Colombia.* Bol. Ofic. Sanit. Pan. Am. 10, 485, 1989.

71. **ORTIZ A.** : *Desarrollo del sistema de registro de informacion sobre drogas en Mexico.* Bol. Ofic. Sanit. Pan Am, 107, 523, 1989.

72. **MEDINA ME., TAPIA CR., RASCON ML., SOLACHE G., OTERO BR., LEZCANO F., MARINO MC.** : *Situacion epidemiologica del abuso de drogas en*

Mexico. Bol. Ofic. Sanit. Pan Am. 107, 475, 1989.

73. NEGRETE JC. : *El alcohol y las drogas como problema de salud en America Latina.* Bol. Ofic. Sanit. Pan Am. 81, 158, 1976.

74. de la QUINTANA M. : *Situacion Nacional de la Farmacodependencia en Bolivia y los problemas derivados de la misma.* Conv. Hipolito Unanue, Quite, 1986.

75. ALFARO E., *Abuso de drogas en Costa Rica. Recopilacion de varios estudios.* Bol. Ofic. Sanit. Pan Am. 107, 504, 1989.

76. BERGONZOLI G., RICO O., RAMIREZ A., PAZ MI., RAMIREZ J., RIVAS JC., SALINAS A., RODRIGUEZ O., SALAZAR O., RINCON N. : *Uso de drogas entre estudiantes de Cali, Colombia.* Bol. Ofic. Sanit. Pan Am. 106, 22, 1989.

77. AGUILAR E. : *Prevalencia del uso indebido de alcohol, tabaco y drogas en la problacion ecuatoriana.* Bol. Ofic. Sanit. Pan Am. 107, 510, 1989.

78. ORTIZ A., ROMERO M., SORIANO A. : *Recent trends in drug use in Mexico city. In epidemiol. trends in drug abuse,* NIDA, Rockville, 1990.

79. FERRANDO D. : *Uso de drogas en las ciudades del Peru. Encuesta en Hogares.* CEDRO, Lima, 1990.

80. MEDINA ME., RASCON ML., OTERO MC., TAPIA R., SOLACHE G., LAZCANO F. : *Drug abuse in Northern Mexico. Result from a National survey, in epidemilogical trends in drug abuse,* NIDA, Rockville, 1990.

81. AGUILAR E. : *Prevalence of the improper use of alcohol, tobacco and drugs in the Ecuadorian population, in epidemiological trends in drug abuse,* NIDA, Rockville, 1990.

82. JERI FR. : *Some recents facts about drug abuse in Peru, in epidemiological trends in drug abuse,* NIDA, Rockville, 1990.

83. MARIATEGUI J. : *Epidemiologia de la farmacodependencia en el Peru.* Rev. Neuropsiquiat. 41, 29, 1978.

84. NAHAS GG. : *Keep of the grass. A scientific enquiry into the biological effects of the marihuana,* pp 80, Pergamon Press, New York, 1979.

85. KOZEL N. (editor) : *Epidemiologic trends in drug abuse. Proceedings of the community epidemiology work group,* NIDA, Rockville, 1991.

86. JERI FR., Del POZO T., SANCHEZ CC., FERNANDEZ M. : *El sindrome de la pasta de coca. Observaciones en un grupo de 158 pacientes del area de Lima.* Rev. Sanid. Minist. Int. 39, 1, 1978.

87. PONCE M. : *Consumo de drogas en une problacion universitaria de Lima,* Minist. Salud, 1973.

TRENDS AND POLICIES REGARDING CANNABIS AND OTHER NARCOTICS IN SWEDEN.

Peter Allebeck.

Associate Professor, Department of Community Medicine, Karolinska Institute at Huddinge University Hospital, Sweden.

ABSTRACT

Sweden has by tradition had a restrictive drug policy, in which cannabis has been treated on the same level as "hard" drugs. The drug situation among young persons is now better than in previous years, with only a few percent of young persons reporting having tried drugs. However, we still have large groups of heavy drug abusers in older age groups. As long as these groups continue to be an important focus for import and dealing with narcotics, there is a great risk of further spread of drug abuse and recruitement of new groups. Preventive activities targeted to young people is still an important task, and new methods have to be elaborated, suited for youth of the nineties. Of great concern at present is the harmonization towards the European Community. Scientists and health care workers have warned against the effects on Swedish alcohol and drug policy -and thus the health of the population- that might result if Sweden would have to change laws and social policy, and close down all frontier controls.

KEY WORDS

"Narcotics", Illicit drugs, Prevalence, Sweden, Drug policy.

SWEDISH DRUG POLICY IN GENERAL

According to Swedish law, Narcotics[1] are those substances that are classified as such and put on the list of Narcotics by the National Board of Health and Welfare. The list, which is regularly updated and revised, comprises various forms and preparations of cannabis, CNS stimulants, opiates, hallucinogens and psychotropic drugs (1).

Sweden has by tradition had a restrictive drug policy. An exception to this was a shorter period with a permissive drug policy during 1965-67 (2). During this period a number of doctors were allowed to prescribe drugs to addicts. Thus more than 200 patients were able to inject themselves with legally obtained drugs, most of them with central stimulants but some of them with opiates. Since this policy increased the amount of drugs on the black market, and of criminal activity among the addicts, there was a return to a restrictive drug policy by the end of the sixties, which has been maintained, and reinforced, since then.

The penalties for narcotics offence are among the most severe within the Swedish criminal code. Several years of imprisonment is the normal penalty for narcotics felony (3). For a long time the policy has been to protect the "victims" of the drug dealers, i.e. that "ordinary" drug abusers are more in need of treatment than punishment, and use of narcotics thus was not considered criminal. During the eighties it became clear that this distinction was impossible to maintain, and that use of drugs almost invariably implies not only buying and selling drugs, but also other criminal activities. Thus, by an amendment to the Narcotic Drugs Act, also use of narcotics was criminalized in 1989. In practice, however, this part of the law is very difficult to apply, and to my knowledge no sentence has been pronounced for merely use of narcotics. The amendment, however, was important in that it indicated a changed attitude within society towards use of narcotics.

In Swedish law and social policy, cannabis has always been treated on the same level as other narcotics. During the sixties and the beginning of the seventess, some groups in society -some artists, "hippies" and young people in general- advocated a liberal policy with regard to cannabis, claiming that alcohol was more detrimental to health than cannabis. No politician or other person in an official position, however, took a stand in this direction or took any action to decriminalize cannabis.

---

[1] *The term "narcotics" is used here in its legal designation: e.g. substances interdicted by law (except for medical purpose) because of their strong potential to induce dependence and impair persistently information processing by the brain. Ed.note.*

Early in 1990, it turned out that some persons close to the recently appointed new Swedish prime minister (conservative party) had advocated a more liberal policy towards cannabis and also towards cocaine, e.g. permission to sell and buy cannabis in certain stores. They had expressed these ideas as part of a general neo-liberalist ideology. The prime minister was quick to react on this issue. he immediately rejected the ideas expressed by his co-workers, and instead declared publicly that some amendments to the Criminal Code on Narcotics were to be expected, giving the police more authority to fight the drug market.

TRENDS IN PREVALENCE OF DRUG USE

According to surveys carried out in schools (at age 16) and among military conscripts (at age 18-19) there has been a steady decline in the proportion of young persons reporting having ever tried narcotics (4). From a maximum around 1971, when about 15% reported having tried drugs, it has declined and has remained at about 4% during the last years. Cannabis is the most commonly used drug, more than 90% of young persons who report ever having tried drugs have used cannabis. There are no recent surveys on heavy drug use in Sweden. Indicators such as number of offences against the Narcotic Drug Act, number of persons in institutional care, and cases of hepatitis, suggest a decrease in the number of heavy drug abusers during the eighties. In particular, there have been important changes in drug use in different age groups. The abuse of cannabis, which was previously viewed primarily as a youth phenomenon, is now extending to older age groups. The proportion of cannabis convictions in the age groups 25-39 years increased from 22% in 1975 to 56% in 1989. The development is similar for CNS stimulants and opiates.

Furthermore, although there is a decrease in the number of young persons who are initiated into heavy drug use, a large cohort of persons heavily exposed during the sixties and the seventees have carried their drug habits with them in older ages. Many of these have severe, long standing addiction, resistant to treatment and rehabilitation. With time and age of HIV/AIDS in these groups has prompted new initiatives for treatment and care (methadone maintenance, needle exchange programmes, etc.), but little new initiatives for prevention.

REFERENCES

1. <u>Facts on narcotics and narcotics abuse.</u> National Board of Health and Welfare. Stockholm, 1979.

2. Bejerot N. Drug abuse and drug policy. <u>Acta Psychiatr.Scand.,</u> suppl.256, Copenhagen, 1975.

3. Narkotikaboken. <u>Socialstyrelsen redovisar</u>, 1988:2, National Board of Health and Welfare. Stockholm, 1988.

4. Report 91. <u>Trends in alcohol and drug use in Sweden.</u> CAN, The Swedish Council for Information on Alcohol and Other Drugs. Stockholm, 1991.

CANNABIS CONSUMPTION IN THE FRENCH POPULATION (12 to 44 years) IN 1992.

Lionel Gaillaud

Impact Médecin, Paris.

ABSTRACT

A sample of 1167 French persons aged 12 to 44 (representative of 24.5 million people) were queried regarding their smoking of cannabis (hashish). One third of the sample was offered cannabis and two thirds were not. Approximately, 4.8 million had smoked hash and there were one million regular or occasional users. The majority were male, young (65% less than 25 years old) city dwellers and college graduate (30 %). Smokers are mainly motivated by pleasure (73 %) or desire to "feel better" (10 %). Parental reaction to hash smoking by their children varies from concern (24 %), strong disapproval (15%), understanding (25 %) indifference (7 %) or approval (6 %). 64% of presents users consume alcohol, 54% are regular tobacco smokers and 2% have used heroin.

KEYWORDS

Cannabis use; Hashish use; French population (12 - 44 year olds); Prevalence; Other drugs; Socio-cultural profile; Motivation; Parents.

The French Polling organization SOFRES made a study in May 92 regarding the use of hashish in the French population aged 12-44. This poll, the first of its kind in France, was performed for the "Fondation Toxicomanie-Prevention Jeunesse" and sponsored by the Fondation de France.

A national sample of 1167 persons aged 12 to 44 (representing 24.5 million people) were interviewed face to face in their homes. This inquiry was set up according to the quota method (in relation to sex, age and occupation of the head of the family by social and professional category). This poll was self-administered: the persons queried filled in the inquiry questionaire by themselves.

I. USE OF HASHISH AMONG 12 TO 44 YEAR OLDS:
prevalence and motivation (Table 1 and 2)

The poll of the SOFRES indicates that approximately one third of this sample representative of 7.5 million people were offered hash and that 4.8 million have smoked the drug (61% of the sample were never offered the drug). At present, one million French people are current hashish users : 250 000 smoke regularly and 750 000 sporadically. Approximately 4 700 000 persons (one out of 5 in this particular age group) who have tried hash at least once : 3.8 million are former hash smokers, 1, 8 million have smoked "more than once" and 2 million "only once".

2,7 million have received offers to smoke but refused and 15 million between the ages 12-44 were never offered hash and never smoked. Two million people refused to answer this question.

In a limited analysis among 12 to 17 year old: 17% of them were subject to offers, 6% accept and half of them (3%) smoke from time to time. For those smokers who define their use more precisely, two thirds of those who stopped smoking hash used it very rarely (monthly or yearly). For those who have continued smoking 30% smoke weekly , 22% smoke daily, 14% several times a day.

Among all reasons for smoking, pleasure is the first and main motivation reported for hashish consumption then comes curiosity, and desire to transgress what is forbidden(6 %). 10% use hash to "feel good". In the case of former smokers curiosity is the first reason invoked (56% against 22% present day smokers). For present day smokers, pleasure is the first reason (73%).

TABLE 1

FREQUENCY OF USE

	Former and present smokers (%)	Present smokers (%)	Former smokers (%)
Several times a day	12	12	12
Once a day	3	4	2
2 or 3 times a week	8	6	9
Once a week	9	16	5
2 or 3 times a month	9	15	6
Once a month	14	18	12
Once every 2 or 3 months	20	12	24
Once a year	14	4	20
No answer	11	13	10

TABLE 2

REASONS FOR SMOKING

	All smokers (present and former (%)	Present smokers (%)	Former smokers (%)
For pleasure	52	73	41
"It makes me feel better"	10	11	10
By curiosity	44	22	56
Because it's forbidden	6	2	8
To follow the crowd	4	0	7
No answer	0	0	0

People interviewed may have given several answers.

TABLE 3

SOCIO-CULTURAL PROFILES OF HASHISH USERS

	Present smokers (%)	Former smokers (%)	Never smoked (%)	No answer (%)
SEX				
Men	4	19	68	9
Women	4	12	77	7
AGE				
< 25 years	6	14	70	10
25 - 34 years	4	21	69	6
35 - 44 years	1	11	80	8
OCCUPATION OF HEAD OF HOUSEHOLD				
Farmer	4	4	73	19
Skilled workers, small businessmen	2	16	69	13
Executives professions	5	18	70	7
Managers	4	22	66	8
White collar	4	13	74	9
Blue collar	2	9	83	6
Retired	10	32	52	6
PROFESSIONAL SITUATION OF INTERVIEWEE				
Employed	3	18	72	7
Unemployed, looking for 1st job	8	13	70	9
Student	6	13	71	10
Housewife, no occupation	2	15	78	5
EDUCATION				
No degree	1	7	82	10
Primary	0	3	90	7
High school	4	13	75	8
College	1	14	75	10
University	7	23	64	6
TOWN POPULATION				
< 2,000	3	9	80	8
>2,000 - <20,000	4	14	74	8
>20,000 - <100,000	0	15	77	8
100,000 or >	5	16	71	8
Paris area	7	27	56	10
REGIONS				
North	2	9	80	9
West	3	13	76	8
South West	4	7	81	8
South East	6	21	67	6
Center	3	19	74	4
East	2	8	80	10
Paris area	6	25	59	10

II. SOCIO-CULTURAL PROFILE OF HASHISH USERS (Table 3)

Of those who use or have used hashish 23% are men and 16% are women.

Beyond 35 years, people appear less confronted to hashish. A high proportion of the 25/34 year olds have stopped smoking. Among the present day smokers, 61% are under 25 years and 30% are between 25 to 34 years. 19% of high school and university students have smoked hash and 6% continue to do so. The education of the user has an influence upon the accessibility to hash : 30% among university graduates, 7% among those with no degree.

Lastly, hash users are most numerous in the Paris region (31%), then in the South-East (27%) . In towns with less than 2000 inhabitants 12% have smoked or smoke hash.

III. RELATIONSHIP WITH THE PARENTS (Table 4 and 5)

41% of present day users have spoken to their parents about their hash use (against 25% former smokers) . When children tell their parents that they have smoked Hash ; 29% of the parents are worried and 25% understand , 6% approve.

IV. CHRONOLOGY AND SETTING OF HASH USE   (Table 6 and 7)

When asked "how old were you when first subject to a hashish offer ?". 18% answer "at 15 years or less" (27% of which smoke presently), and 28% "between 16 and 17 years" (29% of which smoke presently), that is to say before their coming of age (18 years in France).

The use of hashish happens above all in one's home with friends, then in parties, in discotheques or in concerts and in school, whereas the offer happens mostly at parties, then at school, concerts, discos and in the street.

In 88% of the cases, the person who makes the offer is a friend. In 2% of the cases, he also offers harder drugs such as cocaine and heroin. Smoking hashish appears to facilitate contact with people likely to push harder drugs. This contact happens to 43% of those who continue to smoke hashish. It reaches 47% of those who use hash regularly, and to 2% of those who have rejected all offers.

TABLE 4

PARENTAL AWARENESS

	Former and present smokers (%)	Present smokers (%)	Former smokers (%)
Mentioned to parents	31	41	25
Parents became aware of it	12	16	11
No communication	56	43	63
No answer	1	0	1

TABLE 5

PARENTAL REACTION TO THE KNOLEDGE THAT THEIR CHILDREN SMOKE HASH.

Approve	6
Understand	25
Are worried	29
Feel guilty	0
Condemn	15
Are indifferent	7
Don't discuss it	9
No answer	9

TABLE 6

HOW OLD WERE YOU WHEN FIRST OFFERED HASH?

	Former and present smokers (%)	Present smokers (%)	Former smokers (%)
15 or less	18	27	13
16-17	28	29	32
18-19	24	17	23
20 or more	26	19	29
No answer	4	8	3
Average age	18.4	17.5	18.6

TABLE 7

WHERE WERE YOU FIRST OFFERED HASH?

	Former and present smokers (%)*	Present smokers (%)	Former smokers (%)
At a party	39	33	44
At a concert or night club	11	13	12
In school	18	4	17
In a cafe	2	0	3
In the street	9	18	4
In other occasions	12	19	12
No answer	9	13	8

* 88% were offered hash by a friend and 9% by a dealer, 3% no answer

TABLE 8

WHERE YOU EVER OFFERED COCAINE OR HEROIN?

	Former and present smokers (%)*	Present smokers (%)	Former smokers (%)
By the same person who offered hash	2	3	2
By someone else	15	43	16
No	80	52	80
No answer	3	2	2

V. EXPOSURE AND OPINION ON OTHER DRUGS (Table 8)

Among the present hashish users, 64% use alcohol "often" or "from time to time" and 26% "rarely". 54% are regular tobacco users. 4% take drugs without medical purpose "often" or "from time to time". Lastly, 2% of hashish users say they use cocaine "rarely" and 2% take heroin (0,5% "often", 0,5% "from time to time" and 1% "rarely").

81% of those interviewed know that the use of hashish is forbidden, but 12% think it is tolerated and 2% think it is licit.

Moreover, 81% of the people interviewed are against legalization, 14% are for legalization. These numbers become respectively 41% and 57% when the people interviewed were hashish smokers or former smokers, and 71% and 25% for present day smokers. Regarding a possible legalization of "hard" drugs 80% of present day smokers are against it, as well as 88% of former users and 95% of those who have never used hashish.

82% of those who have never smoked hash believe in a unavoidable escalation toward "hard" drugs. 70% of those who have stopped smoking hash reject this belief as well as 87% of present day smokers.

78 % of present day smokers reject the opinion according to which "Smoking hash is a way of opposing oneself to society". Among the non-users, 74% believe that hashish is used by people who feel unhappy.

Progression of police arrests for use and possession
of cannabis in France, 1970-1990.

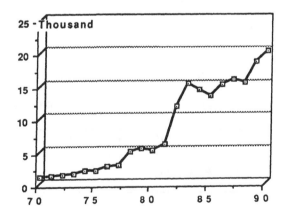

(From F.R. Ingold, 1992)

The Epidemiology of Adolescent Drug Use in France and Israel*

DENISE B. KANDEL, ISRAEL ADLER, AND MYRIAM SUDIT**

Abstract – Based on samples of adolescents residing in urban areas in France and in Israel, cross-
– cultural comparisons of adolescent use of alcoholic beverages, cigarettes, and illicit drugs are reported. Lifetime and current prevalences of use of all substances are higher in France than in Israel.
The relative ranking of the prevalence of use of the various drugs is identical in the two countries,
and is similar to that found in American samples. In both countries, drug use is more prevalent
among males than females, and among older than younger adolescents. There are no differences
among different socio-economic groups. Religiosity affects of non — alcoholic substances and the
amounts of alcoholic beverages consumed in Israel. Differences in the prevalence of substance use
across culture decrease as overall prevalence of use increases(*Public Health 1981 ; 71 : 256-265.*).

Drug use in adolescence has been extensively studied from epidemiological. sociological and psychological perspectives in recent years. The number of studies reflects the growing social concern with
substance use and its presumed causes and consequences. While such concern is growing in many
countries, few epidemiological cross-cultural data are currently available. This paper presents data
from two surveys of adolescents in France and in Israel.

Analyses carried out earlier in the United States indicated that the uses of legal and illegal
drugs*** are closely interrelated[1,2] and that involvement in illicit drugs is preceded by the use of legal
drugs[3]. Alcohol was found to play a crucial step in the developmental sequence of adolescent drug
involvement from non-use to use of illegal substances with beer and wine as the entry drugs, followed by cigarettes and / or hard liquor. Some individuals progress to marihuana : and almost no adolescent progresses to other illicit drugs without having used marihuana. An important issue, therefore, is whether differences also in the use of other drugs in cultures where wide differences exist in
the use of alcoholic beverages. France and Israel were selected for comparative analysis because of
the apparent striking differences in patterns of alcohol consumption in the two countries. Although
no systematic cross-cultural epidemiological data exist on general populations self-reported patterns
of alcohol consumption, such as sales volume of alcohol and indicators of alcohol-related disabilities.
Israel has one of the lowest rates in the Western world; France has one the highest rates[4,5]. In 1972,
France had an apparent consumption of 23.43 l. of absolute alcohol per capita, as compared with
3.25 l. in Israel in 1974, and 10.,48 in the United States in 1975[6].

For both Israel and France, there exist almost no epidemiological data on drug use (whether tobacco, alcohol or illicit drugs) either by adolescents or by adults[7-9]. To the best of our knowledge, to

---

* Cet article publié dans l'American Journal of Public Health (**71** (1981) 256-65) est réimprimé avec
la permission des auteurs auxquels nous adressons notre reconnaisance.

** Address reprint requests to Dr. Denise B. Kandel, Department of Psychiatry and School of Public
Health, Columbia University, 722 West 168th Street, New York, NY 10032. Dr. Kandel is also with
the New York State Department of Mental Hygiene ; Dr. Adler is affiliated with Columbia University SPH and with Hebrew University of Jerusalem ; Ms. Sudit is with Columbia University SPH.
This paper, submitted to the Journal July 8, 1980, was revised and accepted for publication October
6, 1980.

*** The use of certain substances is illegal throughout the western world, and includes drugs used for
recreational purposes, such as marijuana, LSD or heroin, or psychoactive drugs, such as the tranquilizers, stimulants and barbiturates, when used without medical prescription. Alcoholic beverages
and cigarettes can be legally used bu adults and are usually classified as "legal" substances, even
when restrictions exist for adolescents. The legal age for alcohol use in France is 18. There are no age
restrictions on alcohol use in Israel.

date *only two studies of adolescent drug use have been carried out in France.* In 1971, Davidson [10,11] examined smoking, drinking, and the use of medically prescribed psychoactive drugs among 2,339 high school students aged 16-19 in three regions of France. A survey of drinking practices was conducted on a representative sample of 1,535 French youths 15 — 19 years old by the Institut de Recherches Scientifiques Economiques et Sociales sur les Boissons (IREB)[12] a private association of manufacturers of liquor interested in promoting safe drinking. None of these studies asked about the use of marihuana or other illicit drugs*.

In Israel, two adolescent studies have reported about the use of cigarettes and hashish. Peled questioned 3,000 students, randomly selected in 1971 from 20 high schools in the four major urban centers of Israel[13, 14]. In 1973, Shoham, et al, contacted 416 students in four high schools in different parts of Israel[15]. In neither study were questions asked about drinking. A still unpublished study by Shuval and Javetz inquired in 1979 about the use of legal and illegal drugs by over 6,000 seventh through twelfth grade students.

Three epidemiological studies of drinking practices among adults were carried out in France in 1955-1959[16]. Since then, however, among adults only one departmental survey[17] and an unpublished national study by IREB[18] have been carried out, despite the fact that alcoholism is one of the foremost public health problems in France (see also reference 19). Similarly, despite the long standing interest in the low rates of alcoholism among Jews, there is but one recent limited survey of drinking in the general population in Israel[20].

A major goal of the current research was to obtain epidemiological data on substance use by adolescents in France and in Israel.

METHODS

SAMPLES

Two adolescent household surveys of urban youth 14 to 18 years of age were carried out. The French sample includes 499 French adolescents representative of 14 through 18 years old youths in the Parisian metropolitan area (Paris and suburbs) interviewed in Spring 1977. The field work was carried out by the Institut Français d'Opinion Publique. The Israeli sample included 609 Israeli adolescents, 14 through 18 years of age, residing in the four major urban centers in Israel (Jerusalem, Tel Aviv, Haïfa, and Beer Sheba) interviewed in Spring 1979. The field work was carried out by the Israel Institute of Applied Social Research, in Jerusalem. Both samples are urban, and as such not selected to be representative of the countries as a whole. The completion rate in the Israeli sample is about 90 per cent. The French sample was collected as a quota sample, based on respondents age, sex , and student status. Various demographic characteristics of the sample (including parents occupations) were similar to those in the the target population. In both samples, data were collected with structured interview schedules administered in households. The schedules are based on an instrument developed for an earlier American adolescent study[22]. Efforts were made to include identical questions in both instruments, although modifications were introduced to incorporate country-specific conditions. The list of alcoholic beverages included certain beverages available in one country but not the other. Religiosity was measured by a different question in each country. Translation and back-translation were carried out to ensure standardization across cultures. Pre-tests of 18 and 30 interviews were carried out in France and in Israel, respectively.

Both samples are almost equally distributed on sex (53 per cent males in the French sample and 49 per cent in the Israeli). However, the age distributions differ somewhat. The French sample includes a higher proportion of older adolescents (18 years old) than the Israeli sample : 26 per cent vs 13 per

---

* A new study by Davidson[43] appeared after the present article was accepted for publication. The survey was carried out in 1978-79 on 2,088 high school students drawn from the same three regions in France as in the earlier 1971 survey. Questions were asked about the use of tobacco, alcoholic beverages and "drugs". Seven percent of the high school students reported any lifetime experience with any illicit drug.

cent. Since adolescent drug behavior is strongly age-related[23], the data were age-adjusted. All observations were weighted so as to create a uniform distribution for each category on the age variable in each country. All sample sizes in the Tables are the unweighted totals.

OPERATIONAL DEFINITIONS

We focus upon three dimensions of drug use :1) ever use, 2) current use, i.e., use in the month prior to the interview, and 3) lifetime frequency of use for each drug. With the exceptions noted below, the same questions were asked for all drugs in both surveys. The drugs inquired about are listed in Table 1.

The pertinent items in the Israeli and the French questionnaires were :1)«How frequently have you ever used... ?» with categories ranging from "never used" to "used 60 or more times" for alcohol and marihuana, and to "used ten or more times" for other illicit drugs ; and 2) "How frequently have you used... in the past month ?" with categories ranging from "never used in past month" to " used more than once daily." Because a pilot study in Israel had revealed very low rates of lifetime prevalence of illicit drug use, the Israeli questionnaire did not contain questions regarding the current use of these drugs. The smoking question was phrased somewhat differently and asked : "How often do you smoke cigarettes ?" with categories ranging from "never smoked," "only smoked once or twice," "used to smoke but stopped," "smoke occasionally," "smoke less than a pack a day," and "smoke a pack a day or more." The last three categories were combined to arrive at estimates of current smokers.

TABLE 1

Percentage of Adolescents Who Used Various Drugs Ever and In the Last 30
Days in France (1977) and Israel (1979)

	FRANCE			ISRAEL		
Substance	*Ever* %	*Last 30 Days* %	*Ratio*	*Ever* %	*Last 30 Days* %	*Ratio*
Cider^A	84	35	42	—	—	—
Beer	80	54	68	70	27	39
Wine^B	79	54	68	63	27	43
Hard Liquor	75	48	64	52	22	42
Cigarettes	82	64	78	44	16	36
Marihuana/Hashish	23	11	48	3	—	—
LSD	4	2	50	*	—	—
Amphetamines	4	3	75	*	—	—
Barbiturates	6	3	50	2	—	—
Tranquillizers	6	3	50	4	—	—
Heroin	2	1	50	*	—	—
Any Illicit Drug	26	15	58	8	—	—
Total N	(474)	474)		(525)	(554)	

^ANot asked in Israel.
^BIncludes only non-ritualistic use.
*None or less than one per cent.

Although the same drugs were asked about in each country, adjustments were made in the alcoholic beverages to reflect the range of beverages consumed in each country. In France, questions were included about cider ; and "apéritifs" and "digestifs" were specified among distilled spirits. In Israel, "white beer" was specified to avoid confusion with a non-alcoholic beverage known as "black beer." we also distinguished ritualistic and non-ritualistic uses of wine in Israel. Our reporting will be confined to non-ritualistic use.

Both measures of use were dichotomized into "user" and "non-user" categories. Comparing responses to these two items provides a measure of reliability. As in American samples [24], we find a very high degree of correspondence between the two sets of responses. The French sample includes only one case (0,2 per cent) with discrepant responses to the two items on any of the drugs. In the Israel sample, the rate of inconsistent responses ranges from 1,5 per cent for hard liquor to 7,2 per cent for wine. *

RESULTS

The data pertain to two major topics : 1) epidemiological trends in lifetime and current prevalences, and extent of life-time experience with various classes of drugs : and 2) the socio-demographic correlates of patterns of use.

LIFETIME AND CURRENT PREVALENCES

From the rates of reported lifetime prevalence of drug use for the total age-standardized samples in Table 1, several trends are readily apparent. First, for all drugs, the rates of reported use in France exceed those in Israel. These differences are relatively small in the use of beer and wine, are larger with respect to hard liquor and cigarettes, and are highest in the use of illicit drugs, except tranquilizers. With one exception, the relative magnitude of these differences is inversely related to the overall prevalence of use of the various drugs : the higher the reported rate of use. the smaller the difference between the rates reported in each country. For cigarettes, however, the difference is somewhat higher than one would expect on the basis of its overall prevalence.

The relative prevalence of use of the various substances is very similar in the two countries. Beer and wine** are reported to have been used by more young people than hard liquor, and the reported rate of use of all legal drugs exceeds that of all illegal drugs. The only difference between the two countries pertains to cigarettes. Whereas in France the use of cigarettes is higher than that of all alcoholic beverages except cider, in Israel the rate of cigarette use is lower than that of all alcoholic beverages. In this respect, the order of the various drugs in Israel resembles that in the United States[23,25] (see also Table 6). The lifetime prevalence of use of the illegal drugs in Israel is exceedingly small and is nil for three drug classes : psychedelics, amphetamines, and heroin***. In France, the use of marihuana is almost four times as great as use of the next illicit drug, replicating trends in American surveys[23]. Combining the use of one or more illicit drugs into a single variable, "ever use of any illicit drug", clearly confirms the patterns noted above across the two countries and within each country. In both countries, the use of any illicit drug is less prevalent than the use of any of the licit substances, and the differences are larger in Israel than in France.

In line with definitions used in American studies, "current" use was defined as any use during the past 30 days preceding the surveys.

For all drugs, the proportions of youth who have used the drug in the past 30 days is much smaller than the proportion having ever used it. The relative rankings in current prevalence (see Table 1).

---

* In line with analyses of inconsistent responses in earlier American studies, respondents who reported currently using a given drug were classified as having ever used it.

** In Israel, the proportion having used wine for ritualistic purposes (83 per cent) is higher than the proportion using non-ritualistically.

*** The percentages (3 per cent) of Israeli adolescents reporting having used hashish and/or marijuana in 1979 is similar ta the percentages of users (5 per cent and 4 per cent) identified in two earlier Israeli studies carried out in 1971[14] and 1973[15].

TABLE 2
Lifetime Frequency of Use of Various Drugs in Total Sample and among Users
in France (1977) and Israel (1979)

Drug/Substance	*Percentage of Adolescent Reporting Each Frequency of Use*						
	Never %	1-2 times %	3-9 times %	10-39 times %	40-59 times %	60 + times %	N
FRANCE (1977)							
Cider	15	18	18	26	7	16	(480)
	—	21	21	31	8	20	(406)
Beer	19	17	21	16	8	19	(482
	—	21	25	20	10	24	(390)
Wine[A]	21	17	17	19	9	17	(481)
	—	22	22	24	11	22	(381)
Hard Liquor	25	22	18	19	6	11	(489)
	—	29	24	25	8	15	(366)
Marihuana / Hashish	78	9	4	4	*	4	(495)
	—	39	24	18	2	17	(109)
LSD[B]	96	2	1	1	—	—	(464)
	—	53	26	21	—	—	(19)
Amphetamines[B]	96	2	1	1	—	—	(479)
	—	59	18	23	—	—	(17)
Barbiturates[B]	94	4	1	1	—	—	(474)
	—	67	19	15	—	—	(27)
Tranquilizers[B]	94	4	1	1	—	—	(478)
	—	70	17	13	—	—	(30)
Heroin[B]98	*	1	*	—	—	(469)	
	—	22	56	22	—	—	(9)
ISRAEL (1979)							
Beer	30	29	18	11	4	8	(541)
	—	45	28	18	6	12	(345)
Wine[A]	44	31	14	7	2	4	(571)
	—	54	24	12	3	6	(320)
Hard Liquor	51	30	12	4	1	2	(566)
	—	62	25	8	3	4	(276)
Marihuana / Hashish	97	2	*	1	*	1	(654)
	—	47	11	16	*	21	(19)
Barbiturates[B]	98	2	*	*	—	—	(544)
	—	82	18	;	—	—	(11)
Tranquilizers[B]	96	3	1	*	—	—	(542)
	—	74	22	4	—	—	(23)

[A]includes only non-ritualistic use.
[B]Used 10 times or more was last category.
*None or less than one per cent.

The French data illustrate also that current use of illicit drugs is much lower than current use of the licit substances.

The cross-cultural differences observed between Israel and France in the lifetime prevalence of all drugs persist and are accentuated in the measures of current use. On all drugs in which a comparison is possible, the reported rates of current use are at least twice as high in the French sample as they are in the Israeli (see the ratios of current to ever users in Table 1)

The lifetime and current prevalence data presented above include in the user category adolescents who have experimented with a given drug once or twice, as well as those who have used it more frequently. The number of times French and Israeli youths have ever used each of the various drugs is presented in Table 2, with response distributions among the total samples in each country, as well as the conditional distributions among those who ever used each substance. Since the information regarding ever use was supplemented by information on current use, as described above, the percentages of "never use" in Table 2 are not always the complements of the proportions of ever use reported in Table 1.

In France, about one-fourth of those who have ever used the various alcoholic beverages have used them only once or twice : between one-sixth and one-fourth have used these substances 60 or more times. These result stand in sharp contrast to those observed in the Israeli sample, where one observes a much lower percentage of heavy involvement in alcoholic substances (see Table 2). Similar cross-cultural differences appear with respect to cigarette smoking : 63 per cent of the adolescents who ever smoked are no longer smoking in Israel as compared to 24 per cent in France, 3 per cent vs 12 per cent are currently smoking a pack or more a day, and 12 per cent vs 35 per cent smoke less than a pack of cigarettes a day.

In France, the use of marihuana is more experimental than the use of legal drugs. A very low percentage of adolescents report the use of marihuana 40 or more times. The use of illicit drugs other than marihuana is rare and the most experimental of all. Among Israeli adolescents, use of marihuana and / or hashish is less experimental than the use of wine and hard liquor. However, the size of the groups reporting use of other illicit drugs is very small, so that any inferences based on these subsamples must be very tentative.

These results suggest that , for those drugs for which prevalence rates make a comparison possible (i.e., the various alcoholic beverages, cigarettes and marihuana), not only do French adolescents experiment with these drugs more extensively than the Israelis, but also once they start experimenting with a particular substance, they are more likely than the Israelis to remain users. A similar conclusion was reached in a discussion of results on an earlier United States sample : "Substances with the highest overall rates of ever use are also those which are used most recently and most regularly, except for a reversal between marihuana and hard liquor"[26].

DEMOGRAPHIC CORRELATES

Age. In both countries, for almost every single drug, rates of use increase monotonically with age (see Table 3) While slight reversals in these trends occur, the highest rates of use are consistently observed in the oldest cohort and the lowest rates in the youngest one.

The cross-cultural differences in prevalence of drug use observed in the total youth samples prevale in all age groups. Furthermore, the differences in rates of use of illicit drugs between Israeli and French adolescents increase with age. Whereas in the younger cohorts (14 and 15 year olds), the reported rates of ever use are about 2.4 times larger in France than in Israel, the rates are 3.5 and 4.0 times larger in the older cohorts (17 and 18 year olds, respectively). The rate by which use of any illicit drug increases with age in the adolescent years is much larger in France than in Israel.

Sex. Consistents sex differences exist with regard to use of all drugs (Table 4). With two exceptions, in both countries, the reported rates of lifetime experience with all drugs are higher for males than for females. The exceptions involve the use of barbiturates and tranquillizers, where sex differences either completely disappear (tranquillizers in France, barbiturates in Israel). Consistent with this trend is the finding that in France the sex differences in the use of barbiturates are smaller than for any of the other illicit drugs, except the tranquillizers.

The same cross-cultural trends appear for the sex differences as were described for the age differences. The sex differences are greater in Israel than in France, and they are inversely related to the overall prevalence of use of the various substances.

Father's Education. Father's education was used as a surrogate for the socioeconomic status of the

TABLE 3

Percentage of Adolescents Who Used Various Drugs by Age in France (1977) and Israel (1979)

Substance	FRANCE					ISRAEL				
	14 %	15 %	16 %	17 %	18 %	14 %	15 %	16 %	17 %	18 %
Cider^A	81	81	84	88	89	—	—	—	—	—
Beer	70	78	85	89	83	64	72	73	70	73
Wine^B	75	70	74	87	90	48	59	71	66	72
Hard Liquor	58	67	76	84	91	38	49	55	55	64
Cigarettes	63	83	83	93	89	25	38	39	55	62
Marijuana / Hashish	8	17	21	33	31	*	1	1	7	8
LSD	1	*	4	4	5	*	*	*	*	*
Amphetamines	1	6	4	4	8	*	*	*	*	*
Barbiturates	3	6	5	4	11	2	2	2	2	1
Tranquilizers	6	3	6	7	10	4	7	3	5	4
Heroin	1	2	2	1	3	*	*	*	*	*
Any Illicit Drug	12	21	25	35	40	5	9	6	10	10
Total N	(83)	(69)	(105)	(100)	(122)	(84)	(126)	(140)	(137)	(74)

^A Not asked in Israel.
^B Includes only non-ritulistic use.
* None or less than one per cent.

TABLE 4

Percentage of Adolescents Who Ever Used Various Drugs by Sex in France
(1977) and Israel (1979)

Substance	FRANCE			ISRAEL		
	Male %	Female %	Ratio	Male %	Female %	Ratio
Cider^A	87	82	.94	—	—	—
Beer	85	77	.91	79	62	.78
Wine^B	81	77	.95	72	55	.76
Hard Liquor	79	70	.89	61	44	.72
Cigarettes	86	79	.92	52	36	.69
Marihuana/Hashish	26	18	.69	5	2	.40
LSD	5	3	.60	*	*	—
Amphetamines	5	2	.60	*	*	—
Barbiturates	6	5	.83	2	2	1.00
Tranquillizers	6	6	1.00	3	5	1.67
Heroin	3	1	.33	*	*	—
Any Illicit Drug	29	24	.83	9	7	.78
Total N	(246)	(228)		(252)	(266)	

^ANot asked in Israel.
^BIncludes only non-ritualistic use.
*None or less than one per cent.

respondent's family of origin and was dichotomized according to whether or not the father had completed high school. There is no consistent relationship between adolescent drug use and father's education in either country. Whatever differences are observed between the two educational levels are extremely small (data not presented).

*Religiosity** — Much has been written about the role of religious affiliation and religiosity in modulating drinking patterns and alcoholism rates[21,27-31]. The drinking behavior of Jews in particular has been the subject of intense analyses. One popular thesis is that Jews as a distinct sociocultural group are characterized both by low rates of abstinence from alcohol and low rates of alcoholism and problem drinking[21,30,31]. In addition, there may be variations among Jews that are associated with religious orthodoxy, the more orthodox exhibiting lower rates of abstention and of alcoholism than the less orthodox. These findings first reported by Snyder,[21] however, were not replicated by Knupfer

---

* Analyses were also carried out among Israeli adolescents of different ethnic backgrounds. There is an ethnic differentiation in the types of substance used rather than the overall prevalence of use. Youths with fathers born in Israel are characterized by relatively high rates of use of wine and liquor ; those with fathers from Asia/Africa by beer ; and those with fathers drom Europe/America are slightly higher in the use of cigarettes than the other two groups. The proportions having consumed beer, wine, hard liquor or cigarettes are respectively : 64 per cent, 71 per cent, 59 per cent and 41 per cent, in the Israel group ; and 74 per cent, 57 per cent, 49 per cent and 43 per cent in the Asia/Africa group ; and 70 per cents, 63 per cent, 51 per cent and 46 per cent in the Europe/America group

and Room[30]. In neither study was a distinction established between drinking for ritualistic and for non-ritualistic purposes.

Because France and Israel each are characterized by a single religion, we could only investigate the role of religiosity on substance use by adolescents in each country. The cross-cultural consumption patterns discussed above illustrate that as a group Israeli adolescents have much lower rates of use of all substances than the French, except as regards the drinking of wine for ritualistic purposes. If such drinking is taken into account, the lifetime experience with wine is more prevalent among Israelis than among French adolescents.

The lifetime and current rates of use of substances by religiosity are presented in Table 5. In the Israeli sample, religiosity is indexed by type of school attended;* in France, by frequency of attendance at religious services. French adolescents who report attending services more than once a month are classified as religious. The data clearly demonstrate the inhibiting effects of religiosity on the use of all drugs in France.

A more complex pattern is displayed among Israeli adolescents. Religiosity appears to inhibit the use of cigarettes, hashish/marihuana and other illicit drugs, as well as beer. The differences are smaller as regards the drinking of wine for non-ritualistic purposes, and they are completely reversed as regards the use of distilled spirits : a higher proportion of religious than of non-religious Israeli adolescents report having consumed hard liquor ever in their lives or within the last month. The effect of religiosity in Israel manifests itself most strongly not in the frequency of drinking experiences but in the amounts of alcohol consumed at any one sitting. These differences appear for each of the three alcoholic substances, including hard liquor. Religious youths consistently consume fewer alcoholic beverages than the non-religious. On any one occasion, 12 per cent of the non-religious as compared to 5 per cent of the religious report having consumed more than one can of beer : 25 per cent vs 14 per cent drank more than one glass of wine : 12 per cent vs 6 per cent drank more than one glass of liquor.

Thus, within the broader cross-cultural differences in which Jews as a group are less likely to report use of any kinds of substances than the French, including alcoholic beverages, religiosity appears to modulate drinking experiences along the lines suggested by previous investigators : among Jews, religiosity is associated with lower rates of abstinence of those alcoholic substances that are associated with religious observances, such as wine and hard liquor, but lower consumption of any alcoholic beverages at any one sitting. In both countries, religiosity is associated with reduced experimentation with illicit substances.

DISCUSSION

Patterns of drug use among adolescents in France and Israel display both differences as well as close similarities. The overall order in the lifetime and current prevalences of use of the legal and illegal drugs is identical in both countries. The legal drugs are used by a larger proportion of the youth population than the illegal drugs, marihuana is used much more frequently than any of the other illicit drugs, and females are more likely than males to use tranquillizers and sedatives. However, while all four legal drugs have been tried by similar proportions of the French adolescents, the Israeli adolescents are much less likely to have smoked or to have drunk hard liquor than they are to have drunk beer or wine.

The most striking cross-cultural differences, however, appear in the overall lifetime and current use prevalences of use of all drugs, in the extent of use, and in the age-and-sex-specific rates of use in Israel and in France. French youths uniformly report higher lifetime and current use of all the substances investigated in the study, as well as more extensive drug involvement, than the Israeli youths.

In comparing cross-cultural rates of use of various drugs, two alternate consumption patterns could emerge. The high rate of use of one substance, especially alcohol, in a culture could indicate a general positive attitude about the taking of drugs that would be reflected in all aspects of drug tak-

---

* The Israeli educational system, which is under government control, consists of two parallel subsystems : a religious and a non-religious one. Students are free to choose the type of school attended in any place in the country ; those in religious schools attend services daily in school.

TABLE 5

Percentage of Respondents Who Have Ever Used and Used in Past 30 Days Various Drugs, by Religiosity[A] in France (1977)

Substance	FRANCE Ever Use Non-Religious %	FRANCE Ever Use Religious %	FRANCE Past Month Non-Religious %	FRANCE Past Month Religious %	ISRAEL Ever Use Non-Religious %	ISRAEL Ever Use Religious %	ISRAEL Past Month[C] Non-Religious %	ISRAEL Past Month[C] Religious %
Beer	81	64	56	32	73	61	28	25
Wine[B]	80	73	55	41	58	53	27	28
Hard Liquor	77	50	49	31	48	56	18	33
Cigarettes[D]	83	70	43	21	46	37	7	3
Marijuana / Hashish	24	2	12	2	5	*	—	—
LSD	5	*	3	*	*	*	—	—
Amphetamines	4		3	*	*	*	—	—
Barbiturates	6	5	3	3	*	*	—	—
Tranquilizers	7	5	3	3	*	*	—	—
Heroin	2	*	2	*	*	*	—	—
Any Illicit Drug	29	9	—	—	18	2	—	—
Total N	(355)	(85)	(342)	(83)	(350)	(100)	(357)	(94)

[A] Religiosity was measured in France by frequency of church attendance (more than once a month), in Israel by type of school attended.
[B] Includes only non-ritualistic use.
[C] Current use of illicit drugs not asked about in Israel.
[D] Due to different categories, current use not necessarily in past month.
* None or less than one per cent.

ing, such that rates of use of all drugs would be uniformly low or uniformly high. On the other hand, in a culture characterized by a low rate of use of a particular substance, there could be a process of compensation such that the low rate of use of that substance would be paralled by a high rate of use of another substance. If only the French and Israeli data were available for consideration, the first interpretation would appear to be supported. For every substance that was investigated, the French adolescents have higher prevalence of use than the Israelis. Thus, there appears to be an "addition" rather than a "substitution" effect. Sulkunen first developed these contrasting concepts to interpret the uses of various alcoholic beverages in different societies.[4]. In particular, he drew upon the results of the classic social experiment in Norwegian rural areas, where the introduction of new low *alcoholic* beverages constituted additional opportunities for use that became superimposed on former drinking patterns rather than replaced them.[32] The same general principles may be extended to other substances than the different alcoholic beverages.

However, the introduction of the United States as a third comparative case illustrates the complexity of the phenomenon. France, Israel, and the United States represent three especially interesting cases for cross-cultural comparisons. Each country falls into a different category in the various typologies of drinking cultures that have been developed. According to Pittman's typology of cultures based on attitudes about drinking,[33] Israel is classified as "permissive,". France as "over-permissive," and the United States as "ambivalent" In the typology suggested by Sulkunen, based on the preferred alcoholic beverage,[4] Israel is characterized as a distilled spirits culture, France as a wine-consuming country, and the United States as a beer culture. Our data suggest that each country, in turn, appears also to display different patterns of adolescent substance use.

American data are introduced for comparison from a study carried out in 1978 by the New York State Division of Substance Abuse Services on a sample of New York State high school students,[25] in an historical period comparable to the French and Israeli studies (see Table 6). The New York State-wide American sample, however, includes adolescents from a variety of ecological areas.

In the same period, French adolescents exhibit as high (or even higher) rates of use of licit drugs as the Americans but much lower rates of illicit use. The American sample has much higher rates of illicit drugs than either of the other two countries. The national differences are striking. For example, 64 per cent of American adolescents had used marihuana in 1978 as compared to 23 per cent of the French in 1977, and 3 per cent of the Israelis in 1979. Thus, all cultures cannot be unambiguously ordered with respect to overall prevalence of drug use. Countries may be at various stages of evolution as far as specific drug use patterns are concerned. Because of the much higher prevalence of the use of legal drugs in France in Israel, the prospects for further large increases in the use of illicit use would seem to be much more favorable in the former than in the latter.

These data also suggest that the same ethnic or cultural group may behave quite differently in different social contexts (see also Room[9] for a similar point). While Israeli adolescents have the lowest overall prevalence of use of all substances among all three nations, in the United States, Jewish adolescents have the highest rates of use of illicit drugs[1, 22], and Jews have the lowest abstaining rates for alcohol[27,30,31].

Irrespective of overall levels of use, in every culture the uses of legal and illegal drugs are highly interrelated. In France, as in the United States,[1, 2] marihuana users are more likely than non-marihuana users to have used each of the legal drugs (see Table 7). In addition, there is a direct and striking relationship between degree of involvement in legal drugs and involvement with marihuana. Thus, the proportion of French adolescents who report having used hard liquor at least 40 times in their lifetime ranges from 13 per cent among those who have never used marihuana to 86 per cent among those who have used it 40 or more times.

We conclude that despite wide marginal differences in the use of drugs in various cultures, uniformities appear in the relationship of drug use to sociodemographic variables : drug use increases through the adolescent years, there are more male than female adolescents using each type of drug, and there is no consistent association between drug use and the adolescent's socioeconomic status[34]. These uniformities, appearing in societies characterized by widely divergent patterns of use, indicate that drug use shares certain basic features that are not affected by cultural specifics.

Trends appearing in these epidemiological findings suggest important policy implications. There may be a systematic relationship between overall prevalence of drug use in a culture and certain related aspects of drug behavior.In particular, overall prevalence of the use of drugs in a culture appears to be associated with two social processes : 1) a greater and more persistent involvement in drugs, as reflected in the proportion of adolescents who remain current users among those who ever

TABLE 6

Percentage of Adolescents Who Used Various Drugs Ever and in the Last 30 Days, in France (1977), Israel (1979), and New York State (1978)

Substances	FRANCE (1977) Ever %	Last 30 Days %	Ratio	ISRAEL (1979) Ever %	Last 30 Days[E] %	Ratio	N.Y. STATE (1978)[A] Ever %	Last 30 Days %	Ratio
Cider[A]	84	35	.42	—	—	—	—	—	—
Beer[B]	80	54	.68	70	27	.39	96	57	.75
Wine[C]	79	54	.68	63	27	.43		38	.56
Hard Liquor	75	48	.64	52	22	.42		41	.70
Cigarettes	82	64	.78	44	16	.36	82	40	.36
Marihuana / Hashish	23	11	.48	3	*	—	64	45	.44
LSD[D]	4	2	.50	*	—	—	11	4	.38
Amphetamines	4	3	.75	2	—	—	18	8	.47
Barbiturates	6	3	.50	4	—	—	13	5	.50
Tranquillizers	6	3	.50	*	—	—	15	7	
Heroin	2	1	.50	—	—	—	2	1	
Any Illicit	26	15	.58	8	—	—			
Total N	(474)	(474)		(525)	(554)		(25,000)		

[A] From N.Y. State Division of Sustance Abuse Services[25] and special tabulations.
[B] Not asked in Israel.
[C] Includes only non-ritualistic use.
[D] Includes all hallucinogens for New York State (1978) Data.
[E] Current use illicit drugs other than marihuana not asked about in Israel.
[F] Due to different categories, current use not necessarily in past month.
* None or less than one per cent.

TABLE 7

Percentage of Adolescents Who Ever Used Various Licit Drugs by Ever Use of
Marihuana, France (1977)

Substance	Ever Used Marijuana %	Never Used Marijuana %
Cider	90	83
Beer	92	78
Wine	89	77
Hard Liquor	91	70
Cigarettes	99	78
Total N	(116)	(379)

started experimenting with a drug (Table 6), and in the frequency of lifetime use among users of each drug (Table 2) ; and 2) in addition, increased overall prevalence appears to be associated with a spread of the phenomenon throughout all groups in society, such that group differences in drug experiences are attenuated, as witnessed by the decreased sex and age differences in drug use patterns in France as compared to Israel.

The implications of these findings are reminiscent of the distribution model of alcoholism. Originally proposed by Lederman[35] ,the distribution model assumes that the number of alcoholics or heavy drinkers in a society is related to the overall consumption of alcohol in that society and increases multiplicatively with per capita consumption levels[36,37]. Although this model is not without its critics[38], it seems to be quite robust[39-41]. It must be stressed that the data presented in this paper do not deal with the epidemiology of alcoholism, problem drinking, addiction, or problem drug use. It is dubious that such cases can even be captured in any large numbers in epidemiological surveys of general populations[42]. The measure of drug involvement that we analyzed, i.e, lifetime frequencies of drug experience, is a gross measure of involvement. However, those cross — cultural data do suggest a relatively conservative position with regard to accessibility and availability of substances. Both for legal and illegal drugs, persistence and degrees of involvement may be directly related to the overall prevalence of consumption levels in the society. Much remains to be learned about the factors, especially the sociocultural factors, that may explain these broad societal differences in overall consumption patterns.

ACKNOWLEDGEMENTS

This research was supported by PHS research grant DA 01097-06, Scope E from the National Institute on Drug Abuse and the Center for Socio-Cultural Research on Drug Use, Columbia University. We would like to thank Douglas S. Lipton, New York State Division of Abuse Services, Albany, New York. for providing special tabulations from the statewide high school survey, "Periodic Assessment of Drug Use Among Youth, 1978

REFERENCES

1. JOHNSTON L., *Student Drug Abuse*, Ann Arbor : Institut for Social Research, 1973.

2. SINGLE E., D. Kandel and R. Faust, Patterns of multiple drug use in high school, *J. Health Soc. Behav.* **15** (1977), 344-57.

3. KANDEL D., Stages in adolescent involvement in drug use, *Science* **190** (1975), 912-14.

4. SULKUNEN P., "Drinking patterns and the level of alcohol consomption : An international

overview", *in* : Research Advances in Alcohol and Drug Problems, *Gibbons et al.* eds, vol. 3, New York, John Wiley & Sons, 1976, pp. 223-81.

5. World Health Organization Statistics Report : Trends in Mortality from Cirrhosis of the Liver, 1950-71, Geneva, WHO **29**, 52 (1976).

6. KELLER M. and G. GURIOLI, Statistics on Consumption of Alcohol and on Alcoholism, *Journal of Studies on Alcohol.* (1976).

7. MERCER G.W. and R.G. SMART, "The epidemiology of psychoactive and hallucinogenic drug use", *in* : *Research Advances in Alcohol and Drug Problems*, Gibbons R., Y. Israel, H. Kalant *et al.* eds, vol. 1, New York, Wiley, 1974, p. 330.

8. MOSER J., Problems and Programmes Related to Alcohol and Drug Dependence in 33 countries, Geneva, WHO, 1974, Table 8.

9. Room R., "Measurements and distribution of drinking problems in general populations", *in* : *Alcohol Related Distabilities*, Edwards G. *et al.* eds, Geneva, WHO, 1977.

10. DAVIDSON F., M. CHOQUET and M. DEPAGNE, *Les lycéens devant la drogue et les autres produits psychotropes*, Institut National de la Santé et de la Recherche Médicale, Paris, 1973.

11. DAVIDSON F., Etude de l'usage de boissons alcoolisées par les lycéens, paper presented at Meeting of Haut comité d'étude et d'information sur l'alcoolisme, April 25, 1974.

12. Institut de recherche scientifiques économiques et sociales sur les boissons : Les consommations de boissons par les jeunes, Paris, IREB, September 1976.

13. PELED T., The Structure of Motivation for the Use of Haschich : Image and Personal Experiences of High School Students *in* : *Israel, Paper presented at the 2nd International Symposium on Drug Abuse, Jerusalem, Israel, May 1972.*

14. *PELED T. and H. SCHIMMERLING, "The Drug Culture Among the Youth of Israel : The Case of High School Students," in Israel Studies on Criminology 1972-1973*, Shoham S. ed, vol II, Jerusalem, Academic Press, 1973.

15. SHOHAM S., N. GEVA, D. KLIGER *et al.*, *Drug abuse among Israeli youth : epidemiological pilot study, Bulletin on Narcotics* **26** (1974), 9-28.

16. SADOVA R., G. LOLI and M. SILVERMAN, *Drinking in French Culture*, New Haven, College & University Press, 1965.

17. TUYINS A., G. PEQUIGNOT, O. JENSEN *et al.*, La consommation individuelle de boissons alcoolisées et de tabac dans un échantillon de la population en Ille-et-Vilaine, *Revue de l'Alcoolisme* **2** (1975), 105-51.

18. Institut de recherches économiques et sociales sur les boissons, *La Consommation quotidienne de boissons par les Français*, Paris, IREB, December 1974.

19. ROYER R.J. and J. LEVI eds, *Anglo-French Symposium on Alcoholism*, Paris, INSERM, 1976.

20. KANDEL D.B. and M. STUDIT, Drinking Practices Among Urban Adults in Israel : A Cross-Cultural Comparison, Colombia University, 1980 (unpublished manuscript).

21. SNYDER C.R., *Alcohol and the Jews : A Cultural Study of Drinking and Sobriety*, Carbondale, IL, Southern Illinois University Press, 1958 – 8.

22. KANDEL D., E. SINGLE and R. KESSLER, The epidemiology of drug use among New York State high school students : distribution, trends and change in rates of use, *A.m. J. Public Health* **66** (1976), 43-53.

23. JOHNSTON L.D., J.G. BACHMAN and P. O'MALLEY, *Highlights from drugs and the Class of 78 : Behaviors, Attitudes and Recent National Trends*, Rockville, MD, National Institute on Drug Abuse, 1979.

24. SINGLE E., D. KANDEL and B. JOHNSON, The Reliability and validity of drug use response in a large scale longitudinal survey, *J. Drug Issues* **5** (1975), 426-43.

25. New York State Division of Substance Abuse Services, Substance Use among New York State Public and Parochial Students in Grades 7 through 12, Albany, NY : NYS, 1978.

26. KANDEL D., E. SINGLE and R. KESSLER, The epidemiology of drug use among New York State high school students : distribution, trends and change in rates of use, *Am. J. Public Health* **66** (1976), 46.

27. KELLER M., The great Jewish drink mystery, *Brit. J. Addict* **64** (1970), 187-296.

28. SKOLNICK J.H., Religious affiliation and drinking behavior, *Quarterly Journal of studies on Alcohol* **19** (1958), 452-70.

29. SNYDER C.R., "The rarity of alcoholism among Jews : Is it biologically or socio-culturally determin", *in Genetic Diseases among Ashkenazi Jews*, R.M. Goodman and A.G. Motulsky eds, New York, Raven Press, 1978.

30. KNUPFER G. and R. ROOM, Drinking patterns and attitudes or Irish, Jewish and White Protestant American men, *Quarterly Journal of Studies on Alcohol* **28**, 676-99.

31. CAHALAN D. and R. ROOM, *Problem Drinking among American Men : Monograph N° 7*, New Brunswick, NJ, Rutgers Center of Alcohol Studies, 1974.

32. MÄKELA K., Consumption level and cultural drinking patterns as determinants of alcohol problems, *J. Drug Issues* **5** (1975), 344.

33. PITTMAN D.J. ed., *Alcoholism*, New York, Harper & Row, 1967.

34. KANDEL D., "Drug and drinking behavior among youth", *in : Annual Review of Sociology 6, J. Coleman, A. Inkeles and N. Smelser eds, 1980, p. 235.*

35. LEDERMAN S., Alcool, alcoolisme et alcoolisation : Données scientifiques de caractère physiologique, économique et social, Institut National d'Etudes Démographiques, Travaux et Documents, cahier N° 29, Paris, PUF, 1956.

36. BRUNN K., G. EDWARDS, M. LUMIO *et al., Alcohol Control Policies in Public Health Perspective*, The Finish Foundation for Alcohol Studies, vol. 25, New Brunswick, NJ, Rutgers Center for Alcohol Studies, 1975.

37. DELINT J. and W. SCHMIDT, The distribution of alcohol consumption in Ontario, *Quarterly Journal of Studies on Alcohol* **29** (1968), 968-73.

38. PARKER D.A. and M.S. HARMAN, The distribution of consumption model of prevention of alcohol problems, *J. Studies on Alcohol* **39** (1978), 377-99.

39. The Lederman Curve : Report of a Symposium held in London January 6-7, 1977, at the invitation of the Alcohol Education Centre under the chairmanship or the late Professor D.D. Reid, 1977.

40. SCHMIDT W. and R.E. POPHAM, The single distribution theory of alcohol consumption, *J. Studies on Alcohol* **39** (1978), 400-19.

41. FITZGERALD J.L. and H.A. MULFORD, Distribution of alcohol consumption and problem drinking : comparison of sales records and survey data, *J. Studies on alcohol* **39** (1978), 879-93.

42. ROOM R., "Amount of drinking and alcoholism," *in : Proceedings of the 28th International Congress on Alcohol and Alcoholism*, M. Keller and M. Majchrowicz eds, vol 1, Abstracts, Washington, D.C., 1968.

43. DAVIDSON F. and M. CHOQUET, *Les Lycéens et les drogues licites et illicites*, Paris, INSERM, 1980.

STAGE IN ADOLESCENT INVOLVEMENT IN DRUG USE

(Kandel D., Science, 190:912, 1975)

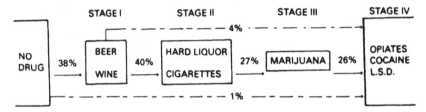

This diagram illustrates the successive stages in adolescent drug use observed in two follow-up surveys ("longitudinal") of 5,468 New York State High School students between fall 1971 and spring 1972, and of 985 seniors 5 months after graduation. Students start use of legal drugs, beer or wine, and go on to the smoking of cigarettes and drinking of hard liquor. While 27 percent of students who smoke and drink progress to marijuana within a 5 to 6 month follow-up period, only 2 percent of those who did not drink or smoke previously do so. *Marijuana in turn is a critical step on the way to other illicit drugs*: while 26 percent of marijuana users will experiment with L.S.D., amphetamines and opiates, only 1 percent of non-drug users and 4 percent of legal drug users do so. This sequence is found in each of the 4 years in high school and in the year after graduation

A 1980 study confirmed the statistical progression of marijuana to heroin and cocaine, "The linkage between marijuana use and later heroin or cocaine use is *ten times* greater than the evidence of linkage between cigarette smoking and lung cancer" (Clayton, R. and Voss, H. *U.S. Jour. of Drug and Alcohol Dependence*, Jan. 1982).

DRUG REFORM: THE DUTCH EXPERIENCE[1].

Richard H. Schwartz

Department of Pediatrics, Georgetown University,
Medical School, Washington D.C., USA

ABSTRACT

An account is given of the governmental approach to the management of the abuse of controlled substances in the Netherlands. Use and possession of 30 gr of cannabis, hash and marihuana is legal and the drug is openly sale in "coffee shops" (300 in Amsterdam alone). This policy which prevails since 1978 has been associated with a progressive increase in cannabis use among 15 to 19 years old : from 4.2% in 1984 to over 8% in 1988-89. While heroin use and traffic is illegal, the Dutch authorities have adopted a policy of "harm reduction" towards heroin addicts: Free distribution of syringes and needles, methadone distribution. Results of this policy have been ambiguous and not prevented the start of cocaine use.

KEY WORDS

Dutch drug policy; Cannabis; Coffee shops; Heroin, Methadone.

INTRODUCTION

The Netherlands has a population of nearly 15 million, much of it concentrated in four cities : Amsterdam, Rotterdam, The Hague, and Utrecht. By law, the cost for medical and unemployment insurance is picked up by the state. Every Dutch citizen is entitled to free medical, pharmaceutical, dental, and hospital treatment. Unemployment is estimated to be 15%, and all unemployed adults receive social security without prerequisiities.

---

[1] *Adapted with permission from Hospital PracticeMay 30, 1991.*

In reaction to what they perceive to be harsh and repressive antidrug laws elsewhere, the Dutch have approached drug abuse as a problem of public health, not one of law enforcement or criminal justice. Coordinated by the State Ministry of Health, the Dutch policy includes

1) "normalization" of the use of cannabis -i.e., legal indulgence of personal use or sale of marijuana or hashish- and de facto decriminalization of personal use of other drug, as heroin and cocaine;

2) widespread distribution of methadone to registered addicts, even those who continue to use heroin, cocaine, or amphetamines;

and 3) syringe and needle-exchange programs. The exchange programs is part of a comprehensive AIDS-prevention effort that also provides for condom distribution, education about safe sex practices and syringe and needle exchanging programs, and counseling for those who request it.

Is the Dutch approach working, and can it be applied in other more populated countries with less generous social and medical free benefits? Some proponents of what is termed drug reform think that the answer is yes to both questions. According to this view, experience in the Netherlands presents a paradigm -a realistic, pragmatic, and sensible middle ground between outright legalization of all drugs of abuse and the general repressive approach prevailing in France and the U.S.

A general evaluation of the Dutch experience with the limited data available, especially in the area of pharmacology and toxicology will attempt to answer these important questions.

THE DUTCH LEGISLATION OF ILLICIT CONTROLLED SUBSTANCES

Although the Netherlands ratified the 1961 Single Convention on Narcotics Drugs, which criminalizes the possession of narcotics, cocaine, and cannabis, it has not yet ratified the 1971 Convention on Psychotropic Substances. The cornerstone of present-day drug policy in the Netherlands, *Backgrounds and Risks of Drug Use*, published in 1972 by the "Narcotics Working Party", concluded that potentially harmful effects from drugs of abuse may be categorized primarily according to the drugs' pharmacologic properties and addictive potential and the purposes for which they are used.

In 1976, the 1919 Dutch Opium Act was amended to distinguish "hard drugs" with unacceptable risks - such as opiates, cocaine, amphetamines, hallucinogenic agents, and "hash oil" (highly concentrated cannabis resin)- from natural hemp products (marijuana and hashish), or "soft drugs". The latter were considered by the Ministry of Health no more dangerous than the paper in which they were wrapped for smoking.

The maximum penalty for trafficking in heroin or cocaine was increased from four years' to 12 years' imprisonment. On the other hand, possession of up to 30 gm of cannabis for personal use was reduced from an offense to a misdemeanor; maximum penalties were set at one month's detention or a fine of 5,000 Dutch florins (about $2,500). The Ministry of Welfare, Health, and Cultural Affairs commented: "The act reflects the view that criminal law plays only a minor part in preventing individual drug abuse. Although the risks to society must, of course, be taken into account, every possible effort must be made to ensure that drug users are not caused more harm by criminal proceedings than by the use of the drug itself."

In practice, penalties for possession of cannabis are virtually always ignored, since the intent of the law is de facto decriminalization of cannabis. Indeed, sale of limited quantities of hashish to persons 16 or older at coffee shops and youth centers is officially sanctioned. This is intended to direct the market for soft drugs into legitimate channels, and so keep hashish sales out of the hands of criminals. There has been no successful effort to generate tax revenues from the sale of cannabis products in the Netherlands.

Efforts to legislate taxes of hashish sales were successfully challenged by hashish users. They argued that, since the government had signed the Single Convention on Narcotic Drugs, it would be illegal to sell an "illegal" substance. Similarly, any drugstore may sell syringes and needles. Possession of these instruments, even when used to inject illegal drugs, is not forbidden.

CANNABIS CONSUMPTION IN HOLLAND

After 1976, with the change in legal climate regarding possession and personal use of natural cannabis products, coffee shops and youth centers where anyone 16 and older could purchase cannabis proliferated. Hashish coffee shops which offer a selection of freshly brewed coffee and of the most popular brand of beers, as well as selected varieties of hashish for smoking, are permitted to advertise. In early 1980, there were fewer than 30 hashish coffee shops in Amsterdam. By the end of the decade, there were approximately 300.

In Dutch towns that have no hashish coffee shops, dealers make purchases in the city and sell hashish from their homes.

Advocates of drug reform in the United States often quote an impressively low 4.2% lifetime prevalence of cannabis use by Dutch youth. Such a figure should be revised as a result of a more recent survey made by the Ministry of Health between 1988 and 1989 at 45 Youth Health Care Centers across the country. A total of 8019 youths aged 10 years and older were surveyed. Most were enrolled in secondary schools. According to survey results, lifetime prevalence of use of illicit drugs by Dutch students increased with age : lifetime use of any illicit drug was 16.5 % for students aged 15 to 16 years and 23.5 % for students aged 17 to 19 years. Use of all drugs was higher among boys than among girls. Lifetime, annual, and monthly use of cannabis is shown in the Table. Nationally, lifetime use and current use (within the last 30 days) of cannabis by Dutch students aged 15 years and older had almost doubled since 1984. These data do not provide information on use by Amsterdam adolescents compared with youths nationwide. In any of the so-called hashish "coffee shops", of which there are 300 in Amsterdam alone, cannabis can be openly purchased by anyone aged 16 years and older. Such shops appear to do a thriving business not only in Amsterdam but also in other cities throughout the Netherlands.

TABLE 1

Cannabis Use by Students in the Netherlands, 1988*

	Age 15-16 y.	Age 17-19y.
Lifetime use	10.8	17.3
Used cannabis in last 12 months	8.5	9.1
Used cannabis in last 30 days	5.2	4.5

* Values are percentages of student surveyed.

In the United States, in 1988 (the same year as the latest available Dutch study), the prevalence of lifetime and current use of cannabis was approximately triple that of the Netherlands. While this contrast is striking, it must be viewed in the context of the overall recent pattern of use in the United States has undergone a significant and consistent decline since 1978, when its use by high school students was at its highest level. Between 1984 and 1988, while the use of cannabis increased almost 100% among upper-high school students in the Netherlands, lifetime use of cannabis by American high school seniors decreased 8 percentage points, from 55% to 47%. During the same time, current use (within the last 30 days from the time of the survey) of cannabis in this country decreased 7 percentage points, from 25% to 18%.

USE OF OPIATES

In the Netherlands, as in Great Britain, as much as 60% heroin users smoke the drug in special pipes or by a method called chasing the dragon. In the latter, granulated heroin is placed in a folded piece of aluminum foil, which is heated by a taper. White smoke containing vaporized heroin moves up and down the foil with the movement of the molten powder, and thus resembles the undulating tail of the dragon in Chinese mythology. The fumes are inhaled through a small bamboo tube or a cylinder of rolled paper.

Forty percent of hard-drug users inject heroin or cocaine, or both. In part because of more lenient treatment of opiate users under Dutch laws, some 10% to 20% of heroin addicts in the Netherlands are nationals of other European countries, mainly from neighboring Germany. Drug dealing is a major source of income for these "addict visitors".

In 1977, the number of heroin addicts in the Netherlands was estimated at 5,000. Since the beginning of the 1980s, the number has increased to 15,000 to 20,000 (1 per 750,000). Amsterdam, the capital city of 640,000, has an estimated 4,000 to 7,000 addicts, 60% to 80% of whom receive municipal assistance, which includes methadone maintenance programs. Thirty percent of the Dutch addict population lives in the other three big cities.

USE OF COCAINE

In 1987, a study reported 4% of 19 year old having use cocaine and 13% of 34 year old. Among 386 Dutch intravenous addicts interviewed between 1985 and 1989, 43% had used cocaine, usually by smoking the free-base form, in the previous six months. However, the number of cocaine-dependant Dutch adults is still small.

THE METHADONE PROGRAMS AND METHADONE BUSES

In 1978, city authorities in The Hague started a program to distribute methadone from mobile buses to reduce the risks associated with L.V. heroin uses. In the first four months of the distribution program, the number of clients increased from 70 to 400; one year later, this number was 1,500. In 1982, health authorities in Amsterdam started a similar program. By 1988, methadone syrup was dispensed daily to 6,300 Dutch heroin users, approximately one third of the country's opiate addicts.

Most minimal-requirement ("low-threshold") methadone programs are run in cooperation with a number of drug-free therapeutic programs. As a result, the number of addicts entering drug-free treatment in Amsterdam has more than doubled since the introduction of methadone and needle-exchange programs.

Methadone Buses. The Amsterdam methadone buses have regular routes, fixed stops of approximately one hour each at eight different locations near drug trafficking areas, and can serve up to 100 addicts each day. Methadone dispensed only to registered clients who have been examined by a physician and carefully selected as individuals which might benefit from and comply with the program. Methadone not given to unknown transients, and no take-home doses are permitted.

The condition for participation in methadone program is quarterly contact with a physician, who writes a prescription for enough methadone until the next visit. Dealing drugs, using drugs, and aggressive behavior are not permitted in the vicinity of the methadone buses. There is no requirement for urine test however, and no mandatory contact with drug counselors.

Such low-threshold programs have benefits. They facilitate contact with some addicts who might otherwise by difficult to reach. On request these programs provide information about specific drugs treatment of adverse drug reactions, direction to the nearest treatment center, and a sterile needle- and syringe- exchange system. Condoms are stocked on the buses and dispensed free of charge. Mobile unit personnels provide crisis intervention and referral to intensive treatment services.

The low-threshold programs have had limited success in reducing crime and prostitution. Addicts seemingly use methadone bot to abolish drug addiction but as a safety net when they are unable to secure heroin or cocaine.

Drug free treatment. The Dutch approach to drug addiction is greatly weakened by an under emphasis on rehabilitation and by a limited number of drug abuse treatment facilities. There are no recent data on vocational rehabilitation or structured aftercare for those who complete treatment at the facilities. Important issues -relapse, cross-addiction, reintegration into mainstream society, pharmacological and toxicological studies, experimental or clinical, are not reported.

CRIME REDUCTION

Advocates of reform in American drug policy contend that crime will decrease significantly when drugs are made legal. Drug abuse experts in the Netherlands held the same view before that country adopted its permissive drug policy -some predicted a 40% reduction in crime. Evidence of the anticipated reduction, however, has not materialized. On the contrary, some recent Dutch studies suggest that the link between drug abuse and criminal behavior may be overstated.

In Rotterdam, for example, one study found that 92% of 3.793 arrests from 1983 to 1987 involved crimes that were unrelated to drug acquisition. The study also analyzed 90 opiate addicts in Rotterdam and found highly unstable backgrounds; more than 75% had not finished secondary school, and most had lost their jobs.

The investigators concluded that much "drug-related" crime is not related to drugs at all. Most addicts had been criminals well before they became addicts, and drug-seeking behaviors merely caused them to redirect their criminal efforts in a different, but still illegal and anti-social, direction. Of note, Rotterdam's illegal market in heroin is now inseparable from that of cocaine, as most addicts are dependent on both drugs. The Netherlands government has concluded that legalization of heroin would not result in a significant decrease in criminality.

DRUG ABUSE AMONG IMMIGRANTS

In the Netherlands, as elsewhere, persons at the bottom of the socio-economic ladder are more likely to become addicted to heroin than those who are better off. Heroin addicts who belong to ethnic minority groups, particularly first- and second-generation children of immigrants from Surinam (formerly Dutch Guiana) or the Moluccan Islands in Indonesia constitute about 30% of the addict population (but only 5% of the general population). Among the 180,000 persons of Moluccan ethnic background in the Netherlands, 2% were estimated to be addicted to heroin.

THE JUNKIEBOND

In 1980, The *Junkiebond* ("Junkie League") was established as a self-help group to counter a movement by some politicians to coerce addicts into treatment. At present, there are Junkiebond groups in all of the country's largest cities. An umbrella organization, the Federation of Dutch Junkie Leagues, serves as a national forum for exchange of information and political activism. The chief aim is to encourage legislative policies that will foster public acceptance of hard-drug users.

The belief is that users of heroin and other injectable drugs know best what their own problems are. Most of those problems arise, it is claimed, because heroin is illegal, which often forces the addict to obtain the drug by illegal means.

The demands of the Junkiebond include access to narcotics, as well as methadone, with no prescription or contact with municipal health authorities. Addicts, it is asserted, should not have to declare their desire to be drug free and, along with municipal welfare services, should share responsibility for the treatment plan. Whether heroin rather than methadone should be supplied by those services ("heroin maintenance") is not currently under discussion.

Opinions on this issue diverge, even among members of the Junkiebond. Sixty-two percent of the addict-clients of methadone maintenance programs in Utrecht would prefer to receive heroin instead of methadone. However, most would rather receive methadone under a less restrictive system than participate in a strongly regulated heroin maintenance program.

SYRINGE AND NEEDLE EXCHANGE

Addicts represent 26% of persons with AIDS in the United States and 50% of those in Italy, in contrast to only 8% of Dutch AIDS patients reported as of 1989. Between 1988 and 1989, the Dutch AIDS population increased by 56%, from 605 to 1,074 patients. In the past three years, however, the percentage of HIV-antibody-positive persons from the addict population has apparently remained stable at 30%. These prevalence figures are derived from self-reported behaviors of addicts in response to surveys in Amsterdam and other Dutch cities. Follow-up verification of those self-reported data was not available.

Syringe- and needle- exchange programs were initiated in the Netherlands in 1984 and are available in 40 Dutch municipalities. These exchanges are part of an integrated plan that includes distribution of free condoms, education of addicts to avoid unsafe injection practices, and counseling by municipal social workers. At the outset, approximately 1,000 needles and syringes were exchanged each week. In 1989, 820,000 needles and syringes were distributed through the 11 exchange locations in Amsterdam. The return rate was 87%. The program is funded by the Ministry of Welfare, Health, and Cultural Affairs, and a longitudinal outcome study has been under way in Amsterdam since 1986.

Conclusions from the first 27 months of evaluation should be drawn very carefully. There are major methodologic problems. Of the 263 addicts, only 72% (189) had injected narcotic drugs at least once during the half year preceding the survey. Only 55% injected daily. Approximately 70% of the participants returned for three of the required quarterly follow-up visits during the study. With respect to reduction in use of shared syringes and needles, the gains represent compliance from only the most motivated (60%) of the study group, which included very few daily heroin or cocaine injectors (who are a greatest risk of HIV infection).

In 1985, the Amsterdam Municipal Health Service started a longitudinal study to evaluate the impact of AIDS prevention among narcotics addicts. As early finding was that high-risk behaviors of new and younger addicts admitted to the program between 1986 and 1988 were undiminished by education efforts. Older participants who had remained with the original cohort, when reinterviewed two years later, reported that their intravenous drug use and needle sharing decreased by about one third. Although some gains appeared to reside from early surveys of the needle- and syringe- exchange programs in the Netherlands, it is not yet certain that such programs have a positive preventive effect on HIV transmission.

CONCLUSION

The Dutch are in the midst of a unique experiment in drug reform that has received widespread publicity in the media. In a small, rather homogeneous, highly industrialized country such as the Netherlands, drug reform advocates are proud of their successes and of their ability to maintain their independent philosophy while at the same time cooperating with the rest of Europe in enforcing international drug laws. Even the most avid proponent is quick to point out, however, that the Dutch system may not work well in countries with more heterogeneous populations. Notably the Dutch approach has met with little or no success in the underprivileged and immigrant populations.

It has not prevented a steady and significant rise in drug addiction (cannabis, cocaine, opiates) among the youthful population 15 to 19 years old and young adults.

REFERENCES

Engelsman E.L., Dutch policy on the management of drug-related problems. Br.J.Addict, 84:211, 1989.

Van de Wijngaart G.F., A social history of drug use in the Netherlands: Policy outcomes and implications. J. Drug Issues, 18:481, 1988.

Van de Wijngaart G.F., Heroin use in the Netherlands. Am.J.Drug Alcohol Abuse, 14:125, 1988.

Driessen F.M., Van Dam G., Results of a drug use survey among students of secondary education in Amsterdam. Instituut voor Sociale Geneeskunde, Vrije Universiteit, Amsterdam, 1988

Sandvijk P., Westerterp I., Musterd S., Het Gebruik van Legale en Illegale Drugs in Amsterdam (The use of legal and illegal drugs in Amsterdam). University of Amsterdam, the Nethelands, 1988.

Musterd S., Sandvijk P., Westerterp I., Drug use in Amsterdam. University of Amsterdam, the Nethelands, 1988.

Stichting INTRAVAL: Harddrugs & Crimimaliteit in Rotterdam (Hard drugs and criminality in Rotterdam). Stichting INTRAVAL, Groningen, 1989, p.120.

Buning E., AIDS-related interventions among drug users in the Netherlands. Int.J. Drug Policy, 1:10, 1990.

Van den Hoek J.A.R., Van Haastrecht H.J.A., Couthinho R.A., Risk reduction among intravenous drug users in Amsterdam under the influence of AIDS. Am.J.Public Health, 79:1355, 1989.

Van de Wijngaart G.F., Competing perspectives on drug use: The Dutch experience. University of Utrecht, the Netherlands, 1990.

Schwartz R.H., Legalization of drugs of abuse and the pediatrician. A.J.D.C., 145:1153-1158, 1991.

Plomb H.N., Kuipers H., Van Oers M.L., Smoking, Alcohol consumption and the use of drugs by school children from age of 10: Fourth National Youth Health Care Survey on Smoking, Alcohol Consumption and the Use of Drugs in the Netherlands. Amsterdam, the Netherlands: Free University Press, 1991.

Johnston L.D., O'Malley P.M., Bachman J.G., Drug use, drinking and smoking: National survey results from high school, college and young adult population: 1975-1988, Rockville, Md, National Institute on Drug Abuse, 1989.

Schwartz R.H., Drug abuse in the Netherlands, A.J.D.C., 146:788-789, 1992.

THE DUTCH DRUG POLICY: A PHYSICIAN'S COMMENTARY.

Karel Frederik Gunning, M.D.

Groene Wetering 32, 3062 PC Rotterdam.

ABSTRACT

According to this commentary of a practicing physician, the free cultivation availability and sale of cannabis in Dutch coffee shops has been associated by an increase of its social acceptability and use among adolescents as a "soft" drug. The "harm reduction" Dutch policy to contain heroin addiction through distribution of free needles and syringes and through methadone distribution has not prevented the spread of this type of opiate addiction in the country and has not curtailed drug generated criminality.

KEY WORDS

Dutch drug policy; Harm reduction; Cannabis; Heroin; Methadone.

As the Dutch Minister of Well-being, Public Health and Culture has not accepted the invitation to speak at this conference, I am the only Dutchman to give you a Dutch point of view. This is a pity, at it would be very interesting to hear a member of the Dutch government explain their drugs policy, which to many non-Dutch seems perplexing. I will therefore try to explain this policy as far as I understand it myself, but I will also criticise it. I am sure our Dutch officials are fully convinced that their approach is the best, but I am not an admirer of this policy. You may therefore tend to believe that I am prejudiced and that the information I give you will be biassed. Fortunately I can quote from some articles which have appeared recently in several Dutch papers and which show that the rosy picture of Dutch successes in dealing with drug, which is presented by Dutch officials, is not quite in keeping with the true facts. The most revealing is a leading article in the issue of March 7, this year, of a magazine called Elsevier, which is the Dutch equivalent of Time Magazine. I will mostly quote therefrom.

You may remember that for many years Holland was regarded as the black sheep, being too tolerant in its dealing with drugs. Our government took the point of view that only the use of hard drugs, such as heroine, LCD and cocaine, was really dangerous and should be discouraged, but that the products of cannabis, such as hashish and marijuana are rather harmless, and that the fight against these so-called soft drugs has no priority. Our so-called opium-law was even changed, so that the possession of less than 30 grams of cannabis products was no longer a crime but an offence. The main target of our drug policy is to separate cannabis use from the use of other drugs. Therefore our governement allows the establishment of so-called coffee-shops, where an official dealer is permitted to sell cannabis products, provided he does not sell them to minors and does not sell any other hard drugs. The official philosophy that cannabis is harmless and not even causing addiction, makes us believe that the use of cannabis can well become integrated in our nation's life, which means that only excessive use is discouraged and that moderate use is no longer seen as abnormal behaviour. Cannabis is grown today in Dutch greenhouses and seems to be one of our most important horticulture products. Only our adherence to the Single Convention dissuades us from legalising cannabis.

But also the fight against what the governement calls hard drugs is waged with a sympathetic attitudes towards users., who are not seen as criminals, except when caught stealing. Addicts coming to the municipal health authorities are officially registered, so that they can receive a daily dose of methadone, distributed in a so-called methadone-bus, which circulates each day around the town and stops at fixed points. In order to prevent infection, especially AIDS, used syringes can be officially exchanged for sterile syringe and needles. Hopeless addicts are given morphine and psycho-social help is available for users who still have a chance to be cured. Addicts who want to kick the habit can be admitted to a rehabilitation centre, but are free to leave as soon as they like. Even in some prisons drugs can be obtained.

When criticised by their non-Dutch colleagues, our officials used to answer rather apologetically that we must acknowledge the fact that drugs are being used. But in 1987 a new policy was adopted. At a United Nations World Drugs Conference our former Minister of Justice began to tell the world how succesfull we were in not only stabilising but even diminishing the use of drugs. He invited experts from other countries to come and study our policy, which soon became known as the Amsterdam model, or rather the Dutch model, as other towns, such as Rotterdam, are even more tolerant than our capital.

An indeed many countries are sending their experts to Holland. According to the Elsevier-article, just mentioned, the Amsterdam health officials are receiving 150 groups from other countries each year and are sending doctors, social workers and police officers to all corners of the world to tell about our successes.

The Elsevier article takes as an example the German city Frankfurt am Main, a town with 600,000 inhabitants, where so-called "rigid" policy of dealing with drugs had resulted in a number of 9,000 addicts and a drug-related mortality of 143 last year. According to the health autorithies the number of drug-addicts in Amsterdam (300.000 inhabitants) has dropped to between and 6 and 8,000 and the drug-related mortality to 40 per year. Three years ago in Frankfort a red-green coalition was installed and the alderman for health went to Amsterdam to study our policy. In November 1990 at a conference in Frankfurt the cities of Amsterdam, Frankfurt, Hamburg and Zürich agreed on the so-called Frankfurt Resolution, which is almost entirely a copy of the Dutch model.

Yet the article in Elsevier's Magazine contains also doubt about the government's claim of success. The head of the drugs department of the Amsterdam health authorities is quoted as saying that the drug-related mortality figure of 40 per year includes only the victims of overdoses and that the real figure of drug-related deaths per year in between 100 and 150. He is also quoted as fearing that besides the 6 to 8,000 officially registered users of heroine a new group of young addicts is rapidly increasing in numbers. He cites the example of a 22 year old prisoner he saw that morning, who did not consider himself to be an addict, but who takes heroine first thing in the morning and again later in the day both heroine and cocaine. This boy is not registered, like many others. Reading this, one may question whether the real figures of Amsterdam with its tolerant policy are so much more encouraging than the Frankfurt figures which were the result of the former so-called rigid policy. Apparently the Frankfurt figures were more realistic than the Amsterdam ones.

What about the separation of cannabis and the other drugs, which was the aim of establishing coffeeshops ? This separation is no longer maintained in the city of Rotterdam, where you will see near a side-entrance of the Central RailwayStation two containers standing side by side, called "Platform Zero", which are used as barracks where you can get, as the Elsevier article mentions, coffee in the one and where you can take injections and even openly buy heroine in the other without any interference by the police. When a few weeks ago some taxidrivers got infuriated by the drug-users and wanted to set the containers on fire, the police rushed in to protect these so that the heroine-use could go on undisturbed. One of the churches in Rotterdam is used for receiving drug users, a coffee bar on the ground floor and a room where heroine users can quietly inject themselves in a cellar downstairs. In many coffeeshops you may get the addresses where heroine and cocaine can be bought. And minors can ask adult friends to buy the drugs they wants, which means that the restrictions to the sale of cannabis have no effect in reality.

What about the figures given by the former Dutch Minister of Justice in Vienna 1987? He told the conference that the use of drugs in Holland had dimished and mentioned that of the youngsters of 19 years and less 2.5 % had ever used cannabis. But an article on November 18, 1987, in the Dutch newspaper Trouw, which many readers consider rather liberal, gave another picture. According to the Union of Consultation Offices for Alcohol and Drugs the use of cannabis was increasing. The figure used by the Justice Minister was the result obtained from an inquiry amongst families in 1983. But an enquiry made in 1985, whereby pupils were asked to fill in a form in class, had shown a percentage of 4 %, and in the age group between 15 and 24 years the percentage was 12 % nationwide and 24 ù in the big cities. Apparently the Minister had not received these later results.

What about the number of coffeeshops? In 1987 every observer could ascertain the fact that this number was rapidly increasing, both in the big cities and in the smaller towns, especially on the borders with Germany and Belgium. According to police reports today there are in the whole country between 1200 and 1500 coffeeshops, whereas a few years ago there were only 800. There are now 400 coffeeshops in Amsterdam alone, causing unrest among the inhabitants, who recently started a demonstration forcing Amsterdam's Mayor to come and speak about his plan to deal with this problem. In Venlo, a small town near the German border, the 35th coffeeshop was opened in February this year. If the use of drugs is really diminishing in Holland, this almost doubling of the number of coffeeshops in a few years time is hard to explain. Dutchmen are not known to be foolish enough to open more shops when the market is decreasing.

Also the yearly amounts of drugs seized by police hardly support the theory that drug use is diminishing in Holland. Whereas the amounts of heroine have not changed very much between 1986 and 1990, the amount of cannabis seized has doubled from 48 à 110 tons. More recent figures on drug seizures are not yet available. On march 7 this year, the Central Criminal Investigation Department published a statement that the drugs trade in Holland had a turnover of 2.5 billion (=thousand million) guilders a year. A few days later an international, mainly Ghanese, gang, which had its headquarters in Amsterdam, was arrested. It imported drugs from Asia and South America and exported them towards the United Stated, Canada and many countries in Europe. According to the police the gang used 1500 adresses where 10.000 Ghanese were living illegally. These figures seem incredible. But together they do not suggest that our Dutch policy is leading to decreased drug use.

The article in the newspaper Trouw also mentioned the fact that cannabis-users have problems with learning, as memory and the capacity to think logically are impaired. On September 13, 1991, the leftist Magazine HP/de Tijd gave an extensive article on addiction to cannabis, and a similar article appeared on January 16, 1992, in the Magazine Panorama, which is rather wellknown for its extremely liberal views. According to the Consultation Offices for Alcohol and Drugs the number of registered cannabis-users, who have asked for help has increased from 567 in 1988 to 913 in 1990. We can conclude that slowly the truth is dawning even in Holland that cannabis is not the harmless drug it was assumed to be.

Is there a solution?

I was asked to present a Dutch point of view. I do not agree with the tolerant policy of the Dutch government. Neither do I agree with a policy that can be described as rigid. I know little about Sweden, but as far as I understand, the Swedes have worked out a strategy which not only aims at reducing the availability of drugs but also tries to help the users to stop their habit and to dissuade non-users from experimenting with drugs.

To dissuade non-users means providing convincing information. I know by experience both in Morocco and in Holland, that cannabis is in itself a dangerous drug and that its use may lead to the use of heroine and cocaine. I think that all the governments in the world should unite on a program to inform the public, and especially parents and educators.

To help users means providing good treatment. If I am well informed success with the treatment of addicts depends on the duration of this treatment (as far as I Know not less than two years) and the complete absence of drugs during the whole period of treatment.

You may say that the number of users in the whole world is enormous and still increasing, so that it is impossible to treat all users. And you may be right. Whatever we do must be done on a huge scale and will ask an enormous effort. But I think there is no other way. The danger to humanity is too great. We must attack the problem together. And together we should be able to win the battle, if we really see the necessity to win.

CANNABIS AND THE LAW

Georges Lagier and Jean Michaud

ABSTRACT

In France, cannabis, its flowering tops (marihuana), its resin (hashish) and its derivates, are classified among controlled dangerous substances.

Their use is forbidden (except for scientific research), and should be penalized.

Accordingly, French law is in strict agreement with the 1961 International Convention on "Narcotic Drugs" ("stupefiants"). In this convention cannabis is listed on Schedules I and IV. French law is also in agreement with the 1971 International Convention on psychotropic substances, which lists tetrahydrocannabinols among Schedule II drugs, only to be used for scientific or medical purposes.

Implementation of this law has been difficult, because of its supplementation by a new law of December 31, 1970, which considers an addict as a patient in need of treatment. A magistrate may enjoin an addict to accept treatment instead of a criminal sentence. This process is called therapeutic injunction, or compulsory referral of the addict into a treatment center. However, in fact, compliance with this injunction is not always systematically carried out by the addict. For such a scheme to be effective, a very close cooperation between the judicial system and the addiction treatment center must exist.

KEY WORDS

Cannabis, tetrahydrocannabinol, narcotics, psychotropic drugs, legal aspects, criminal law, therapeutic injunction

I. THE LAW

Georges Lagier, M.D.
President National Commission on Psychoactive Drugs, Paris, France.

In France, cannabis flowering tops (marihuana) and its resin (hashish) are listed among substances classified as narcotics, in French "stupefiants." So are their active ingredients, tetrahydrocannabinols, their esters, their ethers , their salts and the salts of their derivates.

The use of these substances is forbidden in France, except in the case of exceptional derogation by the Health Department when the drug is used for scientific research.

According to these laws, French law conforms to the 1961 International Convention on Illicit Drugs, which classifies cannabis and its resin in Schedule I and IV.

French legislation also conforms to the International Convention on Psychotropic Drugs of 1971, which requires a rigorous control on the use of tetrahydrocannabinols. Since 1991, Δ-9-tetrahydrocannibol (Δ-9-THC) and its stereoisomers (2 racemic and 4 stereoisomers, including dronabinol) have been transferred from Schedule I to Schedule II in this convention. This change of class allows their therapeutic use in countries, like the United States and Canada, who have approved their use for medical purposes.

Dronabinol (No CAS 1972-08-3; (-)-trans-delta-THC; Marinol® and Deltanyne®) is the only product for which a therapeutic use as an anti-emetic for cancer chemotherapy has been reported. Therefore, all the stereoisomers have been changed in class because of complex medico-legal considerations.

The French delegation to the United Nations unsuccessfully opposed this change of classification for Δ-9-tetrahydrocannabinol and its stereoisomers from Schedule I to Schedule II in the 1971 convention. According to the French specialists, the therapeutic application of dronabinol was too limited and inferior to other medications, and increased the risk of diversion of the drug to recreational use. No medication containing Δ-9-THC is sold in France or continental Europe.

II. IMPLEMENTING THE LAW

Jean Michaud
Magistrate, Cour de Cassation, Paris, France.

The drug problem is one of the most difficult to settle according to the French criminal laws. I want to examine the complexity of this problem before stating the solutions given by the law concerning the drug user.

The law of December 31, 1970 and the Public Health code, defines the judicial process in that field. The article of the code, L.628, punishes with imprisonment and fine the person who uses illegal substances or plants, classified as "stupefiants." This point is very unusual in French law, since it raises the possibility of an infringement of individual liberty. Some critics have claimed that such legislation is unconstitutional. According to The Declaration of Human Rights: Article 4: Liberty consists in doing what is not prejudicial to others. Article 5: Law can only enforce acts that are not prejudicial to society. But does the drug user only harm himself? There must be serious reasons to curtail our human rights by repressing individual freedom. In short, we may state that there is a philosophical justification for the refusal to penalize the drug addict.

The use of drugs leads to several social dangers: spread by users and dealers, it incites various crimes against which society has to protect itself. The use of drugs also leads to the alienation of individual liberty.

Should the drug user be punished as stated in article L.628 of the penal code? This is repression. But there is an awareness of the special problems of the addict. The drug user appears more and more to be a medical patient in need of treatment.

In practice, there is a very thin line between law and medicine. Holding the violation to be more of a symptom than a crime, some experts advocate that the culprit should be treated rather than punished. They feel that the penalty can only worsen the problem. There is a general tendency today, to acknowledge that treatment should sometimes be favored over punishment.

Indeed, for the punishment to be effective, two conditions are required: 1) the awareness of being harmful towards others, and 2) the possibility to avoid relapsing even if it is induced by the threat of punishment.

In that case, the only awareness is one of self-destruction (and the claim of having a right to it). The idea of relapse does not hold out against the overwhelming drive which takes hold of the addict in a state of withdrawal, or when he is tempted to take his drug of choice. Medicine and justice are attempting to act in a complementary way.

The law delineates the different categories of illegal drugs according to Article L.628, substances or plants classified as "stupefiants."

We are free to consider the notion of dependency. Fighting against dependency justifies circumventing the individual liberties of the user. This does not solve the problem entirely. There is still a controversy about the comparable damages of legal drugs (alcohol and tobacco) and illegal ones.

The legislator has attempted to protect against the influence of drugs. This action takes place before any judicial decision. The Public Prosecutor can inform the medical authorities about an addict's case, in order to recommend therapy or medical assistance. This kind of prevention is very specific, and the Public Prosecutor acts outside the prosecution system. He commits the addict to the health services.

The judicial authorities do not interfere in two other cases: when the addicts are reported by social services, or when they go in for treatment on their own accord.

The public prosecutor performs its function in other cases, when it takes the prosecution initiative.

But even in that case, compared to the traditional procedure, there is a possibility of mitigation. If the addict agrees to treatment, the public prosecutor can dismiss the case. But if the addict stops treatment, or if there is a relapse, the judicial system follows its normal course. The case is referred to the examining magistrate and then might continue on to judgement. But again, at those two steps, different opportunities are offered to the accused: the examining magistrate can order treatment for drug addiction, continued by surveillance measures. If the process is completed, the investigation ends without prosecution (even though an offense occurred). If the case goes to court, two alternatives are again given to the defendant: first, a treatment for drug addiction, followed by an exemption from penalty if it succeeds, or second, if the treatment is refused, a probationary suspended prison sentence depending on the acceptance of treatment.

It seems that every effort should be made to spare the addict from criminal sentences, and to offer him the medical assistance he needs to overcome his condition.

In the case of the drug dealer, one has to resort to criminal law. The point is not any more to repress personal behavior without external consequences, but to fight and punish social delinquency. The seller of illegal drugs is harmful to others just as any other criminal. The serious consequences of those crimes lead the legislator to be much more severe: sentences were raised up to twenty years. Prosecutors were released from some prosecuting restraints by extension of preventive detention, by allowing night searches on personal premises and by encouraging informers.

There are two opposite tendencies in French criminal law. The first one looks for social reinsertion of the drug addict, stopping the degradation of his condition. In that perspective, cooperation between medicine and justice is necessary. The second is a reinforcement of repression. It is necessary to stop the drug traffic which is at the same time a way of protecting the drug user.

But this blend of medicine and justice, therapy and penalty, is not always effecive. If the addict does not accept treatment, or if he relapses, the judge will have to interfere again. This means that an unsuccessful treatment could lead to a penalty. What is the basis for a penalty which would have been avoided, and consequently would not have been deserved in case the addict was cured, or on the way to being cured?

This shows the ambiguity of the law. It can be explained by the fact that if the law tries to protect the person, and fails, then it has to protect society.

Have these attempts been successful in curbing drug addiction in France? Not really, according to the available 1991 statistics from law enforcement and treatment centers like Marmotan. Then what should be done? Some recommend that so-called "soft" drugs be legalized. Others recommend reinforcement of repressive measures. It is difficult to make new proposals in this field at a time when permissiveness is considered better than restraint.

We can at least consider that the French law, with its medico-judicial approach, is on the right course. We have to hope that it will be effective in the long run. Times are gone when the names of Baudelaire and Cocteau gave drugs an artistic legitimacy. Drugs are no more a presumed stimulant of creativity, they damage the brain.

Doctors and jurists, more than others, must pursue their complementary tasks in spite of all obstacles.

CANNABIS : POINT OF VIEW OF W.H.O.

Juhana Idanpaan-Heikkila

Division of Drug Management and Policies, World Health Organization, Geneva, Switzerland.

The World Health Organization (WHO) is an intergovernmental organization within the United Nations system. A total of 168 countries are united in WHO, working together for the attainment by all people of the highest possible level of health. A target towards this goal is the attainment by all people of the world, by the Year 2000, of a level of health that will permit to lead a socially and economically productive life, popularly known as "Health for All by the Year 2000".

WHO AND DRUG ABUSE PROBLEM

WHO has recognized that health problems related to drug abuse are of major public and political concern in a large number of countries. It is estimated that there is a total of 48 million drug abusers in the world, including some 30 million cannabis users[1]. Although a dramatic escalation in the abuse of cocaine and heroine has occurred during the last decade, cannabis still continues to be the most widely abused drug.

WHO's Programme on Substance Abuse (PSA) was established and its Strategy Document[2] published in 1990. This new programme emphasizes the crucial importance of demand reduction as part of a balanced approach to combat the drug problem. WHO works in close collaboration with governments, non-governmental organizations and all relevant UN agencies such as the United Nations International Drug Control Programme (UNDCP) and the International Narcotics Control Board 5INCB), both located in Vienna, Austria.

EXTENT OF CANNABIS PROBLEM

According to the UN statistics, and despite increased expenditure on law enforcement, the production, traffic and abuse of cannabis continue to be widespread[3-5].

The quantity of both herb and plants reported by weight increased in 1986-1987 but declined very sharply in 1989. However, the annual seizures of cannabis resin (around 400 tons) have remained almost unchanged during 1989-1990.

According to the INCB[5] cannabis is mainly produced in som African countries such as Morocco, Sudan, Nigeria, Ghana, Rwanda, Zaire and Zambia. Other important production areas are situated in Afghanistan, Pakistan, Bekaja Valley in Lebanon, Nepal, Mexico, Colombia, Jamaica and Belize. Also in the former USSR drug-related problems continue to develop in increasing proportions. The drugs frequently abused are mostly of local origin., namely opium and cannabis. Cannabis grows wild on approximately 4 million hectares in Kazakhstan, on 1.5 million hectares in the far-eastern provinces as well as on vast lands of the lower Volga river basin, in nothern Caucasus and in southern regions of the Ukraine. While many illicit poppy fields are destroyed by law enforcement squads, efforts to eliminate growth of wild cannabis have little success.

In recent years, increasing cultivation of cannabis has been found in the USA and the authorities have taken measures in order to control this important source of domestic consumption, including identification of the distributors of the equipments used by illicit growers.

Data from some Western European countries and the United States seem to indicate that the number of abusers of cannabis is levelling off and in some instances declining[5]. Meanwhile, cannabis abuse is beginning to spread to several Central and Eastern European countries. There are signs of rapid development of organized traffic in cannabis in Czechoslovakia and the former USSR. According to the INCB, in December 1990, some 1.8 tons of cannabis resin seized in Germany were reported to have been transported overland via USSR from Afghanistan[5].

HARMFUL EFFECTS OF CANNABIS ABUSE

A number of speakers in this meeting have emphasized the harmful consequences of cannabis abuse. In order to avoid repetition, I would only like to summarize the point of view of WHO as follows:

- Based on discussions in WHO Scientific Groups, existing International treaties on drug abuse control and resolutions of the UN Economic and Social Council, the World Health Organization considers that cannabis has both acute and chronic harmful effects on human health.

- WHO recommends that all governments should intensify their efforts to reduce the demand for cannabis and cannabis resin and limit the licit use of cannabis products to medical and scientific research only.

- WHO further recommends that scientific research, especially long-term investigations on the effect of cannabis abuse in human organism, should be continued.

- Furthermore, WHO urges all governments to maintain or adopt appropriate preventive measures towards the harmful health and other consequences of cannabis use.

LEGALISATION NOT AN OPTION

Based on the UN Conventions[6-7] and supporting the view of the INCB[4-5] and the Council of Europe, WHO does not support the view which continues to be voiced advocating the legalisation of the possession and use of some or all drugs for non-medical purposes. Such an approach might be interpreted by potential abusers as sanctioning and approving drug use. This can be expected to interfere with programmes to reduce the demand for drugs and to promote life free from drug abuse.

Prohibition of the non-medical possession and use of narcotic drugs and psychotropic substances represent specific obligations of the Parties under the UN single Convention on Narcotic Drugs (1961), the 1971 Convention on Psychotropic Substances and the 1988 UN Convention against *Illicit Traffic in Narcotic Drugs and Psychotropic Substances*.

REFERENCES

1. **WHO Eight General Programme of Work** covering the period 1990-1995, page 124, WHO, Genneva, 1987.

2. **WHO Programme on Substance Abuse**, Strategy Document, WHO/PSA/90.1, Geneva, 1990.

3. **UN ECOSOC**, Situation and trends in drug abuse and illicit traffic. Document E/CN.7/1991/20.

4. **International Narcotics Control Board**. Report of the International Narcotics Control Board for 1990. United Nations. Publication n° E.90.XI.3, New York, 1990.

5. **International Narcotics Control Board**. Report of the International Narcotics Control Board for 1991. United Nations. Publication n° E.91.XI.4, New York, 1991.

6. **United Nations**, Single Convention on Narcotic Drugs, 1961. United Nations, Publication n° E.77.XI.3, New York, 1977.

7. Convention on Psychotropic Substances, 1971. **United Nations**, Publication n° E.78.XI.3, New York, 1977.

Section III

DETECTION, IDENTIFICATION

AND TESTING

1. Identification and detection

DETECTION, IDENTIFICATION AND MEASUREMENT OF CANNABINOIDS AND THEIR METABOLITES IN BIOLOGICAL FLUIDS

Pierre Levillain

Hopital Fernand Widal - Laboratoire de Toxicologie

ABSTRACT

Identification of cannabis use requires a screening test, (immunoassay) for 11 THC carboxylic acid in urine. If the concentration is higher than the cut off value, a confirmation test must be used, gas chromatography/mass spectrometry. The conjunction of a positive immunoassay, a proper chromatographic retention time and at least three mass ions with appropriate intensities constitutes definitive proof of the presence of 11 THC carboxylic acid in urine and of cannabis use. Quantitative analysis of cannabinoids and their metabolites in blood plasma may be required in case of intoxication or for determining time of ingestion. Methods used for drug identification can produce erroneous results. Endogenous substances and other drugs can interfere in the determination, giving cross reactions with immunoassay reagents or identical retention times in chromatographic procedures. They can also increase the detection limits of the various methods. Cannabinoids and their metabolites must be extracted and separated from the biological background before analysis, and during these steps loss of material can occur. Such errors can produce false positive or negative results. Because the use of cannabis is illegal, the consequence of a positive test is important. Its identification must be reliable and interpretation of the urine testing result must be clear. For this purpose, the analyst must have a good knowledge of all the methods used in terms of specificity, sensitivity, accuracy and precision. He must define the limit of detection and be sure that this limit is below the cut off value for the drug. The results must be supported by intralaboratory control data and by external proficiency testing. Standard quality controls practiced in clinical laboratories must be applied to analyses for drugs of abuse.

KEY WORDS

Cannabis, immunoassay, gas chromatography, mass spectrometry, quality control,11 THC carboxylic acid

INTRODUCTION

The low levels of cannabinoids in body fluids make their detection and quantification difficult: Δ-9-tetrahydrocannabinol blood concentration never exceeds a few hundred ng per ml in man. Furthermore, the rapid diffusion and metabolization of these compounds results in a fast decrease of their plasma concentration. In addition, a large number of metabolites have been isolated and complicates laboratory analysis. A rapid survey of pharmacokinetics and metabolism of cannabinoids is helpful in addressing problems raised by their detection and analysis. Use of cannabis is forbidden by law and raises the possibility of legal action against users. Consequently, results of drug testing must be as error free as possible. This requires an answer to three questions. What metabolite must be detected? In what fluid? What methods must be used?

I - PHARMACOKINETICS AND METABOLISM

Cannabinoids in plasma

Among the numerous terpenic compounds of cannabis, three are currently found in smokers blood: cannabidiol, cannabinol and Δ-9-tetrahydrocannabinol (THC). The maximum concentration of THC (the most active compound) is reached 10 to 20 minutes after the beginning of inhalation and is currently about 100 ng/ml. This level is influenced by the way of smoking (puff duration, retention of smoke in lungs, etc.) as well as individual sensitivity. But the lipophily of THC allows it to diffuse easily in various organs, especially the brain, where it is responsible for the characteristic euphoria, heart, with an increase in cardiac rhythm, liver, where it is metabolized, and lipids, where it can be accumulated. This fast diffusion step involves a deep fall in blood concentration; the amount of THC is only about 10 ng/ml after one hour and around 1 ng/ml after 6 hours. At last THC is slowly eliminated from organs over several days, which results in blood plasma concentrations ranging from 1 to 0.1 ng/ml. Cannabinol and cannabidiol possesses a pharmacokinetic behaviour similar to THC. Because of their low polarity, cannabinoids are slightly soluble in water, so that in blood plasma they are mainly adsorbed on lipoproteins, which makes their recovery from this fluid more difficult. Garret and Hunt found that the detection sensitivity for THC in dog plasma could decrease 10 fold after eating.

The main advantage of drug testing on blood lies in that the specimen is taken by health professionals, which eliminates sample falsification. But the small amount of specimen, the low level of drug and its fast elimination, as well as the difficulty in separating THC from endogenous lipids makes it difficult to use this fluid for a large screening of cannabis users. Clean up methods, necessary to obtain partial purification of THC from lipids largely increases the complexity of the laboratory work.

Metabolism

The liver plays an important part in THC detoxification and elimination. Cytochrome P450 is involved in hydroxylation which occurs quickly in C_{11} (11 hydroxy THC can be detected in blood 10 minutes after an I.V. injection of THC) but also, more slowly, in C_8 or in different carbon atoms of the pentyl side chain. Further oxidation can result in the formation of a carboxylic group in C_{11} or in the side chain, or of a ketone in C_8. Some of the acids are found in free form, but most are conjugated as glucuronide ester. Eighty different metabolites of THC have been isolated. All of them are more soluble in water than the parent molecule, which allows them to be eliminated in gall (70%) or urine.

The major metabolite is 11 THC carboxylic acid. It is quickly eliminated in urine as a result of metabolization. But the molecule can be found for several days at low levels in urine, because of the retention of THC in different tissues: duration of excretion can largely vary (from three days for casual smokers to thirty days for heavy marijuana users).

Urine appears to be the best fluid for identification of cannabis users; the long period of excretion of 11 THC carboxylic acid makes its detection easy. In addition, there is less interaction between cannabinoids and urine components than in blood, so extraction is easier. And lastly, large volumes can be readily collected noninvasively.

Many drugs or metabolites can be found in saliva, which has been proposed for detection of THC. But there are some disadvantages in using this fluid: difficulty in obtaining the specimen, and small sample size causes difficulties of confirmation and quantification.

II - ANALYTICAL METHODS

Generally, testing urine for drugs of abuse includes a screening test, used to rapidly, easily and inexpensively separate the positive specimens from those that produce negative results. But because these screening tests are subject to interferences, those specimens that yield a positive screening require an additional confirmation procedure, which may be slower and more expensive, but must provide a more specific result. In some cases, cannabinoids and their metabolites can be quantified.

<u>Screening procedure</u>

<u>Thin Layer chromatography</u> largely is a matter of controversy. It may be a relatively inexpensive method for screening because multiple specimens can be applied to one thin layer plate, but most authors consider this technique to be obsolete and inadequate for drug screening mainly because of its lack of sensitivity and selectivity. Curiously, new papers regularly appear in scientific reviews about this technique. Indeed, sensitivity can be increased by extracting cannabinoids and their metabolites from large samples of urine and concentrating the extract into a small volume before chromatography. But it needs to hydrolyze glucuronides of carboxylic acids before extraction, which makes analyses more complex. Also, staining the plates with Fast Blue B increases the color intensity of cannabinoid spots. Detection limit as low a 2 ng/ml have been determined, which favorably compares with immunoassays. Lack of selectivity results from the nonreproducibility of Rf and from substances that interfere with the detection. But separation of interfering compounds and reproducibility of Rf can be largely improved by using over pressured layers for migration. Differences in the methods used can explain the varying results of blind comparison studies between thin layer chromatograpy and other methods. In conclusion, thin layer chromatography can provide good results only at the expense of simplicity of use; the manipulations must be rigorously standardized and only executed by well trained technicians. Thin layer chromatography is also used for cleaning up samples for more sensitive techniques such as mass spectrometry or RIA.

<u>Immunoassay</u> is by far the most commonly used technique for screening 11 THC carboxylic acid in urine. Advantages of this method are well known: it needs only small samples, allows direct detection without preliminary treatment of sample and its sensitivity is high

(ng amounts of drugs can be detected). Furthermore, the reaction can easily be automated, which increases precision and speed of analysis. But immunoassay suffers from the lack of specificity of antibodies, as different molecules (mainly other cannabinoids) can react with them. This possibility of cross reactions requires that positive results be confirmed by more specific methods.

Several means are used to detect the immunochemical reaction. The most popular is the Emit or enzyme multiplied immunoassay technique, which offers two advantages. It is a homogenous assay, avoiding separation steps, and the final detection is based on spectrophotometric measurement, which makes it easy to adapt this technique to classical automates used in biochemical laboratories.

The Emit screening test compares the absorbance obtained for each specimen with the value measured for a calibrator. This standard contains a specified concentration of 11 THC carboxylic acid that is considered as the limit for positive results (cut off value). The comparison allows us to determine whether the specimen level is lower or higher than the limit. Evaluation of the Emit test (with a cut off value of 100 ng/ml) with radioimmunoassay and gas chromatography/mass spectrometry has shown a 4% rate of false positive and a 10% rate of false negative. Anything that modifies enzyme activity (for ex temperature change or presence of enzyme inhibitor) can produce erroneous results. Also, false positive tests have been obtained with urines containing high concentrations of antiinflammatory drugs (ibuprofen, nafroxen and fenoprofen), ethacrynic acid, promethazine and riboflavine. These false reactions have been eliminated by use of monoclonal antibodies. On the contrary, melanin and indole compounds do not interfere. In the same way, endogenous malate deshydrogenase can perturb a test when this enzyme is used for labeling 11 THC carboxylic acid, so that glucose 6 phosphate deshydrogenase nowadays is currently used as a marker.

Different procedures have been successfully applied for the analysis of cannabinoids in blood by the Emit method. They involve protein precipitation and drug extraction by an organic solvent, evaporation of the solvent to dryness and dissolution of residue in buffer before analysis by immunoassay.

The cannabinoid radioimmunoassay Abuscreen is also frequently used for urine screening. The marker is a ^{125}I radiolabelled cannabinoid which needs a counter for detection. The method involves several steps, including physical separation before radioactivity measurement, which is less favorable for high volume tests. Comparison of positive and negative results with GC-MS are good, but metabolites quantitations are rather different since RIA detect several metabolites while GC-MS detect only one.

Fluorescence polarization assay is also commercially available, but needs a specific apparatus for detection, which limits application of the method.

For all immunoassays, choice of the cut off value is important. If it is too low, many positive responses can be found when the metabolite is not present or at a concentration below the limit value. Furthermore, passive inhalation of marijuana smoke can be detected. But, if it is too high, some positive specimen will be found negative. The U.S. National Institute for Drug Abuse recommends a cut off value of 100 ng/ml.

Adulterants (salts, liquid soap, detergents, lemon juice, household cleaners, etc.) can easily produce false positive or negative results when used at relatively low levels. This must be kept in mind during urine sampling as drug users can easily adulterate specimens.

Confirmation and quantification of cannabinoids and their metabolites

Confirmation of positive screening test and quantification of cannabinoids and their metabolites are best performed by physical methods which are more specific than the previous one.

They generally involve a pretreatment of biological fluids. In urines, glucuronide esters of THC metabolites are hydrolyzed by enzymic reaction or in basic media to obtain free carboxylic acids. After urine adjustment to low pH (3 to 5), the metabolites are extracted by organic solvents or adsorbed on a solid phase, washed to eliminate interfering substances and eluted by an organic solvent which is next evaporated. After dissolution in the proper solvent, the residue is ready for analysis. In blood plasma, the cannabinoids are first separated from lipoprotein by an organic solvent (methanol, acetone...) in neutral or acidic medium.

Excellent separation can be obtained for cannabinoids and their metabolites by high performance liquid chromatography (HPLC). Reversed phase columns are nowadays the most often used for this purpose, either in urine or in blood and saliva. The main problem is detection. Owing to the small molecular absorptivity of these compounds, U.V. detection lacks sensitivity. Furthermore, detection occurs at wavelengths (mainly 210 to 225 nm) where interference from biological compounds can occur. The detection limit can be largely improved with suitable derivatization: post column reaction with Fast Blue B allows the detection of cannabinoids at 490 nm. A very sensitive fluorimetric detection can be used with dansylderivatives of cannabinoids. Transformation of these molecules to highly fluorescent compounds by U.V. irradiation is also possible. But the necessary reaction steps make the analysis more complex.

Importance of electrochemical detection for cannabinoids regularly increases in HPLC. Their phenolic group is easily oxidized at a carbon electrode, either in the amperometric or coulometric mode, which increases sensitivity for THC or THC carboxylic acid, both in blood and urine.

Results of HPLC separation can be quantified by comparison with standard curves or better, with an internal standard. But identification is based on the retention time of the substance, which cannot be considered as an absolute proof of identity. That is why HPLC has found application in cleaning up extracts or separating metabolites before mass spectrometry.

The most widely used analysis method for cannabinoids is based on gas chromatography. Most applications are concerned with identification of 11 THC carboxylic acid in urine, but a few are related to the determination of cannabinoids in blood and saliva. Good separation can be achieved with the free neutral cannabinoids (THC, cannabinol...), but better results are obtained after derivatization. Derivatization is unavoidable with acidic cannabinoids (mainly 11 THC carboxylic acid) as the free compounds are less volatile and decarboxylate by heating. Methyl or trimethylsilyl derivatives are currently used, but labelling with fluorinated compounds (pentafluorobenzoate, pentafluoropropanoate, bromopentafluorotoluene, etc.) allows a more sensitive detection with electron capture based methods.

Conventional packed columns afford good separations, but capillary columns are finding increasing applications. The stationary phases most commonly used for coating these columns have low or medium polarity (OV 1 or OV 17 for instance).

In gas chromatography, detection is easier than in HPLC. Good sensitivity is obtained by using flame ionization methods (detection limit around 10 ng/ml) but sensitivity is largely increased with electron capture detectors (detection limit below 1 ng/ml), which are currently used. Fourier transform infrared spectrometers can now be interfaced with gas chromatographs, which allows identification of cannabinoids in cannabis extracts, but this method has not yet been used in biological fluids.

Combination of gas chromatography with <u>mass spectrometry</u> provides the best level of confidence for identification of cannabinoids and their metabolites. It is also highly sensitive, so it allows the detection of quantities in the pg range, although sensitivity is largely lower in biological media because of perturbations induced by endogenous products. Although instrumentation design and cost have been improved recently, the main drawbacks are the expense of the equipment and the necessary clean up steps of samples before injection.

Several ionization methods (chemical ionization, electron impact with different energies...) can be used, that results in a variable number of mass ions. Detection can be operated in different ways which produces spectral data of different types. In the selected ion mode only a few selected ion mass, corresponding to fragments of the drug to be determined are measured, which increases sensitivity but involves a loss of selectivity. The full scan mode, on the contrary, provides a record of the complete mass spectrum, but with a loss in sensitivity. Identification as well as quantification are best performed by comparison with an internal standard such as a stable isotope of the cannabinoids or of their metabolites. The usual standards are deuterolabelled cannabinoids, but also $\Delta 8$ isomers, not found in natural samples.

REFERENCES

- Garrett, E.R. and Hunt, C.A. Pharmacokinetics of Δ^9-tetrahydrocannabinol in dogs. J. Pharm. Sci., 66:395-407.

- Harvey, D. Chemistry metabolism and pharmacokinetcs of the cannabinoids. In: Marihuana in Science and Medicine (G.G. Nahas, ed), pp. 37-39, Raven Press, New York, 1984.

- Drug abuse in the work place - Clinical Chemistry, 11 B (supplement issue), 1987.

DETECTION OF ILLEGAL DRUGS IN BODY FLUIDS:
INTERPRETATION OF RESULTS

Paul Lafargue

Institut de Recherche Criminelle de la Gendarmerie
93111 Rosny-sous-Bois, France

ABSTRACT

The metabolic pathways of delta-9-THC, cocaine and opiates are summarized, and their main urinary metabolites described. For delta-9-THC: 11 nordelta-9-THC carboxylic acid. For cocaine: benzoylecgonine and methylester of ecgonine. For opiates: monoacetylmorphine (heroin), morphine, codeine. All opiates have the same urinary catabolites: morphine and codeine. The latter compound is a therapeutic agent. Detection of urinary cannabinoid metabolites (exceeding 20 ng/ml) as a result of passive exposure to cannabis smoke have been reached experimentally under conditions which rarely prevail in real life situations. Opiate and cocaine urinary metabolites may be detected 4 to 5 days after cessation of their consumption. Cannabinoid urinary metabolites may be detected 3 to 5 days after occasional smoking, and up to 77 days after chronic use (5 cigarettes a day).

KEY WORDS

Cannabis, cocaine, opiates, urinary metabolites detection, codeine, morphine, 11 nordelta-9-THC carboxylic acid, benzoylecgonine, methylester of ecgonine

Prior to the analysis of analytical results it is necessary to give a short review of the metabolism of the main illegal drugs.

I. REVIEW OF METABOLIC PROCESSES

A. CANNABINOID DERIVATIVES

Approximately one hundred molecules can be isolated from extracts of Cannabis sativa. The main psychoactive structure is delta 9-tetrahydrocannabinol or Δ9-THC.

Preparations offered to consumers contain highly variable amounts of active principle (1 to 40%) depending on the origin of the plant, on the quantity of added products and on their form : leaves, flower tips, haschisch, resin,

The metabolism of Δ9-THC is relatively simple. It is summarized in table 1.

The active principle is mostly transformed in the liver :

   -oxidation of the methyl group, in position 9, into primary
    alcohol function
   -oxidation of the alcohol function into carboxylic acid
   -conjugation of the acid function by glucoronic acid
   -urinary excretion of the conjugated derivative

B. COCAINE

Of the many chemical structures isolated from leaves of Erythroxylon coca, the most important one is an alcaloid : cocaine.

It is sold on the market in various forms, from white powder to mixtures containing various percentages of added products.

Cocaine can be administered in different ways :

   -sniffed
   -injected (shoot)
   -ingested (chewed by Incas)
   -inhaled as smoke : more recently, with the advent of crack.

Although more complex than that of Δ9-THC the metabolism of cocaine is relatively simple as shown in table II. The active principle is mostly transformed in the liver and in the plasma. Under the action of esterases two products are formed : (table 2).
   -benzoylecgonine (35 to 54%)
   -methylester of ecgonine (30 to 50%)

TABLE I

Diagram of the metabolism of delta-9-Tetrahydrocannabinol, active principle of cannabinoids.

ACIDE 11 NOR △9-THC
CARBOXYLIQUE

ACIDE 11 NOR △9-THC
9 CARBOXYLIQUE LIÉ A
L'ACIDE GLUCURONIQUE

TABLE II

Diagram of the metabolism of cocaine.

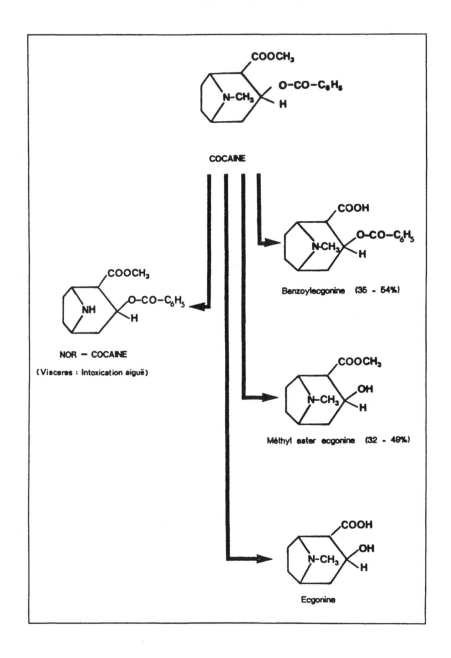

Aside from these two major metabolic pathways two substances can be found :
   -ecgonine   (1 to 10%)
   -nor-cocaine. The presence of a N-demethyled derivate of cocaine is the sign of an acute intoxication by this alcaloid.

C. MORPHIN DERIVATIVES

Many alcaloids are derived from Papaver somniferum but only two of them can induce drug abuse :
   -morphine   (5 to 20%), main compound
   -codeine or methylmorphine (0.5 to 1.5%)

Two neighboring structures should be associated with these natural molecules :   codethyline   (ethylmorphine)   and   pholcodine (morpholinethylmorphine) also used as anti-cough molecules.

When not used for therapeutic purposes heroin is a semi synthetic substance obtained by simple acetylation of the alcohol and phenol functions of morphine. It is considered as one of the hardest drugs because it becomes addictive after it has been used for some time. Drug addicts most often use it intravenously ("shoot") but it can also be smoked or sniffed.

The major metabolic pathways are summarized in Table 3. They show :
   - a significant glucoconjugation of the alcohol and phenol functions (approx. 70%),
   - a N-demethylation (approx. 20%) with the production of nor derivates (nor-morphine, nor-codeine...),
   - a dealkylation (loss of ethyl and methyl radicals) inducing the endogenous formation of morphine in the case of codethyline,and even more so of codeine, which complicates the interpretation of analytical results obtained on biological samples.

It seems of interest to mention that bioactivation of codeine in human hepatocytes is obtained through the action of polymorphous mono-oxygenase, better known as cytochromous dbl/bufl, which also catalyzes the hydroxylation of debrisoquine.

The O-demethylation of codeine into morphine shows that there are very active metabolizers (which create problems in the interpretation of results) and subjects with a deficit in polymorphous mono-oxygenase, whose urine never shows any morphine.

TABLE III

Diagram of the metabolism of morphine derivatives.

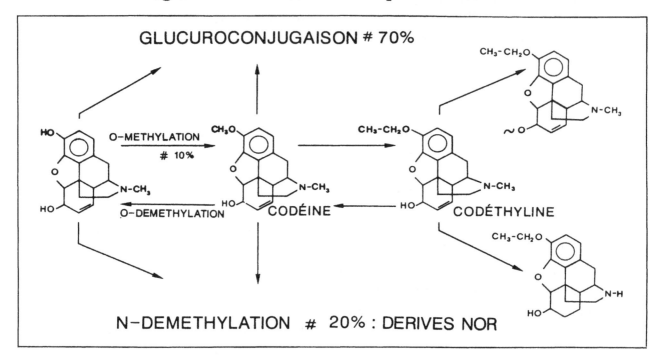

TABLE IV

Interpretation of results.

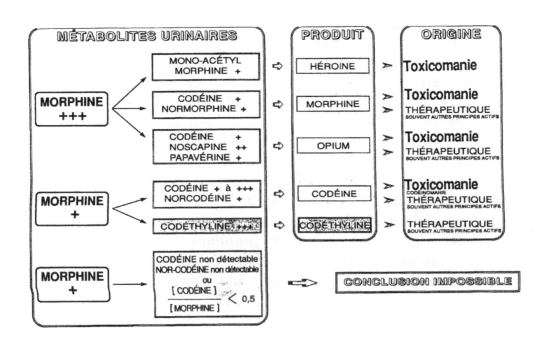

FIGURE 1

Kinetic of urinary excretion of codeine and morphine after absorption of codeine.

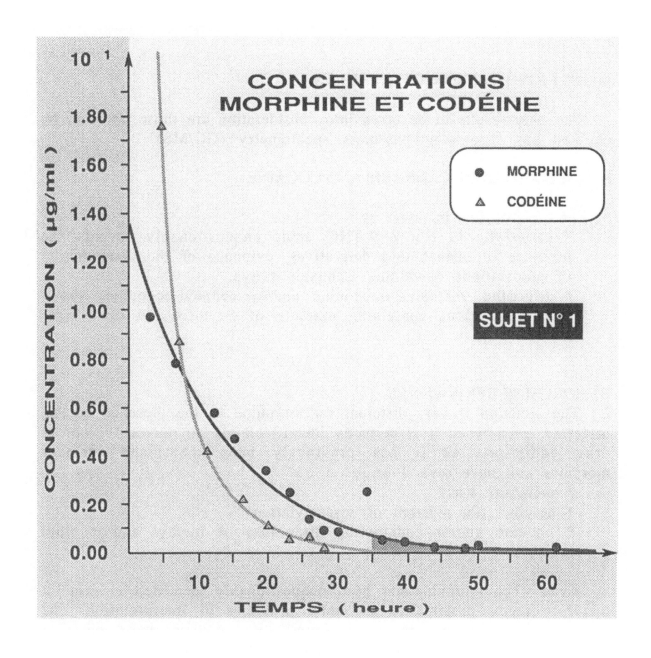

Figure 1 shows the kinetic of the urinary excretion of codeine and its demethylated metabolite, morphine, in an adult, after oral absorption of 1 mg codeine phosphate per kilogram of body mass.

Results show that :
* the morphine concentration exceeds the concentration of codeine after the 13th hour
* codeine can no longer be identified after the 30th hour
* only morphine is present between the 30th and the 65th hour.

II INTERPRETATION OF RESULTS

The only results to be taken into consideration are those obtained by coupled gas chromatography/mass spectometry (GC/MS).

A. DERIVATIVES OF CANNABIS AND COCAINE

The identification in urines of :
* carboxylic 11 Nor Δ 9-THC acid, unquestionably reveals the presence of cannabinoid derivatives, evidence of the consumption of preparations containing Cannabis sativa.
* ecgonine, methylesterecgonine and/or benzoyl-ecgonine, show that preparations containing extracts of Erythroxylon coca were used.

B. MORPHINE DERIVATIVES

The situation is very different for morphine as the presence of this molecule, even when it is perfectly identifiable, is no univocal proof of drug addiction. As it has previously been mentioned, urinary morphine can have several origins :
* morphine itself
* heroin, after reaction of desacetylation
* codeine and/or codethyline after loss of methyl and/or ethyl radicals.

Even when morphine has been unquestionably identified in urine it therefore seems possible to encounter problems of interpretation. The reason for this situation is that a number of compounds have the same urinary catabolite, morphine. This must constantly be remembered by all those who may be involved in fitness examinations or who may impose sanctions. Table 4 summarizes all the various possible cases that must be considered.

TABLE V

Passive smokers. Experimental conditions.

PASSIVE SMOKERS

Cone E.J., Johnson R.E., Clin.Pharmacol.Ther., 1986, n°3, 247-256.

<u>EXPERIMENTAL CONDITIONS</u>

Number of subjects	: 5
Volume of room	: 12.5 m3 (5.2 m2 x 2.4 m)
Ventilation	: None
Length of exposure	: 1 hour (08:30-09:30) / 6 days
Number of "smoked cigarettes"	: 4 to 16

"Goggles were worn during exposure to minimize eye irritation from smoke"

<u>RESULTS</u>

(Total of urinary samples collected during 6 days)

"4 cigarettes"
- 20 ng/ml : 4.6% of urinary samples
- 75 ng/ml : No sample

"16 cigarettes"
- 20 ng/ml : 35.2%
- 75 ng/ml : 2.2 % of urinary samples
- 100 ng/ml : 1.0%

- NIDA standard cigarette = 27 mg of delta-9-THC
- Malboro cigarette (not light) = 1.09 mg of nicotine

FIGURE 2

Positive test reading time.

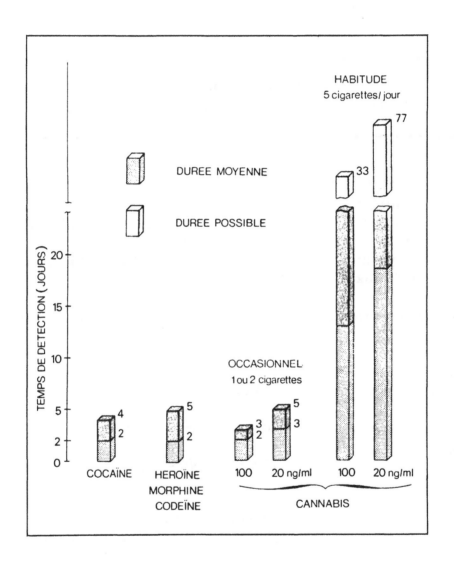

C. PASSIVE SMOKERS

Aside from problems relative to morphine derivatives, we should also discuss the problem of passive smokers in the case of Cannabis. Table 5 describes the experimental conditions implemented by CONE and JOHNSON to show that cannabinoid derivatives can be present in urines of subjects who passively inhaled smoke. These results are at the origin of a heated debate still unsettled. However, results we obtained on several thousand assays call for certain restrictions to the conclusions of these authors. The volume of the room where the experiment took place was only 12.5 cubic metres, which, for a distance to the ceiling of 2.40 meters, amounts to a groung surface of 5.20 square meters. Placing five individuals in this room and making them breathe one hour a day for six days at least the equivalent of four cigarettes of haschisch containing 27 mg Δ 9-THC (very high quality cigarettes) with no ventilation seems rather unrealistic. The authors have the honesty to report that subjects had to wear goggles to minimize eye irritation from the smoke. And even under these conditions only 4.6% of all urine samples showed a positive reaction at the 20 ng.ml-1 threshold after passive inhalation of four cigarettes,, and none showed a positive reaction at the 75 ng.ml-1 threshold. This means that under the most common conditions of use (party, piano bar, night- club...), the threshold of 20 ng.ml-1, which is the smallest available in immunology, is sufficiently high to preclude any confusion with the tricky and irritating problem of passive smokers.

D. POSITIVE TEST READING TIME

Positive reading times for urine tested for illegal drugs are shown in Figure 2. They indicate that a subject can be proved to have been a Cannabis smoker more than two months after he (she) has stopped using it.

REFERENCE

CONE E. J. and JOHNSON R. E.
Contact highs and urinary cannabinoid excretion after passive exposure to marijuana smoke
Clin. Pharmacol. Ther. 40 : 7-256, 1986

SALIVARY THC FOLLOWING CANNABIS SMOKING CORRELATES WITH SUBJECTIVE INTOXICATION AND HEART RATE [1]

David B. Menkes*, Richard C. Howerd*, George F.S. Spears*, and Eric R. Cairns**

*Departments of Psychological Medicine, Social and Preventive Medicine, and the Neuroscience Centre, Otago Medical School, Dunedin, New Zealand.
**Chemistry Division, Department of Scientific and Industrial Research, Private Bag, Petone, New Zealand.

ABSTRACT

A cannabis smoking trial was conducted using paid volunteers. Subjective intoxication, measured using a visual analogue scale, was compared with heart rate and with salivary delta-9-tetrahydrocannabinol (THC) levels at various times after smoking a cigarette containing 11 mg THC. Subjective intoxication and heart rate elevation were significantly correlated with the log of salivary THC. Salivary THC levels are a sensitive index of recent cannabis smoking, and appear more closely linked with the effects of intoxication than do either blood or urine cannabinoid levels.

KEY WORDS

Human, cannabis, THC, saliva, intoxication.

---

[1] Reprinted with permission from Psychopharmacology, 103:277-279, 1991.

INTRODUCTION

Cannabis intoxication is mainly due to THC, one of the many pharmacologically active cannabinoids absorbed after smoking or ingestion (Harvey 1987). Partly because of its extreme lipid solubility, blood THC levels correlate only modestly with cannabis intoxication (Ohlsson et al. 1980; McBay 1988). Urinary levels of THC metabolites are present for weeks or months after exposure, and thus offer an even weaker index of current intake or intoxication. Recently, the detection of THC in saliva has been demonstrated to offer a simple, non-invasive index of current exposure (Gross et al. 1985; Thompson and Cone 1987). This study reports the correlation between subjective intoxication, heart rate, and salivary THC.

MATERIALS AND METHODS

Thirteen paid volunteers were used. All were male, ranging in age from 22 to 36 (mean 25.5 years) with average height = 184 cm and average weight = 84 kg. All were experienced cannabis smokers, using the drug 2-12 times per month, and had no other substance use other than moderate alcohol intake. One week's abstinence from cannabis was required prior to the study. All subjects were screened and judged fit on the basis of physical examination, brief psychiatric interview, ECG and routine laboratory studies (haematology screen, urea and electrolytes, liver function tests). Urine screening confirmed the presence of THC metabolites in all but one subject.

Subjects individually smoked one cannabis cigarette (500 mg) containing 11 ± 1 mg THC. Subjective intoxication was measured at baseline and at various intervals after smoking by asking participants to place a mark on a 80 mm linear visual analogue scale between "completely straight" and "the most stoned I have ever been (on cannabis alone)." McBay (1988) and others have indicated that subjective intoxication is well correlated with other behavioural measures, such as cognitive task impairment. Supine heart rate was estimate from the radial pulse over 30 s.

Pilot studies showed that expectorated saliva stimulated with chewing gum (1 min, Wrigley's Juicy Fruit) was substantially easier to collect and analyze reproducibly than either unstimulated saliva or that collected with a dental cotton roll or cotton bud-on-a-stick. Samples, collected in silanised screw-cap vials (8 ml) containing sodium fluoride (40 mg) as a preservative, were stored in the dark at 4° C until analysis. Food and drink were allowed, but monitoring showed these to have little effect on THC time curves.

Salivary THC was measured using a modification of the GC-mass spectrometric technique of McBurney et al. (1986). An internal standard (12 ng of the heptyl analogue of delta-8-THC in 5 µl ethanol) was added to 200 µl 8 M urea in silanised glass tubes. Samples of 20-200 µl saliva were added, vortexed and then extracted with 4 ml pentane by rotating for 30 min on a turntable. After centrifugation, pentane was recovered under nitrogen at 50° C. Extracts were then derivatised by adding 50 µl hexane and 30 µl pentafluoropropionic anhydride (Pierce) and incubating at 60° C for 15 min. After evaporation again under nitrogen at 50° C, residues were dissolved in 40 µl heptane and 4 µl volumes were injected into the GLC. A Hewlett-Packard GLC (5890) with autoinjector (7673) and mass selective detector (5970), operating in a single-ion monitoring mode, were used with HP-UX software. Further details of this technique, including comparisons with HPLC, TDx immunoassays, and RIA, are available from one of the authors (E.C.).

Statistical treatment of data included logarithmic transformation of [THC] and subsequent parametric (Pearson product-moment) correlation with intoxication ratings and heart rate. In addition to such correlations for each individual, these were combined into pooled estimates for all subjects. Because inflated type 1 error is commonly overlooked in such repeated measures designs, a MANOVA procedure (SPSS-X) was used to provide an unbiased estimate of the overall correlation across time of log [THC], subjective intoxication, and increase in heart rate (O'Brien and Kaiser 1985).

RESULTS

Baseline testing showed five or seven subjects had detectible salivary THC, ranging from trace (< 0.2) to 3.4 ng/ml (mean 0.36 ng/ml). Cannabis smoking produced marked increases in salivary THC, subjective intoxication and heart rate (Fig. 1). With one exception, these three measures all peaked at 20 min and decayed monotonically thereafter. There was no relationship between baseline measures of [THC], intoxication (mean 2.8 out of 80.0, range 0-7) or heart rate and subsequent effects of cannabis smoking. All subjects reported their "high" was generally similar to their usual social experiences of cannabis smoking.

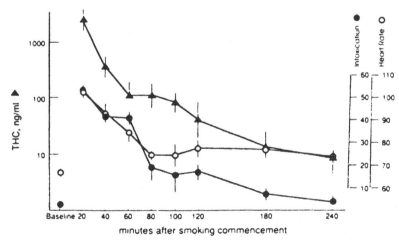

Fig. 1. Effects of cannabis smoking on salivary THC, subjective intoxication and heart rate. Values indicate mean ($\pm$SEM) of 13 observations (baseline, 20, 40 min), 7 observations (60, 120, 180, 240 min) or 5 observations (80, 100 min). One subject was unable after smoking to produce adequate saliva volumes for assay; accordingly [THC] values at 20 and 40 min reflect the mean of 12 observations. Note that THC is expressed on a logarithmic scale; mean baseline THC was off the scale at 0.36 ng/ml

The correlation between log [THC] and intoxication was strongly positive for each of the 12 subjects completing the protocol (range 0.85 - 0.98, n= 4-6 data pairs/subject, all P < 0.02). The MANOVA procedure confirmed an unbiased overall within-subjects, correlation [r= 0.71, Pillai's statistic = 1.50, $F(9, 24)$ = 2.68, P= 0.026], but failed to show a between-subjects effect (r= 0.05, NS). In most cases, subjective intoxication was marked above THC values of 100 ng/ml and minimal below 10 ng/ml.

The correlation between log [THC] and increase in heart rate was also strong but somewhat more variable across 12 subjects [ranging from 0.63 (NS) to 0.98 (P< 0.01), n= 4-6 data pairs/subject]. The MANOVA procedure gave an unbiased estimate of the overall within-subject correlation between these variables of r= 0.55 [Pillai's statistic = 0.90, $F(9, 24)$ = 1.15, P> 0.1]. However, the between-subjects correlation between log [THC] and heart rate was stronger [r=0.69, F(1,10) = 9.11, P = 0.013].

By contrast, the correlation between subjective intoxication and heart rate increase was positive but variable across 13 subjects (range 0.21 - 0.97, n= 6 - 8 data pairs/subject). The MANOVA procedure estimated the overall within-subject correlation between these variables to be r= 0.64 [Pillai's statistic = 1.24, $F(9,24)$ = 1.88, P= 0.10] but failed to show a between-subjects effect (r= 0.03, NS).

DISCUSSION

This study indicates a significant correlation between salivary THC and subjective intoxication in experienced users after smoking a moderate-strength (11 mg THC) cannabis cigarette. This correlation, which remained significant after the conservative MANOVA correction for repeated measures, was predominantly due to a within-subjects effect. This suggest that subjective intoxication is closely associated with salivary THC in a manner which may differ between individuals (cf Ashton et al., 1981). This association is stronger than has been reported for blood THC levels (r= 0.53, Ohlsson et al, 1980), particularly since the latter result was not corrected for repeated observations on each of 11 subjects (see materials and methods).

Under the present conditions, salivary THC also correlated with cannabis-induced tachycardia, but in this case the between-subjects effect was predominant, suggesting a more general physiological effect of THC across subjects. While it is most unlikely that salivary THC is causally related to subjective or cardiac effects of cannabis, its correlation with them suggests that salivary levels may co-vary with THC at relevant sites in the CNS.

THC levels in saliva appear to derive mainly from sequestration of the drug in the mouth during smoking or ingestion (Gross et al, 1985). However, there is also evidence for some transport of radio-labelled THC into saliva after IV injection in monkeys (Just and Weichman, 1974). Interestingly, consumption of food or drink failed to have much effect on salivary THC in this or a previous study (Thompson and Cone, 1987). More work will be required to delineate the kinetics of salivary THC, and to correlate these with behavioural and physiological effects in more naturalistic settings, e.g. with repeated smoking, co-consumption of alcohol, etc.

Acknowledgements. We thank A.R. Zoest for technical assistance, Dr C.T. Hung for discussions, and M.A. Jensen and L.M. Williams for typing the manuscript. This work was supported by the Department of Scientific and Industrial Research, Ministry of Transport, and the Road Traffic Safety Research Council (New Zealand).

REFERENCES

Ashton H., Golding J., Marsh V.R., Millman J.E., Thompson J.W., (1981). The seed and the soil: effect of dosage, personality and starting state on the response to tetrahydrocannabinol in man. *Br.J.Clin.Pharmacol.*, 12:705-720.

Gross S.J., Worthy T.E., Nerder L., Zimmermann E.G., Soares J.R., Lomax P., (1985). The detection of recent cannabis use by saliva delta-9-THC radioimmune quantitation. *J.Anal.Toxicol.*, 2:98-100.

Harvey D.J. (1987). Pharmacology, metabolism, pharmacokinetics and analysis of the cannabinoids. ISI Atlas of Science: Pharmacology 1:208-212.

Just W.W., Weichman M. (1974). Detection of delta-tetrahydrocannabinol in saliva of men by means of thin layer chromatography and mass spectrometry. *J. Chromatogr.* 96:189-194.

McBay A.J. (1988). Interpretation of blood and urine cannabinoid concentrations. *J. Forensic Sci.* 33:875-883.

McBurney L.J., Bobbie V.A., Sepp L.A. (1986). GC/MS and EMIT analyses for delta-9-tetrahydrocannabinol metabolites in plasma and urine of human subjects. *J. Anal. Toxicol.* 10:56-63.

O'Brien R.G., Kaiser M.K. (1985). MANOVA method for analizing repeated measures designs. *Psychol. Bull.* 97:316-333.

Ohlsson A., Lindgren J.E., Wahlen A., Agurell S., Hollister L.E., Gillespie H.K. (1980). Plasma delta-9-tetrahydrocannabinol concentrations and clinical effects after oral and intravenous administration and smoking. *Clin. Pharmacol. Ther.* 28 (3):409-416.

Thompson L.K., Cone E.J. (1987). Determination of delta-9-tetrahydrocannabinol in human blood and saliva by high-performance liquid chromatography with amperometric detection. *J. Chromatogr.* 421:91-97.

2. Testing

TESTING FOR ILLICIT DRUGS IN THE WORKPLACE

Peter B. Bensinger [1]

Former Director, Drug Enforcement Administration, Washington, D.C.

ABSTRACT

Testing for illicit drugs through urinalysis has had a dramatic impact in the United States in reducing illegal drug use on and off the job for employees and for individuals in professional and amateur sports and in military occupations. Drug testing has been increasingly utilized effectively to stem the widespread availability of illegal drugs and the significant threat to safety, performance and health. Illegal drugs can be most effectively identified by urinalysis, although other testing methods under consideration include blood, saliva, eye movement and hair. Drug testing by urinalysis has been conducted accurately and legally in the United States for the armed forces, employees in industry (both union and nonunion), government agencies and in sports. The process utilized has generally been to take a preliminary screen from a urine specimen, either by enzyme immunoassay, fluorescent polarization or radio immunoassay technology, and if the specimen is positive, to confirm with gas chromatography/mass spectrometry. Blood testing, while also accurate for illicit drugs, has been viewed as more invasive, painful and expensive. In addition, some drugs of concern, such as marijuana, are not identifiable in the blood after one day, while detectable in urine over a week or more depending on usage rate.

KEY WORDS

Testing, urinalysis, illicit drugs, workplace, safety, absenteeism, job performance, health

---

[1] *Reprint from "I. Internationales Symposium gegen Drogen in der Schweiz", 19-20 November 1990, Kongresshaus Zürich, VPM Zürich 1991, p.483-486.*

INTRODUCTION

Drug testing is a powerful deterrent to illegal drug use if sanctions apply to individuals found to be positive. This has been true in the military and for private and public employees, and for professional athletes. Drug testing is also a valuable treatment tool and is considered essential for recovering substance abusers returning to work and attempting to stay drug free.

METHODS

The testing process using urinalysis on the preliminary test can provide almost immediate reporting depending on the type of test methodology used. A written printout of the results can be made available. Illegal drugs that are normally screened include marijuana, cocaine, amphetamines, barbiturates, PCP, opiates, benzodiazepines and methaqualone. A number of individuals and organizations have challenged the use of drug testing as an invasion of privacy and as scientifically inaccurate. The United States Supreme Court has found urinalysis drug testing to be legal and accurate, even on a random basis, for individuals tested in safety-sensitive jobs in transportation and law enforcement occupations. The court reviewed the purpose of the testing, scientific accuracy of the methodology, invasiveness of the specimen collection procedures and the consequences of the actions taken on positive results. Hair testing and saliva testing have not been used on a widespread basis in the United States, nor has blood testing for illicit drugs, although blood and breathalyzer testing have been increasingly utilized for the identification of alcohol.

The urinalysis drug test results can report quantitatively and qualitatively. Established cut-off levels are selected for each drug above which any test will be reported as positive. Urinalysis testing costs depend on the volume and the methodology utilized. Employers also may require physicians to review drug test reports to assure that legal prescriptions are appropriately identified and confirmed with the individual doctor.

RESULTS

The use of drug testing in the military was the principal factor driving down illegal drug use by young sailors in the navy, over 40 percent of whom tested positive in 1981, as compared to less than 4 percent testing positive today. In private industry, the use of pre-employment, post-accident and post-treatment drug testing has resulted in dramatic decreases in the number of individuals testing positive and using illegal drugs on and off the job. In the United States, over 85 percent of the largest employers utilize urinalysis drug testing as a means of preventing illegal drug use on and off thejob. The options for drug testing must include questions such as when such tests are performed, what methodology is used, what drugs are included in the test panel, what cut-off levels are selected, where the specimen collection takes place, what laboratory does the testing and what action is taken with the results. The effectiveness of drug testing depends on a well thought out, carefully explained process, regardless of where or how it is used.

Several studies have been conducted that examine work performance of employees using illegal drugs compared with those who have tested negative for such drugs. At the U.S. Postal Service, accident rates, injury, absenteeism and disciplinary problems were found to be far higher for illegal drug users. The U.S. Postal Service, the biggest employer in the United States with 900,000 employees, wanted to examine by means of a drug test whether people who may have used drugs would have a major negative impact in the work place. So they did a test for 4000 workers applying for jobs, since they were taking urine samples anyway. 8.5% tested positive, slightly over 300. And then they went to work for the Postal Service. The Postal Service wanted to know what happened to those who had tested positive for using illegal drugs, when compared to those who had not used illegal drugs. After 8 months, those who had tested positive for drugs, were absent (43%) more often that those who tested negative (32%). Those who had tested positive (and their supervisors did not know it) had termination (because of rule violations, problems on the job) and a turnover rate higher than those who tested negative. After 16 months their termination rate rose to 59%, absenteeism to 60% and the turnover rate to 50%. The Postal Service decided to do pre-employment drug testing to save 62 million dollars a year, and decrease absenteeism, turnover and accidents.

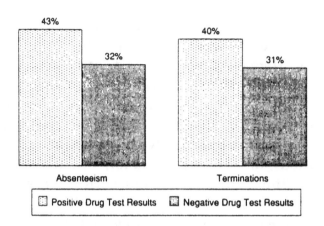

Negative impact of drugs in the workplace
(Postal Service)

The postal Service now requires everyone to pass a pre-employment drug test before being hired, as do nine out of ten of the other largest employers in the United States. Major companies and government agencies use random drug tests for current employees involved in particularly safety-sensitive jobs in the airline, railroad, nuclear power, oil and gas pipeline, maritime, mass transportation and trucking industries, as well as in law enforcement agencies.

COMPANY RESPONSIBILITIES

- Safe work environment
- Clear policy rules
- Conformity with the law
- Protection of key company assets-
     people as well as property
- Requirement to intervene when safety is compromised

Employers in the United States are saying: "We need a safe work environment, clear rules, conformity with the law, protection of our key asset - that is people, not the equipment." It is the people that make the company - and the requirement to intervene when their safety is compromised. Those are the responsibilities of an employer. A policy in which drugs are available or tolerated conflicts with safety, conformity with the law and protection of people.

DRUGS IN THE WORKPLACE - KEY ISSUES

- Clear company policy
- Fitness for duty - condition of employment
- Education and awareness
- Management support
- Labor/Union participation/education
- Drug testing
- Employee assistance
- Supervisory training

In Europe you say: "Well, that is the U.S. experience, here it is different." But let me just say that the jobs are the same: Driving a truck, working in a warehouse, on a fork lift truck, on a construction job, on a railroad, on a bus, in a watch factory. While the language may be German, French, Italian or English, the type of work is the same, and requires reasoning, judgement, coordination and memory. And the drugs are the same: marihuana, hashish, cocaine, amphetamines, heroin. The impact on health, on safety and performance is the same. One of the compelling factors in the United States which has decreased drug use, has been the crusade of the parents movement. It has not been people like me, making speeches. It has not been solely the arrests by the police. It has been the parents and their children, acknowledging the truth of health information as they saw their friends dropping out of society. And finally, the knowledge that people looking for employment will not be able to get any jobs if they use drugs.

```
┌─────────────────────────────────────────────────────────────────┐
│               DRUGS IN THE WORKPLACE EUROPE- USA                  │
│                                                                   │
│   - Same type job assignments                                     │
│        factory                                                    │
│        assembly line                                              │
│         construction worker                                       │
│         heavy machinery operator                                  │
│   - Same drugs                                                    │
│        hashish                                                    │
│        cocaine                                                    │
│         amphetamines                                              │
│   - Same impact on performance                                    │
│                                                                   │
└─────────────────────────────────────────────────────────────────┘
```

Some of the work place issues are: fitness for duty as a condition of employment, union and management working together, employee-assistance, drug-testing and education.

```
┌─────────────────────────────────────────────────────────────────┐
│               THE LIABILITY OF DOING NOTHING                      │
│                                                                   │
│   - Safety of the work environment                                │
│   - Security of plant equipment, company property and assets      │
│   - Health problems of employees                                  │
│                                                                   │
└─────────────────────────────────────────────────────────────────┘
```

Safety of the work environment, security of plant and equipment and health of the employees are jeopardized if one does not stem the tide of drug abuse in the workplace. It is a major issue facing Western industrialized nations. It requires opposing drug legalization.

```
┌─────────────────────────────────────────────────────────────────┐
│               EMPLOYEE ASSISTANCE PROGRAM                         │
│                                                                   │
│   - Voluntary referral                                            │
│   - Management referral                                           │
│   - Management consult                                            │
│   - Treatment follow-up                                           │
│                                                                   │
└─────────────────────────────────────────────────────────────────┘
```

Programs that can help employees who have problems are essential, and have been neglected. The earlier one intervenes, the better the opportunity for individual rehabilitation.

CONCLUSION

Drug testing is no magic wand, nor a panacea, but urinalysis testing for drugs has been found to be a valuable preventive, informational and rehabilitative tool. I believe it will be increasingly used around the world to fight the spread of illegal drug use, and the damage such use can cause, not only to the user but to society as a whole.

SCREENING STUDENTS FOR CANNABIS.

Richard H. Schwartz, M.D.

Department of Pediatrics, Georgetown University, School of Medicine,
Washington, D.C.

ABSTRACT

When warning signs of cannabis abuse occur in student, it is far better to light a candle, by ordering tests for drugs of abuse in urine, than to curse to darkness. When clinically indicated, drug tests may well foreshorten the clinical diagnosis of cannabis abuse by one year or more. In private pediatric practice, such requensts ususally come from the student's parents or guidance counselor in response to marked deterioration in academic performance, serious problems with conduct at home or school, and episodes of observed intoxication or possession of drugs. While obtaining a routine social and medical history, I have found that positive answers to four questions alert me to a possible drug abuse problem. (1) Do you smoke cigarettes every day? (2) How many times have you been drunk and, if more than once, when was the last time? (3) Did you receive more than one D or F grade on your last report card? (4) Are you into partying? Positive responses to several of the above items, especially when there is a strong family history of alcoholism or parental or sibling drug abuse. A series of positive tests offers eloquent testimony as to loss of control over the drug in question. A series of negative tests offers strong support to innocence and causes other than drug use must be sought for. Immuno-assay screening tests for cannabis are today, acceptably accurate, rapidly performed, and informative. It is necessaary for the clinician to understand the specific terminology of drug tests and to be aware of the pitfalls in interpreting results of such tests.

KEYS WORDS

Adolescents, School performance, Drug use, Drug testing, Cannabis, Immuno-assay.

INTRODUCTION

In the United States, peak use of cannabis occurred in 1978 and 1979, when lifetime, annual, monthly, and daily use of the drug by senior high school students was 60.4%, 50.8%, 37%, and 10.7%, respectively.[1] Since then, numerous national and state-wide surveys have documented continuous declines in lifetime, annual, and monthly use of cannabis (marijuana or hashish), as well as most other illicit drugs.[2-4] The 1991 nationwide U.S. data for lifetime, annual, monthly, and daily use of cannabis by senior high school students fell to: 37%, 24%, 14%, and 2%, respectively.[1] Compared with figures from the late 1970's, each of these percentages represents a steep decline, ranging from a 40% decline in lifetime use to an 80% decline in daily use. In the Netherlands, where lifetime, annual, and monthly use of cannabis is approximately one third that of U.S. high school students, cannabis use by Dutch high school students increased by almost 100% between 1984 and 1988.[5] During the same time, there was an 8% decline in lifetime and monthly use by American high school students[1]. One exception to the otherwise encourging trend is that U.S. youth who leave high school prematurely have much higher rates of drug use than those who remain in school.[1,6] Also of concern is that the potency of ordinary and sinsemilla varieties of cannabis has increased since the 1970's. Sinsemilla, an exceptionally potent form of marijuana which is the preferred form of cannabis according to aficionados, contains almost twice the concentration of tetrahydrocannabinol (THC) as does high quality hashish[7]. Moreover, use of cannabis does not offer protection from abuse of other drugs. On the contrary, habitual use of cannabis by students is often followed by use of cocaine and hallucinogenic drugs such as lysergic acid diethylamide (LSD).[1] Adolescents who smoke cannabis at least monthly are likely to be addicted to tobacco and to abuse alcohol.[1]

Concern about marijuana smoking, once considered by many physicians as quite harmless, is well-founded. Acute cannabis toxicity is accompanied by negative effects on learning and memory, as well as psychomotor impairment. The typical effects of cannabis are similar to those of a transient acute brain syndrome; they include deficits in attention span, concentration ability, short-term memory, and ability to synthesize and organize material. Cannabis impairs the ability to apply general principles to specific problems to make judgments in complex tasks and situations, and to make timely decisions.[8] Marijuana has measurable adverse effects on coordination and reaction time that interfere with one's ability to operate a motor vehicle or aircraft.[9] These effects may linger for up to 24 hours after the 3-hour period of acute intoxication ends. There is a clear cause and effect association between use of cannabis, and injury or death from motor vehicle accidents[10,11] and from suicide[12], two of the three major causes of death for American adolescents. Measurable visual and auditory memory deficits persist up to six weeks after cessation of daily use of cannabis.[13,14] Some experienced cannabis smokers learn to overcome these effects and, even when acutely intoxicated, may appear completely normal, unless sophisticated tests of short-term memory or complex motor tasks are administered and measured. Intoxication with cannabis is certainly associated with an increased risk of unsafe sex which may lead to increases in sexually transmitted diseases including AIDS, and to unwanted pregnancy. Antisocial acts such as vandalism and robbery are, according to results of urine tests obtained by our criminal justice system, clearly associated with cannabis use by adolescents. Drug use is not a victimless crime. Adolescents who use drugs regularly adversely effect the well-being of every person living in the home.[15] The normal trusting relationship within the nuclear family usually undergoes steady erosion and deterioration, and mutual suspiciousness and distrust displace family harmony and loyalty.[15] Adolescents who smoke cannabis frequently are much more likely to do poorly in academic subjects, are more likely to be truant or to drop out of school before graduation when compared to their non-cannabis using peers.[5] For older adolescents who are employed, drug use can have an equally deleterious effect on performance. There are reliable statistical data to prove that an employee who uses drugs: is more likely to be involved in accidents and injuries on the job; is absent more than twice as often; and receives sickness benefits many times those of the non-drug user. Drug users are less productive than non users and the quality of their work is lower. Drug users are also more likely to steal from the company, ostensibly to support their drug habait and more likely to leave the company voluntarily or to be dismissed.[16,17,17B]

RISK FACTORS FOR ADOLESCENT CANNABIS USE

Adolescents who smoke cannabis usually do so infrequently, in small quantities, and only in social settings such as parties or rock concerts. Only a minority of adolescents progress rapidly to weekday as well as weekend use, solitary use, and/or uncontrolled use. More than 50% of those who are frequent cannabis smokers began to use the drug frequently by age 15.[1] Risk factors for progressive and out-of-control cannabis use during adolescence include: 1) parent or sibling alcohol or drug abuse; (2) poor academic motivation and performance; (3) attention deficit disorder with or without hyperactivity; (4) poverty; (5) preference for drug-using peers; and (6) unresolved, serious sexual or physical trauma during formative years. Adolescents who use cannabis heavily and often, tend to engage in serious risk-taking behaviors, to show repeated refusal to heed parents and authority figures, and to exhibit little sense of remorse for wrongdoings. In many cases, there is an unfortunate mismatch between the temperament and lifestyle of parent and child.[18] In spite of these common, "red-flag" warning signs, among the middle-class of American adolescent drug abusers, 30% have no detectable risk factors (Schwartz, RH: Unpublished data).

Drug use, it must be understood, has many of the features of an infectious disease because of its social spread to younger siblings and to friends who are neophytes. Detection of student drug use, even at a relatively early stage could possibly alert physicians, parents, and school personnel to a potential serious problem and avoid both its progression and its spread. It is for that reason that drug testing has emerged as an issue in pediatric practice.

OFFICE SCREENING FOR ADOLESCENT DRUG USE

When warning signs of cannabis abuse occur in student, it is far better to light a candle, by ordering tests for drugs of abuse in urine, than to curse the darkness. When clinically indicated, drug tests may well foreshorten the diagnosis of drug abuse by one year or more. In private pediatric practice, such requests usually come from the student's parents or guidance counselor in response to marked deterioration in academic performance, serious problems with conduct at home or school, and episodes of observed intoxication or possession of drugs.[18] A series of positive tests offers eloquent testimony as to loss of control over the drug in question. A series of negative tests offers strong support to innocence and causes other than drug use must be sought for.

How is the physician, non-physician mental health worker, or teacher to suspect a possible drug problem during adolescence? In keeping with the risk factors cited above, there are several areas on which the clinician should focus. Adolescent drug use must be strongly suspected when there is tobacco addiction prior to age 16, alcohol abuse with binge drinking on a regular basis, a sharp decline in academic performance, a relentless pursuit of pleasure at the expense of hard work, untrustworthiness, and association with, and strong and unchangeable preference for, friends who have the same hedonistic philosophy and lifestyle. Adolescent cannabis smokers are convinced that the majority of teenagers in their school smoke marijuana or hashish ("Everyone is doing it"), and most would like to see cannabis legalized.

While obtaining a routine social and medical history, I have found that positive answers to four questions alert me to a possible drug abuse problem. (1) Do you smoke cigarettes every day? (2) How many times have you been drunk and, if more than once, when was the last time? (3) Did you receive more than one D or F grade on your last report card? (4) Are you into partying? Positive responses to several of the above items, especially when there is a strong family history of alcoholism or parental or sibling drug abuse, are much more important than only one or two positive items. Tests for drugs of abuse in urine can be a powerful diagnostic tool when one is confromted with signifi cant behavioral evidence such as delineated above.

ADDITIONAL REASONS TO SCREEN FOR DRUGS OF ABUSE

Drug abuse is often associated with deception, denial, and dishonesty, consequently, drug histories may not be trustworthy[19,20]. Moreover, many adolescent drug abusers do not enter the office of a private pediatrician. Many U.S. emergency medicine specialists therefore, recommend screening tests for drugs of abuse, particularly alcohol, cannabis and cocaine, for reasons such as the following: 1) serious automobile accident, (2) suicide attempt, (3) unexplained seizure, (4) toxic psychosis, (5) drunkenness, (6) symptoms of manic-depressive affective illness, (7) violent and explosive outbursts of temper, (8) antisocial acts, including gang fights, theft, and running away from home, and (9) sexually transmitted diseases, repeated pregnancies, and promiscuity. Screening tests for drugs of abuse are also widely used by adolescent drug abuse treatment programs at the time of admission, after drug users return from holiday, when their behavior is suspect, and during the prolonged recovery phase. Drug tests are also used frequently by the juvenile justice system.[21]

LEGAL AND ETHICAL CONSIDERATIONS

Such testing, however, poses legal and ethical obstacles that compound the traditional concerns about the accuracy and precision of a laboratory test procedure.[22,23] Legal concerns include the right to privacy and the right to resist bodily searches without the searcher providing direct evidence that the accused has performed an illegal act.[24,25] Ethical concerns include violation of informed consent guidelines and questions about sharing test results with parents, school authorities, and insurance companies.[24,26] Screening large populations such as an sports team, or all teenagers who visit physician's offices, is expensive, and in most populations, the percentage of positive tests would be generally be too small to warrent such an effort (Schwartz, RH, unpublished data). Should selected populations of students prove to have a rate of drug use exceeding 10%, screening of large populations may be worthy of consideration, if ethical and legal concerns can be addressed.

Ethical Considerations. In 1989, a policy statement issued by the American Academy of Pediatrics (AAP) Committee on Substance Abuse states that involuntary drug tests should not be ordered for competent older adolescents, even if the parent so wishes.[26] This AAP statement reflects the philosophy of many of the ethicists and adolescent medicine specialists in the Academy. Confidentiality is not unconditional and informed consent prior to testing for drugs of abuse may be waived when there is reason to doubt competency or where medical assessment suggests a high risk of serious damage due to substance abuse.[26-28] The AAP statement, many clinicians believed, made it appear that the right of the teenagers privacy and confidentiality superseded the gravity of the malevolent effects of drug use on the life and well-being of adolescents under their care. I concur. In my opinion, paternalism can be justified when the evil prevented to the student is greater than the wrong caused by the violation of a moral rule. Harm and danger to the adolescent should, however, be the only basis for justified limitations of liberty; annoyance of others won't do. When adolescents younger than age 18 years exhibit poor judgment, cannot make a positive treatment alliance, are acting in a way that poses danger to themselves or to others, do not appear to show concern for his or her condition, and refuse to get an indepth evaluation for a probable drug problem, is it really unethical to obtain a urine specimen for drugs of abuse testing without informed consent?[27,28]

DRUG TESTING

Analytical tests for drugs of abuse in urine consist of two classes of tests: screening tests and confirmation tests. At present, immunoassay screening tests for cannabis (THC) in urine are highly reliable when correctly used and interpreted.[21-23] A positive screening test for THC, when clinically indicated and when followed by a gas chromatography/mass spectrometry confirmation test, offers clear evidence of cannabis use by smoking or oral ingestion. A positive test, does not, however, determine the degree of brain impairment (intoxication), the frequency and extent of drug use, or even the time of last use.

Screening tests detect nanogram quantities of drug metabolites, usually by one of three different immunoassay procedures, at preselected "cut-off" levels of detection, below which the test result will be negative.

SPECIFIC SCREENING METHODS

Thin layer chromatography (TLC): THC, the traditional method for analysis of milligram quantities drug metabolites in urine, will not detect cannabis use unless a cannabis-specific TLC test is ordered and even then it is less sensitive than immunoassay methods.

Immunoassay Screening Tests: Immunoassay tests can detect between five to ten different drugs with a sample size of 20-30 mL of urine. The prototype immunoassay method, is the radioimmunoassay (RIA) method. It has, however, become less popular today because of the cost of equipment, longer turn around time, and the requirement for a license to use radioactive isotopes. The most widely used method in the United States is the EMIT™ (Enzyme Multiplied Immunoassay Test) method, which is based on antigen recognition by cannabinoid-specific antibodies. The endpoint is an invisible color change in nanometers of wave length, proportional to the drug concentration in the sample, measured by a photometer. The Syva Corporation, manufacturer of EMIT™ products, recently announced the addition of three new cannabinoid tests, 20ng, 50ng, and 100ng assays which employ monoclonal antibodies and provide greater overall stability and more consistent results. A third immunoassay test that is gaining popularity is the FPIA (Florescent Polarization Immuno-Assay), trade name ADx™. FPIA is a sensitive, rapid, method, somewhat more expensive in terms of reagent cost, which is based on a reduction in the intensity of polarization of a light beam when antibody recognizes antigen (drug metabolite) in urine. FPIA is the only screening test that permits semiquantitation without additional reagents. ON- TRAK™, a simple-to-perform immunoassay test based on the latex-agglutination procedure, does not require expensive instrumentation, is as accurate as the EMIT (unpublished comparison study) and can be performed with a single drop of urine for a single drug metabolite. The cut-off point for ON-TRAK™ is 100 ng/mL making it too insensitive for adolescent screening use.

Confirmation Tests. Confirmation tests are a higher order of sensitivity, specificity, and accuracy. They are far less influenced by adulterants purposefully added to urine and they are highly specific. In the U.S., the highly sophisticated and costly gas chromatography/mass spectrometry (GC/MS), the industry reference standard, is the only test acceptable for forensic analysis and U.S. government-mandated workplace drug testing. Only 3-4% of cannabinoid-positive test results performed by immunoassay screening methods are not confirmed (i.e. false positives by immunoassay) by GC/MS.

CUT-OFF DETERMINATIONS

Among the controversies involved in drug testing procedures is the appropriate cut-off point, defined as the lowest level of drug metabolite that will cause a test to be a positive test result. These cut-off points are preset by the manufacturer or the medical laboratory, based on national or industry-wide standards. It is imperative that the clinician understand the meaning of cut-off points and apply that knowledge to the proper interpretation of results of screening tests for cannabis.

For cannabis (marijuana and hashish), the cut-off point for the major marijuana metabolite, 9-carboxy tetrahydrocannabinol (THC), has been set at 100 ng/mL for American workplace or armed forces test procedures. Forty-eight hours after smoking a marijuana cigarette, one in four urine specimens while still containing 60-90 ng of THC per mL of urine, will be given a negative result, if the cut-off point is set at 100 ng/mL. For criminal justice officials and at drug treatment facilities, the cut-off point is usually set at 20 ng/mL. In comparison with the 20 ng/mL cut-off point, the 100 ng/mL cut-off point will fail to

detect at least one of every three urine specimens that contains
THC metabolite.[29] By the end of 1992, the cut-off point for THC
testing in the American workplace will be reduced to 50 ng/mL.

DURATION OF EXCRETION OF CANNABIS

        How long will cannabis be detectable in the urine?
Cannabis can be detected for two or three days following casual
use of the crude drug.[22] After daily use of cannabis, about 20%
of cannabis-dependent adolescents will continue to excrete
measurable amounts of THC metabolites for one month or more after
cessation of use.[22] The higher the potency of the cannabis that
is consumed and the longer the time it has been smoked daily, the
longer one can excrete cannabinoid metabolites. Certain factors
can increase the duration of positive results; these include
starvation, vigorous exercise, and dehydration. This can be
explained by mobilization of cannabis from fat stores during
starvation or exercise and increased urinary concentration of THC
during dehydration.

PASSIVE INHALATION

        Passive inhalation of marijuana or hashish smoke during
exposure of innocents who are in an average-sized, ventilated
room or in an open space, will not produce a positive test for
THC in the urine. Some studies have demonstrated positive THC in
urine by passive inhalation; however, all were conducted in tiny,
nonventilated rooms or in the passenger compartment of sealed
automobiles.[22]

PITFALLS OF DRUG TESTING

        Certain pitfalls may be associated with drug testing. One
should not interpret these caveats to mean that drug testing is
inaccurate or ineffective, rather one must understand what the
test really shows and what are the pitfalls in their interpretation. The primary method of deception by the at-risk student is
purposeful adulteration of the specimen.
        Among the clever deceptive techniques practiced to avoid
detection of their drug use, substitution of a drug-free urine
specimen obtained or purchased from a drug-free friend or even
from the family pet. Substitution of tea or apple juice, dilution of the specimen by drinking large quantities of liquids
immediately prior to voiding, ingesting diuretic medication to
cause urinary dilution, adulterating of the voided specimen[30]
with salt, soap, acids or alkali or by adding certain eyedrops
containing benzylalkonium as a preservative,[31] may deceive some
of the immunoassay tests and negate a positive result. Some
cannabis smokers believe that ingestion of megadoses of vitamin C

or cranberry juice shortly before urinating will negate a positive result, this has been shown to be untrue.[32] The specimen should, if possible, be collected soon after awaking, preferably on Monday morning to capture weekend drug use. The urine specimen should feel warm to the touch and the temperature should range between 32.5 and 37° Centigrade shortly after voiding. The urine optimally should be distinctly yellow and the pH should be within physiological range. Specimens must be securely stored in the refrigerator or in the freezer compartment.

SUMMARY

Urine tests for cannabis are accurate, relatively inexpensive, and easy to perform. They are clinically indicated for students whose behavior, conduct, and academic performance suggest the possibility of drug use. Ethical and legal issues must be addressed beforehand. Alteration or adulteration of the urine specimen may occur with "streetwise" cannabis smokers, making vigilance, at the time of specimen collection and storage, imperative.

REFERENCES

1. Johnston LD, O'Malley PM, Bachman JG: <u>Drug Use, Drinking, and Smoking: National Survey Results from High School, College, and Young Adults Populations: 1975-1991</u>. National Institute on Drug Abuse, Rockville, MD, 1992 (In Press).

2. Summary of trend analysis from 1990 National Household Survey on Drug Abuse. <u>NIDA Capsules</u>. National Institute on Drug Abuse, Rockville MD, December, 1990.

3. Report to the Attorney General: <u>Biennial Survey of Drug and Alcohol Use Among California Students in Grades 7,9, and 11: Winter 1989-90,</u> Office of the Attorney General, Crime Prevention Center, Sacramento, CA, 1991.

4. Maryland Department of Health and Mental Hygiene, Alcohol and Drug Abuse Administration and the Juvenile Advisory Council of Maryland, <u>1989-1990 Survey of Substance Abuse Among Maryland Adolescents</u>, Baltimore, MD, October, 1989.

5. Plomb HN, Kuipers H, van Oers ML. Smoking, Alcohol Consumption and the use of Drugs by School Children from the age of 10: Fourth National Youth Health Care Survey on Smoking, Alcohol Consumption and the Use of Drugs in The Netherlands. Vrije Universiteit (Free University) Press, Amsterdam, the Netherlands, 1991.

6. Friedman AS, Glickman N, Utada U: Does drug and alcohol use lead to failure to graduate from high school? <u>J Drug Educat.</u> 1985; 15:353-364.

7. ElSohly MA: Potency Monitoring Project, Report #39, July 1,-September 30, 1991, Research Institute of Pharmaceutical Sciences, School of Pharmacy, University of Mississippi, University City, MS.

8. American Pychiatric Association, Position Statement on Psychoactive Substance Use and Dependence: Update on Marijuana and Cocaine. <u>Am J Psychiatry.</u> 1987; 144:698-702.

9. Leirer VO, Yesavage JA, Morrow DG: Marijuana carry-over effects on aircraft polot performance. <u>Aviat Space Envirom. Med.</u> 1991; 62:221-7.

10. Soderstrom C, Trifillis A, Shankar B, et al: Marijuana and alcohol use among 1023 trauma patients. <u>Arch Surgery.</u> 1988; 123:733-7.

11. Sloan EP, Zalenski RJ, Smith RF, Sheaff CM, Chen EH, et al: Toxicology screening in urban trauma patients: Drug prevalence and its relationship to trauma severity and management. <u>J Trauma.</u> 1989; 29:1647-53.

12. Berman AL, Schwartz RH: Suicide attempts among adolescent drug users. <u>American Journal of Diseases of Children</u> 1990; 144:310-314.

13. Schwartz RH, Gruenewald PJ, Klitzner M, Fedio P: Short-term memory impairment in cannabis-dependent adolescents. <u>AJDC.</u> 1989; 143:1214-19.

14. Schwartz RH: Heavy marijuana use and recent memory impairment. <u>Psychiatric Annals.</u> 1991; 21:80-82.

15. Jenny L, Schwartz RH: Adolescent drug dependency and the family. <u>Virginia Medical.</u> 1985; 112:711-713.

16. Taggart RW: Results of the drug testing program at Southern Pacific Railroad. NIDA Research Monograph. 1989; 91:97-108.

17. Zwerling C, Ryan J, Orav EJ: The efficacy of pre-employment drug screening for marijuana and cocaine in predicting employment outcome. <u>JAMA.</u> 1990; 264:2639-43.

17B. Normand J, Salyards SD, Mahoney JJ: An evaluation of preemployment drug testing. <u>J Applied Psychol.</u> 1990; 75:629-639.

18. Schwartz RH: What to do when you find one of your patients using marijuana. <u>Postgraduate Medicine.</u> 1989; 86:91-96.

19. Chasnoff IJ, Landress HJ, Barrett ME: The prevalence of illicit drug or alcohol use druring pregnancy and discrepancies in mandatory reporting in Pinnellas County, Florida. New Engl J Med. 1990; 322:1202-6.

20. Silber TJ, Getson P, Ridley et al: Adolescent marijuana use: Concordance between questionnaire and immunoassay for cannabinoid metabolites. J Pediatr. 1987; 111:299-302.

21. Visher C, McFadden K: A comparison of urinalysis technologies for drug testing in criminal justice. National Institute of Justice, Research in Action, Washington, D.C. 1991; June 1-6.

22. Schwartz, RH: urine testing in the detection of drugs of abuse. Arch Int Med. 1988; 148:2407-12.

23. Schwartz, JG, Zollars PR, Okorodudu AO, Carnahan JJ, Wallace JE, Briggs JE: Accuracy of common drug screen tests. Am J Emerg Med. 1991; 9:166-170.

24. King NMP, Cross AW: Moral and legal issues in screening for drug use in adolescents. J.Pediatr. 1987; 111:249-250.

25. Chamberlain TR: Legal issues related to drug testing in the clinical laboratory. Clin Chem. 1988; 34:633-6.

26. Committee on Adolescence, Committee on Bioethics, and Provisional Committee on Substance Abuse: Screening for Drugs of Abuse in children and adolescents. Pediatrics. 1989; 84:396-7.

27. Silber TJ: Justified paternalism in adolescent health care: Cases of anorexia nervosa and substance abuse. J Adolescent Health Care. 1989; 10:449-53.

28. Silber TJ: Parental request for a "drug test": What should the caring pediatrician do? Clinical Pediatrics. 1991;30:643-645.

29. Smith DE, Gutgesell ME, Schwartz RH, et al: Federal guidelines for mairjuana should have lower cut-off levels: A comparison of results from immunoassays and gas chromatography-mass spectrometry. Archives Pathology and Laboratory Medicine. 1989; 113:1299-1300.

30. Warner A: Interference of common household chemicals in immunoassay methods for drugs of abuse. Clin Chem. 1989; 35:648-51.

31. Pearson SD, Ash O, Urry Francis M: Mechanism of false-negative urine cannabinoid immunoassay screens by Visine^{TM} eyedrops. Clin Chemistry. 1989; 35:636-638.

32. Schwartz RH, Bogema S: Ingestion of megadoses of ascorbic acid will not produce "clean" urine from marijuana smokers. <u>Archives of Pathology and Laboratory Medicine</u>. 1988; 112:769.

TESTING FOR DRUGS IN THE MILITARY AND TRANSPORTATION INDUSTRY

Paul J. Mulloy*

United States Naval Academy, McLean, VA 22102

ABSTRACT

For the past ten years, the U.S. Navy has used a ten point program to satisfactorily control alcoholism and illicit drug consumption, which in 1980 involved 27.6% of all uniformed personnel. These ten points comprise:

1) Assessment
2) Policy formulation
3) Policy communication
4) Counseling assistance and treatment
5) Training and education
6) Security
7) External support
8) Alternatives
9) Testing
10) Quality assurance

Success in the armed forces ensured a more productive and effective military, and set an example for other organizations. The regulated transportation industry and postal service experienced similar benefits after initiating a similar policy based on employee assistance programs, education and testing. Along with leadership, testing is a key component, and recent studies correlate testing results to performance prediction.

KEY WORDS

Drug testing, assistance program U.S. Navy transportation, railway, postal service, aviation

* Rear Admiral, U.S. Navy (Ret)

In 1990 the U.N. subcommittee on narcotics reported that the international illicit drug trade approximated $500 billion and was growing.[1]

In addressing events that are occurring with regards to illicit drugs in mass transportation in the United States, it is helpful to review the origin and progress within the federal government. The first most comprehensive program to combat illicit drug use began in 1981 in the military. An anonymous written survey conducted throughout all services in 1980 revealed that 27.6% of all uniformed personnel had used drugs in the past 30 days.[2] The Navy validated the written survey with a urinalysis-based survey of their young (18-24 year old) members. It revealed that 47% of their young enlisted sailors had marijuana in their urine.[3]

The Navy used a Ten Point Program which I have shared with federal agencies, other governments and corporations over the past decade. It is a reference for the program that the Department of Transportation has used for its directive for all its administrations. These are the ten components:

1. *Assessment:* Survey employees to evaluate their attitudes towards illegal drug usage and to provide a foundation for the formulation of an effective drug-free workplace policy.

2. *Policy Formulation:* Work with employees to develop a comprehensive drug-free workplace policy. Seek maximum employee participation while encouraging compliance with on-site drug testing as the cornerstone of an effective drug-free workplace program.

3. *Program and Policy Communication:* Provide each employee with written and verbal instructions on the policy and procedures that will be utilized by the employer to conduct on-site drug tests to achieve the goals set by the drug-free workplace program.

4. *Education and Training:* Emphasize educational and training programs for employees so that each employee is fully aware of 1) the manner in which the drug-free workplace program works and 2) the value the program provides in enhanced safety on the work site and improved productivity of fellow workers.

5. *Counseling, Treatment and Rehabilitation:* Provide access to qualified professional counseling assistance and residential treatment programs for any employee needing addiction intervention assistance.

6. *Security Improvements:* Establish an effective worksite security program to protect workers from exposure to illegal drug activities and protect them from an unsafe working environment created by addictive behavior of a fellow employee.

7. *External Support:* Utilize government and private sector substance abuse programs for employees needing treatment.

8. *Substance Abuse Alternatives:* Develop work site programs which provide alternatives to employees who might otherwise engage in substance abuse behavior.

9. *Testing:* Incorporate commercially available, state-of-the-art urinalysis-based drug testing tools, including on-site devices, to maximize work site safety while assuring maximum protection for the individual rights of employees.

10. *Quality Assurance:* Continuously reevaluate each segment of the drug-free workplace program to assure employees are provided the maximum protection and that the goals of the drug-free workplace program are being achieved.

Success in the Armed Forces not only set an example for other organizations, it ensured a more productive and effective military. The regulated transportation industry has experienced similar benefits since it inaugurated its program which has Policy, Employee Assistance Programs, Education and Testing as bedrock components. Even so, from the beginning I have stressed that waging and winning the War on Drugs is fundamentally a leadership campaign wherein the top executive needs to take charge and be held responsible but aided by experts in the legal, medical, clergy, scientific and law enforcement professions. In gaining a drug-free workplace, I believe in a zero tolerance policy. But it should be one designed to get rid of the abuse more than the abuser. Key provisions should include providing immunity from prosecution for those who come forward for help with their substance abuse problem while applying clearly defined and enforced sanctions for those who continue to violate the law.

Along with leadership, testing is a key component and recent significant studies correlate testing results to performance prediction.

In 1990, Dr. David Blank concluded a controlled urinalysis study of 1,000 Navy recruits.[4] The study comprised 500 recruits who tested positive for marijuana use and 500 who tested negative. After 4 years, 70% of those who tested negative were performing satisfactorily and were eligible for reenlistment. However, only 46% of those who tested positive still remained in the Navy. Significantly, 35% of those discharged were for performance and substance abuse-related reasons.

Another study was conducted since 1987 by the United States Postal Service[5] which tested 4,375 applicants at 21 facilities throughout the country. Those who initially tested positive were absent from work 66% more than non-drug users. Members who tested positive were also fired 69% more frequently than those who tested negative. Those findings resulted in a policy that no applicants who tested positive would be hired.

What these illustrate is the significance of pre-employment testing. However, there is a societal issue which needs examination in the decision of not hiring - or hiring with such identification.

Proceeding to the transportation industry in 1987, an Amtrak train accident resulted in 16 deaths, 174 injuries and $57 million dollars in damages. The engineer at the controls tested positive for marijuana. That accident added impetus to the Department of Transportation promulgating its very comprehensive rules and regulations in 1988 which included a broad spectrum of testing. These included: pre-employment for applicants, reasonable cause, post-accident, periodic and random testing.

In Urban Mass Transportation (now the Federal Transit Administration) the rules affect about 195,000 people and in the Federal Railroad Administration about 90,000. The Federal Highway Administration covers 4,000,000 people and the Federal Aviation Administration is responsible for over 410,000 people. The Department of Transportation requires testing for the so-called NIDA (National Institute of Drug Abuse) 5 substances: Marijuana, Cocaine, Amphetamines, Phencyclidine (PCP) and Opiates. They allow employers to test for more where there is reasonable cause. This year by Congressional direction, requirements are being expanded to include alcohol and split sample specimen collection. Like the military, they also require an extensive laboratory quality control proficiency and inspection regime with the National Institute on Drugs.

Random drug testing in American railroads began a phase-in basis in January 1990. Based on 35,000 tests, a positive rate of 1.04% was reported for 1990.[6] 194 specimens were found positive for cocaine, 175 were found positive for marijuana, 10 for opiates and 2 for PCP. FRA rules also require the collection of blood and urine specimens following certain major train accidents, collisions and employee fatalities.

In addition to random testing, the Federal Railroad administration has also tested for drug use since 1986. Post- accident results are down from 5.6% in 1988 to 1.1% in 1991, for reasonable cause from 5.4% in 1988 to 2.4% in 1990. For marijuana, the number of cases decreased for the following causes: post accident decreased from 25 in 1988 to 6 in 1991, and reasonable cause decreased from 222 in 1988 to 90 in 1990.

In the Federal Transit Administration (Mass Transit) a self-reported survey published this past November[7], employees in safety sensitive positions reported illicit drug use as 29.1% having ever used, 7.2% in past year and 6.9% in the past month. Significantly, employees indicated cocaine as the dominant drug. Almost 7% of sensitive safety employees acknowledged drug use on the day before or on the day of their work duty. Over 10% reported daily or weekly use. Mandated testing of this industry will commence in 1993 under the FTA.

In aviation, the FAA announced the results of its first year of drug testing for employees and applicants in July 1991.[8] 230,621 tests were administered in 1990. Of these, 966 (0.4%) were positive. 46% were detected in pre-employment and none was hired for safety sensitive positions. 520 or 54% of the positives were found in aviation safety security positions mainly through random testing. They were removed from their positions. The majority were maintenance or maintenance applicants. Flight crew members and applicants accounted for 28. Approximately 340,000 aviation employees are subject to drug testing.

These early results for drugs in all the transportation industries are encouraging. As the coverage and reporting of mass transit systems expand, a more comprehensive assessment will be made from a new information system.

However, I do caution that results currently reported throughout the United States are based on what I consider to be high cut-off levels set, for example, at 100 mg/ml for screening of THC marijuana, much higher than needed to counter claims of passive inhalation. The U.S. Navy set this level in 1982 partly to allow service personnel an opportunity to change in a strategy of deterrence more than detection. With the massive publicity and years of expanding the effort, lower cut offs should be instituted as the military is doing this year. As national and industry specific screening results show a downward trend, I recommend in a program of deterrence and detection that the cut off levels be correspondingly lowered. One laboratory director estimated that 60% higher positive finding would result when screening cut off was lowered from 100 mg/ml to 20 mg/ml in a sample size of 1,000 specimens.[9]

Taking the lead, the military in January of this year lowered screening cut offs for marijuana to 50 mg/ml and for cocaine 150 (from 300), confirmations are at 15 and 150 respectively for each substance and the U.S.

Department of Transportation has indicated they will follow suit. Unquestionably in the military and in industry as well as national, the trends are encouraging. The military, as a whole, decreased to 4% since 1981. The Navy is now below 1% usage. But with 49% of the high school seniors surveyed still saying they have used drugs, and the rise in cocaine and heroin, the battle is still on.[10] The latest report on emergency room admissions indicates a rise in 1991 in incidents of heroin and cocaine use. With these indices, without a doubt, testing is still a very necessary deterrent and detection tool. In a survey I had conducted in the Navy, 83% of the young people said testing was the #1 deterrent. Perhaps more significant, 27% said they would resume using drugs if testing were halted![11]

In addition to laboratory testing, with technical improvements today, I strongly recommend use of on-site testing. The screening devices currently approved by the U.S. Food and Drug Administration (FDA) for marketing are 98% presumptive accurate or better; results are obtained in minutes; and they convey significant cost savings in time and money compared to laboratory screening practices by avoiding delays, chain of custody, and expensive laboratory procedures. For employees it reduces anxiety by providing immediate results. It is especially useful in pre-screening applicants for employment. Of course on a presumptive positive test and to safeguard individual rights, I recommend laboratory confirmation with gas chromotography/mass spectrometry.

When one considers the average laboratory takes 24 hours to 72 hours to report urinalysis test results and a pilot of a 747 or express train engineers, school bus driver or dangerous machinery operator may be involved, why not gain a highly accurate, timely, simple presumptive finding to remove hazard and enhance public safety? Procedures to protect confidentially, improper disclosure, access to information and rigorously limit embarrassment can be developed for even the most sensitive positions. Surely to protect the public trust, greatly improve public safety and confidence we can accept such provisions. We do it in many ways for lesser reasons and waiting for the next tragic accident is hardly prevention oriented. On-site testing (with proper safe-guards) helps achieves these objective now. And of course, to industry, the cost avoidance can be significant. In the U.S., laboratory costs for processing a specimen average $23-25, the U.S. Office of Management and Budget cites $50-60; and an official with a major defense contractor told me that their accountants estimate that total costs associated with collecting, processing and reporting a specimen results amounts to $100/specimen. With on-site drug testing, plates costing about $10 to detect four drugs is about ten recruits. Drug Screening Systems, Inc. of Blackwood, New Jersey, a new on-site drug testing company, offers a plate of four drugs called Mach IV for about $12 and reduced cost for orders of 100. This capability should be universally examined and employed. In the U.S., with surveys indicating that less than 20% of the workforce use drugs, that predicates a potential 80% reduction in laboratory submission when on-site testing is employed. In an incremental adaptation, on-site pre-employment screening might be a good place to start.

Without question, the War on Drugs will be a long one. The enemy is drugs but its purveyors are criminally brilliant in all spheres of a global business: shipping, banking, law, accounting, terrorism and interlocking financial and distribution systems. Recently liberated Eastern Europe and former Soviet Union are prey to these enemies who have hard currency to offer those who are in desperate economic straits.

Civilization, as we know it, is at risk from the scourge of illicit drugs. Interdiction efforts are important but Demand Reduction is crucial. Conducting proper surveys help overcome denial and sensitize leaders and policy makers to the issue. Education, rehabilitation and testing are key to ensure there is assistance available for users to counter addiction. The workplace should be the focus to program success and the CEO should personally commit to this leadership effort. The workers will respond and expose the dealers and users from a comprehensive program which has both compassion and enforced sanctions in its policy. Properly implemented, the knowledge and countervailing skills will flow back into the home and community. Emphasis on the workplace will reduce demand and put the thugs (not lords) and gangs (not cartels) out of business. We know the way, the question is do we have the will? In my judgement, we must.

BIBLIOGRAPHY

(1)   United Nations Report of the International Narcotics Control Board for 1990.

(2)   Bray, R. M., (Etalia) <u>1988 Worldwide Survey on Substance Abuse and Health Behaviors Among Military Personnel</u>. Research Triangle Institute, Research Triangle Park, N.C.

(3)   Booz, Allen and Hamilton, Inc. <u>Urinalysis Test Results Analysis</u> (December 8, 1980 Survey) U.S. Navy Bureau of Personnel. Arlington, VA 1981.

(4)   Blank, Ph.D., David <u>Early Employment Testing for Marijuana</u>. Demographic and Employee Relation. Arlington, VA 1990.

(5)   *An Empirical Evaluation of Pre-Employment Drug Testing in U.S. Postal Service.* The Journal of the American Medical Association. 1990.

(6)   U.S. Department of Transportation News Release on the Federal Railway Administration. 1991.

(7)   U.S. Department of Transportation. <u>Substance Abuse in the Transit Industry</u>. 1991.

(8)   Discussion with Yale H. Kaplan Ph.D., Director National Center for Forensic Science; Baltimore, MD. 1992.

(9)   U.S. Department of Transportation News Release on the Federal Aviation Administration. 1991.

(10)   <u>National High School Survey on Drug Abuse; Population Estimates 1990</u>, National Institute of Drug Abuse.

(11)   Booz, Allen and Hamilton, Inc. <u>Relative Effectiveness and Impact of Navy Drug Control Initiatives</u>, Arlington, VA. 1983.

DRUG USE AND DRUG DETECTION IN THE FRENCH ARMED FORCED.

Pierre Juillet.

Médecin-Génral Inspecteur (C.R.), Membre de l'Académie Nationale de Médecine, Paris, France.

ABSTRACT

Over the past two decades drug use and dependence in the French Armed Forces are surveyed by military physicians who carry out special interviews. These interviews are targeted to a sample of population of draftees entering compulsory military service, and to a sample of the military in armed forces units and hospitals. In addition, urinary detection of illicit drugs is performed systematically in the flying personnel of the Air Force and random testing in all other services. Drug use and dependence must be closely monitored among draftees who practically represent every year the whole 18 to 19 year old French male population.

KEY WORDS

Cannabis; Cocaine; Drug detection; French Armed Forces; Epidemiology; Urinary detection.

Following the epidemic of illicit drug consumption in the sixties, the Armed Forces instaured epidemiological studies to determine the extent of illicit drug use in the military according to a system called "APETOX".

THE APETOX SYSTEM

These studies were based on confidential interviews carried out by a military physicians with soldiers in the armed forces, or with subjects during the medical examination which precedes their induction into the armed forces. Three groups of subjects were targeted :
- during the medical examination performed in all 18 to 19 years

old french males before their compulsory military service.
- randomly in military recruits irrespective of status or rank
- in military hospitals.

All this information was anonymously coded and centralized in the "Center for the Processing of Medical Information" in the Armed Forces.

In 1990, 9,016 interviews were processed and analyzed :
- 6,657 originated from 438,068 draftees
- 1,357 originated from soldiers in military units
- 1,001 originated from military hospitals.
95% of the sample came from draftees private soldiers

Results of this survey are summarized in Table I. Data provided by the APETOX epidemiological survey may not correspond to actual drug consumption : some of the draftees might report drug taking in order to avoid the draft. However, the final decision for deferment is taken only after an overall analysis of the history and physical as well as psychological examination of the subject.

SYSTEMATIC DETECTION OF ILLICIT DRUGS IN THE ARMED FORCES

Since 1988, urinary detection of cannabis, cocaine and opiates has been established. Samples are analyzed by the immunoenzymatic method of EMIT and positive results validated by GC-MS system (Table II). In 1990, the cut off point for positive cannabinoid detection was set at 20 ng/ml (down from 100 ng/ml in 1989). This allows for the detection of a "week end" joint on the following tuesday.

CONCLUSION

Illicit drug use does not constitute a major problem among officers and men enlisted beyond their period of compulsive service. However this practice must be closely scrutinized among draftees. In this respect, the Medical Service of the Armed Forces plays an important role year after year in monitoring the physical and mental health of all 18 to 19 years old French males.

TABLE I

Non-Medical use of drugs reported by 9,016 conscripts in 1990 (2.06% of the total number examined: 438,068)

Nature :
- cannabis : 67.5 %
- psychostimulants : 10.5 %
- analgesics, opiates : 8.5 %
- hallucinogens : 4 %
- inhalants : 4 %
- psychedelics : 0.5 % (poppers, ectasy, crack)

Frequency :
- Use of one drug : 48 %
- Use of two drugs : 26 %
- Use of three drugs or more : 16 %

Mode of administration :
- Intravenous injections : 8%
- Addicted : 16 %
- In association with alcohol : 71 %

Profile of drug user :
- Mean age : 16
- Mean age of the first intravenous injection : 17.5
- Personality problems : 50 %
  (major psychic disorders : 0.5 %)
- Professional activity :
  • employed : 37 %
  • unemployed : 45 %
- Delinquent behavior : 20 %
- Broken home (divorce) : 42 %

Answers to proposed therapeutic assistance :
- Accepted : 7 %
- Refused : 22 %
- Pending : 13 %
- Deferred : 63 %

Referrals :
- for assistance in the military : 13 %
- to orientation center : 32.5 %
- to civilian specialized organization : 6 %

TABLE II

Urinary detection for cannabis and cocaine in the French armed forces.

Sampling and analysis by EMIT and GC-MS was performed in 1990 on 23,484 subjects. EMIT-test is considered positive when concentration is 20 ng/ml or more.

Urinary Detection	Cannabis	Cocaine*
Proportionate testing in Army or Air Force units :		
3,300 draftees	13.4 %	--
300 commissioned officers	0	0
300 non commissioned officers	1.3 %	0
Random testing in the 3 armed services :		
7,684 enlisted men	8 %	0
4,563 draftees	12 %	(1 subject)
Testing in Air Force Flying personnel		
11,900	2 °/oo**	--

* *Systematically analyzed only in the Airforce flying personnel*
** *18 non C.O. out of 21.*

GENERAL CONCLUSIONS

GENERAL CONCLUSIONS

An international colloquium on illicit drugs organized with the assistance of the City of Paris was held at the National Academy of Medicine in Paris April 8 - 9, 1992. The program dealt essentially with human effects of cannabis consumption, and the following general conclusions were drawn at the end of the colloquium:

1. The toxicity of cannabis is today well established, experimentally and clinically. This drug adversely effects the central nervous system, the lung, immunity and reproductive function.

2. Epidemiological studies have reported that the use of "hard drugs" rarely occurs among subjects who have never consumed cannabis.

3. Consequently, the participants to this colloquium rebut the distinction made between "soft" and "hard" drugs.

4. The trivialization ("decriminalization") of cannabis use, where it has occurred, has resulted in a considerable increase of its consumption and of its subsequent damaging effects.

5. It is important to foster a campaign of information and prevention bearing both on the legal aspects and the health consequences of cannabis consumption.

Professor Henri Baylon
President, French National
Academy of Medicine
April 10, 1992

APPENDIX

HASHISH AND MENTAL ILLNESS: THE EXPERIENCE AND OBSERVATIONS OF MOREAU [1]

Hélène Peters

Macalaster College, Saint-Paul, Minnesota.

Jacques-Joseph Moreau (1804–1884) was born in Montresor, a small town in the Loire valley. His father, an officer in the armies of Napoleon, fought in many battles, and retired after Waterloo to devote himself to the study of mathematics. Moreau first studied medicine in Tours, near his home town, hence the surname he added to his own when he later went to Paris. His first teacher was a very famous clinician of his time, Bretonneau. The period was still the prescientific era of medicine, which was based primarily on the study of gross anatomy and pathology. Physiology as a science did not exist. One man could still learn everything that was known about medicine, and Moreau attempted to do just that. Psychiatry had hardly emerged as a specialty.

When he was 20 years old he went to Paris to complete his studies and earned his degree of doctor of medicine. At 22 he applied for the position of assistant physician in the largest mental hospital in France, just outside Paris; L'Asile d'Aliénés de Charenton. The hospital director was Jean Etienne Esquirol, one of the first physicians to consider mental patients as sick people who should not be shackled, and should be treated with care, concern, and kindness. Moreau learned a great deal from his master and his associates, including Pinel and Aubert.

At that time a prolonged voyage to a distant and exotic land in the company of a psychiatrist was a customary treatment for mental illness among the wealthy people of France and Europe. Moreau accompanied patients to Egypt and the Orient. During his trips he observed that the Arabs used hashish as an intoxicant as commonly as the Turks or Chinese used opium and the Europeans used alcoholic beverages. Moreau was impressed by the unusual effects of this substance on the mind and on the behavior of Oriental people. Back in Paris in 1840, the year of Esquirol's death, Moreau was appointed physician of the Bicêtre mental hospital. He decided then to experiment systematically with hashish and began the first recorded clinical experiments on the psychopharmacology of *Cannabis* extracts.

Moreau seems to have prepared his own hashish. He used a crude preparation, the fat-extracted material which the Arabs call "dawamese."

> The flowering tops of the plant are boiled in water to which fresh butter has been added. When this concoction has been reduced by evaporation to a syrupy liquid, it

---

[1] condensed from "J.J. Moreau, Hashish and Mental Illness". H. Peters and G.G. Nahas, Editors, Raven Press, New York, 1972.

is strained through a cloth. One thus obtains a butter of greenish color which contains the active ingredient. This extract is never absorbed in its pure form because of its obnoxious and nauseous odor. It is sweetened with sugar and flavored with scented fruit or flower extracts. When freshly prepared the finished product is not too un-pleasant to the taste, but it becomes rancid.

"However," according to Moreau, "this extract keeps its intoxicating properties for as long as ten years."

Moreau might have prepared his own extracts in order to avoid the many additives (ginger, cinnamon, cloves, and powder of cantharides) added by the Arabs. How-ever, he had to start with imported plants of *Cannabis*. His attempts, with his friend Aubert, to grow the plant near Bicêtre yielded a variant, which contained minimal active ingredient, insufficient to produce intoxication.

Moreau claims that one has to ingest an amount of his hashish preparation equal to a walnut (30 g) to obtain any results. Because such a large dose was necessary, Moreau's extract could not contain more than 0.5% delta-9-THC: In 30 g this would amount to 150 mg of the drug, a very large dose. With one-half or one-quarter of this dose, says Moreau, "... one will feel happy and gay, and one might have a few fits of uncontrollable laughter." But it is only with much higher dosage (which Moreau and his pupils ingested) that one obtains the profound effects designated in the Orient by the word "fantasia."

Moreau and his pupils went about their experimentation in a very systematic way. They took many different doses of hashish. The pharmacist of the hospital, after a specially high dosage (3 doses, or about 90 g), presented for 3 days a psychotic episode, with hallucinations, incoherence, and great agitation. All the experimenters took notes during their intoxications, giving detailed accounts of all their symptoms. They observed each other carefully, and exchanged and discussed their impressions, feelings, and thoughts during and after hashish intoxication.

As a result of this experimentation in the clinical pharmacology of hashish, Moreau published in 1845 a book, *Hashish and Mental Alienation*. Alienation was the common medical term to describe mental illness or madness. No better account of the acute effects of *Cannabis* intoxication on mental function has been given since Moreau's own descriptions, which will be summarized here and which earned the author one of the first prizes awarded by the National Academy of Medicine.

"Hashish Intoxication and Mental Illness"

Moreau candidly describes the purpose of his experimentation with hashish.

Most psychiatrists have carefully described the infinite variety of symptoms of the many patients with whom they have lived, but I do not know one who, in speaking of madness, has given us the benefit of his personal experience, or described it from the point of view of his own perceptions and sensations. Only curiosity originally led me to experiment upon myself with hashish. Later, it was difficult to forget the exciting memory of some of the effects I owed to this drug. But, from the very outset, I was motivated by another reason: I had seen in hashish, or more specifically in its effect upon the mental faculties, a powerful and unique method of exploring the genesis of mental illness.

Madness most frequently explodes without the afflicted being warned by any appreciable impairment of the organism, and his physician is unable to relate it to any physical malfunction. This is a first point of similarity between the effects of hashish and mental illness. The cause is obvious, but the origin remains unknown. This is what most often happens when that cause acts directly on the organ of the mind. We shall see that the manifestations of hashish intoxication are completely analogous to those reported by mental patients. Mental patients and eaters of hashish express themselves similarly when they want to convey their experiences; it seems that both groups have been under the same morbid influence.

I have compared the principal characteristics observed in mental illness to the symptoms caused in me by hashish intoxication. The insights provided by my own study gave me a better understanding of mental illness.

Moreau next lists the eight cardinal symptoms which are either observed during hashish intoxication or reported by mental patients. They are, in order of increasing mental disorganization:

Unexplainable feeling of bliss, happiness
Excitement; dissociation of ideas
Errors of time and space appreciation
Development of the sense of hearing; the influence of music
Fixed ideas (delusions)
Damage to the emotions
Irresistible impulses
Illusions; hallucinations

Feeling of Happiness

It is a feeling of physical and mental comfort, of inner satisfaction, of intimate joy; that you seek mainly to understand or analyze that for which you cannot find the cause.

It is really happiness that hashish gives, and by that I mean totally mental joys, not at all sensual as one might be tempted to believe. It is a very curious fact from which one can draw strange conclusions: All joy, all contentment, even though their cause is strictly mental, and our most spiritual, idealistic enjoyments could well be in reality only purely physical sensations, developed in the core of our brain exactly like those produced by hashish.

This is the first hint ever made at the existence of brain reward mechanisms centered in the limbic area. They were described by Heath one hundred years later (1954); they are chemically regulated and may be triggered by dependence-producing drugs.

Freud, nearly a century later (1931), also suggested a similar mechanism of action for "intoxicants," which when present in the blood or tissues directly cause us pleasurable sensations. "The symptoms we have just described also occur frequently at the beginning of mental illness: They are the feelings of happiness, of intimate joy in which the sick find so much hope, so much confidence in the future and which, alas, are only precursory symptoms of more violent insanity."

Moreau gives several examples of mental patients who in the course of their illness experienced great joy with no apparent reason. It was a young woman of fine, observing mind, convalescing from "a mania" secondary to childbirth who told Moreau: "In the night, I woke up with a feeling of comfort I cannot describe. I felt happy as I never had. My happiness, my joy, were overflowing and I felt a need to share them with my surroundings. I waited impatiently for daylight to announce the good news. I was wild with delight; I wanted to embrace everybody, even my servants."

Excitement: Dissociation of Ideas

One of the first noticeable effects of hashish is the gradual weakening of the power that we have to orient our thoughts as we wish. Imperceptibly, we feel ourselves overwhelmed by strange ideas unrelated to those on which we want to concentrate.

The action of hashish weakens the will—the mental power that rules ideas and connects them together. Memory and imagination then predominate.

So long as the disorder has not gone beyond certain bounds, one readily recognizes the mistake in which one is temporarily involved; there occurs an uninterrupted succession of false ideas and true ideas, of dreams and of realities, which constitute a sort of composite state of madness and reason and make a person seem to be mad and rational at the same time.

Moreau compares the dream state of hashish intoxication with the twilight of consciousness of the person who, while still asleep, is aware that he is dreaming and wishes to prolong his dream. Superimposed on the dream state and the free flowing of the stream of consciousness, there is a state of emotional lability and an exacerbation of sensory perceptions.

The train of our ideas can be broken by the slightest thing; we undergo the most contradictory influences. We turn in all directions. By a word, a gesture, our thoughts can be directed in succession to a multitude of diverse subjects with a speed and yet with a clearness that is marvelous. A deep feeling of pride takes hold of you with the growing exaltation of your faculties which you feel are growing in energy and power.

Moreau concludes that this aspect of *Cannabis* intoxication, with its composite mental excitement where "fragments" of the imagination constantly interact with enhanced sensory perception, recalls the symptoms of "manic" madness in all its details.

Under the influence of hashish, the mind can fall into the strangest errors concerning time and space. Time seems at first to drag with a slowness that exasperates. Minutes become hours, hours, days. Soon, with more and more exaggeration, all precise ideas of the duration of time escape us, the past and the present are merged. The speed with which our thoughts follow one another and the resulting dream state explains this phenomenon. Time seems longer than when it is measured by terrestrial clocks because the actions or the facts contained in an interval of time, by virtue of their intensity, extend the limits of this interval.

The phenomenon we have just described may be compared with certain extravagant ideas that are sometimes encountered in psychotics. Some of them believe they are a hundred, a thousand years old. The young women I mentioned earlier believed in the first days of her manic illness that she had no age. She imagined that she had lived in all the historical periods which she could remember. "I accused those who surrounded me of stealing from me the measure of time; for me, it no longer exists. I told them. My days and nights passed in an instant too rapidly for me to be able to carry out all the vast plans that filled my mind. I refused to recognize my mother for the reason that I could not have a mother younger than I was."

The action of hashish could not cause impressions as distorted as those I have just indicated. With the awareness of himself, one easily recognizes the illusion by which psychotics are naturally deceived and one avoids drawing, as they do, extravagant conclusions.

Development of the Sense of Hearing: The Influence of Music

Moreau first recalls, despite their poetic exaggeration, the words that Gautier used to describe this phenomenon: "My hearing was fantastically sharpened. I heard the sound of colors—green, red, blue and yellow sounds in perfectly distinct waves.... An overturned glass echoed through me like thunder..."

"I have observed these effects in several people," adds Moreau. "I have witnessed their cries of joy, their songs, their tears and their lament, their deep depression or their foolish mirth, depending on the harmonic mode in which sound reached their ears."

> Pleasant or unpleasant, happy or sad, the emotions that music creates are only comparable to those one feels in a dream. It is not enough to say that they are more vivid than those of the waking state. Their character is transformed, and it is only upon reaching a hallucinatory state that they assume their full strength and can induce real paroxysms of pleasure or pain. At that moment, the immediate, direct action of the harmonics and the actual auditory sensations are combined with the most varied and fiery emotions which result from the assocations of ideas created by the combination of sounds.
>
> One day, I had taken a strong dose of hashish. When I had reached a sufficiently high state of intoxication, in order to subdue the mettle of my ideas and feelings by imparting to them a single direction, I begged a young lady artist to sit at the piano to play a sad and melancholic tune. She chose a waltz by Weber. From the first notes of this tune, so deeply imprinted with sorrow, I felt a chill go through my entire body. The waltz evoked in me only sad thoughts and distressing memories. The faces of several people surrounding me reflected the sinister mood of my imagination. A sadness, a somber melancholy, a painful anxiety overtook me. The prayer from the opera *Moses* gradually restored calm in my soul. I had that physical and mental feeling of comfort that one experiences upon waking from a bad dream or that one enjoys at the end of a bout of fever.
>
> I had not been deceived for a moment of my illusions, although these illusions had affected me like reality itself.

Moreau explains the causes of the rapture produced by music during hashish intoxication. There is first a physical, "organic," basis to this condition, and *Cannabis* must stimulate the organs associated with hearing. "The overstimulation that hashish causes in the whole nervous system is felt most particularly in the portion of this system concerned with the perception of sounds. Hearing acquires an unbelievable sensitivity."

In the second place, Moreau stresses the power of music to stimulate the imagination and to recall old memories, which confers to harmony its social significance. "To feel music it must be understood, which is to say that sounds must be associated with familiar ideas. Music plays a minimal part in the emotions you feel, memory and imagination do most of the recollection." Stimulation of imagination by hashish also explains the influence of music on the hashish experience. "Memories of mourning and death ally themselves immediately with sad songs, happy thoughts with gay songs, religious memories with religious songs, and these thoughts and memories exert an almost unlimited influence upon one's judgment."

"When conscious thinking is obliterated by hashish, the mind abandons itself to impressons which are no more considered in perspective and assume a greatly exaggerated importance."

Fixed Ideas (Delusions)

"This 'intellectual lesion,' so frequent in mental illness, is also caused by hashish, but only when intoxication has progressed to an advanced stage, a stage which one would seldomly reach deliberately."

Moreau experienced delusions only once, when he was first experimenting with the drug and might have absorbed too large a dose. The idea occurred to him that he had been poisoned by his colleagues, and this idea progressively took hold of his mind, dominating all other thoughts in the most absolute manner. He even accused his colleague, Dr. Aubert-Roche, of being an assassin. The denials he received only strengthened his conviction, and he could not recognize the absurdity of his thought. Another delusion, much more extravagant, followed: He was dead, his soul had left his body, and he was on the verge of being buried.

> In the case of mental illness, delusions are essentially characterized by their total and exclusive domination over the mind, as if they had absorbed the individual's personality. Delusions caused by hashish are similar. They can only exist with their distinctive characteristics when the consciousness and the ego are impaired and involved in the overall disintegration of the mind.
>
> We all have fantastic ideas crossing our mind, of power, love, wealth, immortality; in the waking state—in our state of reflective power and complete independence, of self-power—we consider all of these fantasies playing in our mind as if they were in some way alien to us. The slightest impulse of our will makes them change endlessly. Like images from a kaleidoscope moved by the hand, we clear them away without any effort.
>
> If, as a result of the action of hashish, this intellectual capacity weakens and disappears temporarily, the fantasy, that had only crossed our mind before, is transformed into a conviction, a fixed idea, because our judgment, directed by inner consciousness, cannot combat it, accuse it of error and discard it. With hashish, unless intoxication is excessive, the delusions are very short-lived.

Moreau attributes the appearance and fixation of delusions in the mind to a primary "excitement" of brain function which he "wishes eagerly to explain clearly." His insight leads him to define the cause of delusions in terms which modern psychopharmacologists would not reject. This phenomenon of mental excitement "which precedes delusion I would willingly call a dissolution, *a molecular disintegration of intelligence*, if I dared explain it as I feel it. A delusion is the result of this disintegration."

When Moreau examines the circumstances under which delusions develop among mental patients, he believes that these delusions are preceded by "a confusion of all the elements of intelligence and a general disorder of the mind."

I call attention to the observations of physicians who have considerable experience with psychotics. When their patients were in a condition to understand them, the physicians often asked them: "How did your illness begin? How could you put in your head such absurd, extravagant ideas?" Answer: "I became so sad, I was so sick that I was completely confused; I was no longer normal; I had lost my bearings; my ideas were all upside down; I did not know what I was saying or doing, and I imagined that..." This is the start of the delusion which will later dominate all thought processes and will survive the general confusion, the upheaval of the faculties of the mind. Delusions do not always rapidly overwhelm the intelligence of mental patients. Sometimes delusions will have to struggle with the inner consciousness: this is the moment of uncertainty, of indecision, of anxiety, of extreme flight of ideas. These delusions follow patterns of waning and waxing, disappearing and reappearing with still greater intensity.

Subjective observation by the inner consciousness of the mental processes in hashish intoxication has also enabled me to establish that dissociation of ideas and the resulting dream state were the primary source of delusions. Delusions can only be the result of a general upheaval of our mental faculties. A delusion is not an error in thinking. A psychotic does not make a mistake. He functions in an intellectual sphere essentially different from our own. He has a conviction against which neither the reason of someone else, nor his own, can prevail; no more than any reasoning, any thought of the waking state can rectify the reasoning and thinking of the dream state. The same difference exists between the psychotic and the sane man as between the man who dreams and the man who is awake.

Delusions are only separated parts, temporary manifestations of a dream state that extends into the waking state.

By a systematic analysis into the origin of delusions as they appear under the influence of hashish or in the course of mental illness, Moreau became convinced of the primary organic nature of mental disturbances.

Disturbance of the Emotions and "Swinging Moods"

Moreau first describes, by using his own "subjective awareness," the labile emotional state and "swinging moods" associated with *Cannabis* intoxication.

With hashish, the emotions display the same degree of overexcitement as the intellectual faculties. They have the mobility and also the despotism of the ideas. The more one feels incapable of directing his thoughts, the more one loses the power to resist the emotions they create. The violence of these emotions is boundless when the disorder of the intellect has reached the point of incoherence.

The emotional instability induced by hashish explains the importance of set and setting on the drug-induced experience.

Depending on the circumstances, on the objects that you see, and the words that you hear, you will experience the most vivid feelings of happiness or sorrow, the most contradictory passions with unusual violence. If something frightens us, we are soon assailed by fears, unexplainable anxieties which cast a dark shroud on all our surroundings. One day in the midst of a strong hashish intoxication my ears were suddenly struck by the sounds of bells. This was hardly an hallucination but, being in a bad mood, I associated to this sound the idea that it was sounding for a funeral.

Those who use hashish in the Orient, when they want to abandon themselves to the rapture of hashish intoxication, exert an extreme care to eliminate anything that might turn their madness into depression, or might arouse anything other than pleasant and tender feelings. They take advantage of all the practices that the corrupt customs of the Orient place at their disposal. In the depths of their harem, surrounded by their women, under the spell of the music and the suggestive dancing performed by the dancing girls they relish the intoxicating hashish; and aided by their expectations, they are transported into the midst of the countless wonders that the Prophet has gathered in his paradise.

Moreau believes that this unbalanced emotional state cannot be related to any "main injury to the emotions," as many of his contemporaries thought. He attributes the emotional lability of hashish intoxication to the primary "pathological state" of the intellect created by the drug.

So long as the association of ideas is regular, so long as the uncontrollable speed of perception does not disturb the mind, emotions—gay or sad, hateful or kind—are in no ferment. They remain under the control of the will. The dream state which is the necessary consequence of hashish intoxication unleashes all of the power of emotions.

Irresistible Impulses

Impulses, those instinctive urges which build up in us almost without our knowledge, acquire under the influence of hashish an extraordinary driving power, one that is irresistible if the toxic action is very strong. Impulses are like passions; they draw their strength from the excitement, the mental disturbance that prevents regular, free association of ideas. They have their periods of waning and waxing and are only irresistable at the height of the excitement.

As Moreau recalls, "Seeing an open window in my room I got the idea that if I wanted I could throw myself from that window. Though I did not think I would commit such an act, I asked that the window be closed; I was afraid I might get the idea of jumping out the window. Deep down in my fear, I felt a growing impulse, and I had an intimate feeling that I might have followed it with a stronger 'excitement'."

Moreau observed very similar manifestations in mental patients:

The actions of psychotics are not always irresistible, although in no case can they be held responsible. Often these actions are only the logical consequence of false convictions. At other times, a particular disposition of his intelligence will thrust the patient, without any possible resistance, to all his impulses. He will then act without knowing or realizing what he is doing. These patients then act mechanically, as if they obeyed a dream, according to their favorite expression.

Illusions and Hallucinations

Illusions and hallucinations, two serious manifestations of mental illness, derive from the same basic mental change of cerebral excitement "which carries the seed of all mental pathology as the trunk of a tree, its branches, its leaves, and its flowers are contained in the grain."

Progressively, as "excitement" grows, our mind shuts itself off from external impressions to concentrate more and more on subjective ones; as this kind of metamorphosis takes place, we are drawn away from real life to be thrown into a world where the only reality is the one created by our memories and our imagination; progressively, one becomes the toy, first of simple illusions and then of true hallucinations which are like the remote sounds, the first lights, which are coming to us from an imaginary and fantastic world.

Illusions of Hashish Intoxication

When any sort of thing, alive or lifeless, strikes our sight, or when a sound, such as the song of a bird strikes our ears while the "excitement" is still weak, we feel that two distinct phenomena are occurring in our mind:

1. We have seen, we have heard, clearly and distinctly as we do in the waking state.

2. Then suddenly, as the result of certain similarities of which we may or may not be aware, the image of another object and the sensation of another sound are awakened within us. As a result of these intracerebral impressions, due to the action of memory and imagination, the mind pauses, shortly fusing the two sensations into a single one, covering over the real sensation with the imaginary one and projecting the latter upon the external object.

Therefore, an illusion is made of two component parts:

1. a sensory impression

2. a cerebral sensation (immediately deriving from the former) and which is totally due to the action of the imagination. There lies its essential psychic nature.

The external features of an illusion, with the numerous varied forms that it is likely to assume, will be necessarily borrowed from the particular nature of the mood and habitual thoughts of a person. It is understandable that images or ideas that have made the strongest impression on the mind are the first to be awakened or that the cerebral fibers that vibrate most often are more readily disturbed than others.

Visual illusions.   Moreau subjected visual illusions to analytic study.

We often see the face of a person totally unknown to us but which resembles the face of a familiar person. This resemblance, however slight, is enough to recall from our mind the memory of that person; this memory has all the vividness of the sensory impression, for the mind perceives it in the same manner as it perceives in the dream state.

From that moment, what we have seen with the eyes of the mind is put in the place of what we have seen with the eyes of the body. The creations of our imagination have taken the place of reality. And if all reflective thinking is denied to us by the violence of the cerebral disturbance due to hashish intoxication, the two sensations are fused into a single one, and the error is inevitable.

Moreau gives the following example of a visual illusion experienced under the effects of hashish:

When the visage of the old woman evoked in me the youngest and most attractive face, I felt perfectly that the subjective image which my overactive imagination had made me see in a dream was taking the place of the real image. I told myself there were two explanations to this illusion: (1) in taking hashish, I thought that all sensations had to be pleasant, that I must see everything as if it was beautiful; (2) the image of a pretty woman, by the admiration it creates and the emotion it causes, engraves itself spontaneously and deeply in our mind and consequently may be reproduced with great ease.

Auditory illusions.   According to Moreau, illusions of hearing are infrequently caused by hashish intoxication. Rarely are sounds distorted. However, sounds may be amplified, or give rise to emotions or thoughts which are colored by the prevailing mood. Moreau recalls going to a reception while he was under the influence of hashish.

> Hearing the bells ringing in a nearby church, I asked what it could be. Somebody answered, "Someone has probably died." Immediately, the last word echoed five or six times in my ears, as if each person in the room had, in turn, repeated it in a more and more dismal tone. At that time I was not completely sure of the innocuity of hashish. I feared I had taken too high a dose, which explains the sad nature of this illusion.

Moreau displays great insight as he analyzes the different component parts of auditory illusions, which, like visual ones, include:

> 1. A sensory impression, which is the true physical sensation.
> 2. A second sensation that follows the first immediately, a sensation totally in the mind and purely subjective.
> 3. A temporal error of the mind that confounds the two sensations—or, rather, forgets the first one and only retains the second, from which results the distorted perception.

This is the first definition known to this author of what psychologists today call "temporal disintegration or disorganization" (Melges et al., 1970a,b). The mental condition created by *Cannabis* intoxication is characterized by an impairment of immediate memory and a disintegration of sequential thought which is essential for the proper interpretation of sensory perception.

> As rapidly as these three phases of an illusion succeed each other, the mind perceives them distinctly, not at the very moment the manifestation occurs but immediately after; it is the impression of a dream that remains in the mind and which harkens back to it. If we were to give a faithful account of this impression we would say, "I dreamed that I heard..." a true statement that a psychotic changes to "*I heard,*" because he is deprived of his awareness and necessarily confuses the dream state with the waking state.

Illusions of general sensibility (awareness of body image). Illusions of general sensibility cannot be accounted for by the analytic study to which Moreau subjected auditory and visual ones.

> When I felt my body grow in size, inflate like a balloon, this sensation, however unusual, could not be distinguished from ordinary sensations. It was impossible to distinguish, as in the case of visual or auditory illusions, the actual sensation from the creation of the imagination. Illusions of general sensibility are the result of special sensory alterations as real as those that take place in normal sensations. Only the origin of these changes differs. Contrary to what happens in ordinary sensations, it is not in the peripheral parts of the organs or at the nerve endings that one first instinctively locates the source of the abnormal sensation which forms illusion. This sensation is concentrated entirely in the brain. It is in the central nervous system that it first develops, and then irradiates to the organs.
>
> Thus, when I felt my body inflate, I passed my hands over my body and I was not able to confirm this sensation. My hands informed me that my body had kept its usual size at the very moment that I felt it increase excessively. This contradiction between my touch and my subjective perception created the strangest ambiguous experience in me.

The sensory illusions reported by psychotics are similar, according to Moreau, to those produced by hashish, provided the dose absorbed is large enough to create a cerebral disturbance which obliterates awareness.

> Supposing a degree of intellectual disturbance sufficiently intense to obliterate in me all consciousness, as happens among psychotics, we may well imagine that my illusions might have become the point of departure of delusions similar to those seen in psychotics. I could have believed myself transformed into a bird or a balloon, afraid of being carried away by a gust of wind or punctured by the slightest shock or burned by a spark. If the intoxication was intense enough I might believe I had the power of rising into the air, of flying through space like a bird.
>
> It is exactly what happened to the young man who thought he had been changed into a steam engine piston.
>
> Several persons, after taking hashish, claimed that their brains started boiling and the tops of their head rhythmically rose and dropped as if lifted by sprays of steam. I myself felt a similar sensation. It is one of the sensations which most frightens those not yet accustomed to hashish.

Moreau distinguishes illusions from delusions or fixed ideas, which need not be triggered by sensory stimuli. He also distinguishes illusions from hallucinations. In illusions, the mind is still on the border of a dream state, and the imagination has not yet shaken off its dependence on external stimuli. The illusion is confined within certain limits like the activity of the senses to which it is related: Imagination acts within the limits of the sensory activity; visual and auditory impressions trigger the appearance of the dream and of the resulting illusion.

Hallucinations

By contrast, hallucinations include all the faculties of the mind; their only limits are those that nature has placed upon the activity of mental functions. As a result, all mental manifestations may be "hallucinated," not only those related to the perception of sounds or images. Hallucinations are the most striking manifestations of the cerebral disturbance or "excitement" which is produced by hashish intoxication or by mental illness in which the dream state takes precedence over the conscious state.

> The "hallucinator" hears his own thoughts as he sees and hears the creations of his imagination, as he is moved by the impressions he uncovers in his memory.
>
> As the action of hashish is more keenly felt, one passes imperceptibly from the real world into a fictitious world without losing consciousness of oneself. In a way there exists a sort of fusion between the dream state and the waking state. One dreams while awake.
>
> One evening I was in a salon, with close friends. We played some music which helped greatly to stimulate all of my faculties. There came a moment when all my thoughts and memories carried me back to the Orient. I spoke enthusiastically of the countries where I had traveled. As I was telling of my departure from Cairo to Upper Egypt, I suddenly stopped and shouted, "Here! Here! I am now hearing the song of the sailors rowing on the Nile: Al bedaoui! Al bedaoui!" I repeated this refrain as I had done in the past.
>
> It was an hallucination because I heard clearly and distinctly the songs that in the past had so often struck my ears.

This was the first time I experienced this manifestation in such a distinct and clear-cut manner and despite the disturbance of my thoughts, which were whirling in my mind, I applied myself to study this hallucination as precisely as I could.

This hallucination had only left in my mind the fleeting memory of a dream. I perceived only a difference in the degree of intensity between the memory and the impression itself.

I dreamed I heard: I believed that I was hearing with that full and complete conviction that one has when dreaming. Such was the invariable response that I made to myself when I sought to account for what I had felt. I responded with complete presence of mind to the questions that were asked concerning the songs I said I was hearing. Several times I had the occasion to assure myself that other hallucinators experienced impressions of this type. They invariably described these same impressions: "I dreamed that I saw; I dreamed that I heard; and yet I knew perfectly well where I was, who was around me. It was unbelievable."

Exactly the same thing happens in the dream state; it is our own thoughts that we express and to which we respond when we converse with other people in dreams. These are various impressions that we have formerly received which repeat themselves and which we associate and combine in all sorts of ways. All these manifestations derive essentially from the dream state, and one should not mistake them for the hearing of voices when the intellectual faculties have not undergone a change.

At times it happens that we are suddenly awakened by voices that we had been hearing in a dream. And we are as keenly impressed by these voices as if they were real, to the extent that we have to think for a few moments in order to convince ourselves that we were dreaming.

Let us suppose now that these manifestations recur intermittently and that in the intervals we are perfectly lucid. We will then have an idea of what happens in hashish intoxication and in other mental disturbances where hallucinations occur.

The Primary Organic Nature of Mental Illness

By a systematic analysis of the symptoms described by mental patients considered in the light of observations during his own hashish intoxication, Moreau became convinced of the primarily organic nature of mental illness. This conviction brings him to point out "the little differences between the folly of the 'demented' (schizophrenic) and that of the manics. The difference in the two conditions being more a matter of degree of the same basic lesion on the combined mental faculties." He tries to define this abnormal cerebral condition "as a general state of excitement of mental faculties, a rapid and confused agitation of ideas, a dynamic nervous lesion, a kind of oscillatory movement of nervous activity." This functional lesion cannot exist, according to Moreau, independently of the brain (Fig. 1). It is linked to a completely material and molecular change, however imperceptible its nature, "imperceptible as the changes that take place in the intimate texture of a rope upon which one applies vibrating motions of variable intensity."

The existence of this organic change is revealed to us with complete certainty by subjective observation; but how are we to discover its traces when life is gone from the organs, even supposing that this change can leave traces? Take apart, piece by piece, the keyboard that gives so discordant sounds when touched by inexperienced hands, and you will look in vain for the cause of the disharmony that offended your ears. Similarly in your search for the cause of a psychosis you would look in vain at the inner texture (histology) of the brain which will have presented a functional disturbance for some time.

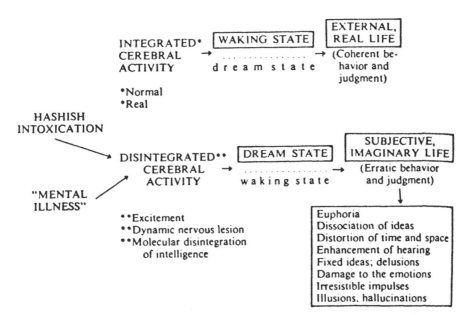

FIG. 1. A schematic representation of Moreau's concept of mental illness, showing its analogy with hashish intoxication.

And what are the causes of this primordial change in brain activity? Moreau mentions the small number of known reasons such as physical, mental, and pathological changes which abruptly shock mental faculties and exaggerate their action. In his opinion,

the psychological reasons that appear so frequently in the development of madness are only secondary and have an occasional value. They do not contain in themselves the power necessary to cause the disease. There is almost always a latent, more or less pronounced, organic predisposition. That is why one sees the most insignificant psychological reason trigger the explosion of the most violent disturbances. Perhaps this explains why, when we see "psychological reasons" triggering so easily mental disturbances, methods of the same nature are so ineffective in curing them.

And finally, Moreau recognizes our ignorance concerning the great number of other possible organic causes of mental illness. He hints at some which will be recognized more than a century later as probable enzymatic changes. "We completely ignore the greater number of causes, those which are hidden, and made (synthetized) in the secret depth of our tissues, which are discharged from one system into another and are transmitted by inheritance."

Moreau's generalizations created great controversies among his contemporaries, who defined insanity as a purely functional disturbance of intelligence and classified the manifestations of mental illness in a strictly symptomatic way according to the different categories of the mind: There were the lesions of the emotions, the lesions of the will, the lesions of the instincts.

To formulate his hypothesis of the primarily organic nature of mental illness, Moreau utilized to the fullest possible extent all tools available to him in 1840: *Cannabis*, the only drug known at the time to profoundly disturb mental function; his own extraordinary insight; and his keen sense of clinical observation. Now, some 140 years later, in spite of all the many investigative tools in neurophysiology, neurochemistry, and psychopharmacology at their disposal, psychiatrists are still pondering the basic nature of mental illness and the validity of Moreau's theory.

And no one is ready to answer Moreau's following query: "What is the lesion of the brain, what is the disposition of cerebral molecules which may be correlated with the false convictions, the erroneous ideas which are common to all of us, from the most learned to the most ignorant?"

Moreau and the "Club des Hachischins"

For Moreau, therefore, the hashish experience was a way to gain insight into mental disease. He advised some of his pupils and friends to share in this extraordinary psychological and emotional experience. One of them was a writer of considerable talent, Théophile Gautier, a member of the Romantic iconoclastic coterie which also included Charles Baudelaire and Alexandre Dumas, which became "Le Club des Hachischins" or "The Club of the Hashish Eaters". Gautier (1846) described his experience at the Hotel Pimodan when, in the company of his literary friends, he absorbed during a lavish meal a potent hashish extract so that he could "taste the joys of Mohammed's heaven". Gautier reveled in the ensuing intoxication, which he describes in great detail and with some poetic license in an article written for *La Revue des Deux Mondes*

> "Hallucination, that strange guest, had set up its dwelling place in me. It seemed that my body had dissolved and become transparent. I saw inside me the hashish Ihad eaten in the form of an emerald which radiated millions of tiny sparks. All around me I heard the shattering and crumbling of multicolored jewels. I still saw my comrades at times but as disfigured half plants half men. I writhed in my corner with laughter. One of the guests addressed me in Italian which hashish in its omnipotence made me hear in Spanish."

Transposition of tongues, though it is referred to in the Bible, is a sign of profound mental disorganization, reported during LSD intoxication. As the hashish intoxication continued, Gautier experienced the waning and waxing of his hallucinations.

> "For several minutes I found myself with all my composure, and quite amazed at what had happened. Then I fell again under the power of hashish. Millions of butterflies, with wings beating like fans, continuously swarmed in a faintly luminous atmosphere. I heard the sounds of colors: green, red, blue and yellow sounds in

successive waves. An overturned glass echoed through me like thunder. My voice appeared so powerful that I dared not speak for fear of breaking the walls and bursting like a bomb. I became entirely disengaged from myself, absent from my body, that odious witness which accompanies you wherever you are. I experienced the particular discontinuous effect of hashish which takes you and leaves you –you mount to heaven and you fall back to earth without transition, as insanity has its moments of lucidity".

Charles Baudelaire (1858), another member of Le Club des Hachischins, also wrote of his experiences with the drug in *Les Paradis Artificiels*. He was fascinated by the sensory effects of hashish and described, as Gautier did, the interchange of sensory modalities: "Sounds have colors and colors are musical. The eyes pierce infinity and the ears perceive the most imperceptible sound in the midst of the sharpest noises". He experienced hallucinations: "External objects take on monstrous appearances and reveal themselves in forms hitherto unknown". Baudelaire also describes the more esoteric aspects of the hashish experience. He depicts "the oriental kif where tumultuous and whirling sensations are replaced by a calm and motionless beatitude, a glorious resignation". And finally comes the ultimate religious experience, the glorious merging of oneself with the cosmos: "You are a king unrecognized," says a voice inside him, "and have become the center of the universe. Everything on earth has been created for me, for me. I have become God." Soon this storm of pride changes to restful beatitude and the universality of man is announced colorfully in a yellow dawn. However, Baudelaire became disenchanted about hashish and gave it up, claiming that, "like all solitary pleasures, it makes the individual useless to men and the society superfluous for the individual. Hashish never reveals to the individual more than he is himself. Moreover, there is a fatal danger in such habits. One who has recourse to poison in order to think, will soon be unable to think without taking poison."

EFFECTS OF THC ON BRAIN AND SOCIAL ORGANIZATION OF ANTS[1]

Peter Waser.

Institute of Pharmacology, University of Zürich, Switzerland.

ABSTRACT

After a single dose of 3H-LSD or THC (100 µg/ml), the maximum brain concentration of LSD reached 150 pg after 12 hours and that of THC 800 pg after 36 hours.

Ants fed sugar water containing 100 µg/ml of LSD or 1 mg/ml of THC (the maximum which could be solubilized in water-tween) presented impairment of social behavior. Behavioral reactions or the performance of individual ants were likewise altered.

KEY WORDS

LSD, THC, Ants, Brain concentration, Social behavior.

---

[1] Reprint from "Advances in Pharmacology and Therapeutics, vol.8, Pergamon Press, Oxford and New York, 1979, pp.299-308.

In man, social factors such as group formations among children and adolescents, broken home situation, educational system and subculture, each factor separately or in combination, are determinants for the development of drug addiction. Group influences are also seen in behavioural manifestations of the drug effects themselves. Altogether, the role of human society in the manifestation of drug abuse is very important.

This is the main reason to test hallucinogenic substances in animals, living together in highly organized communities - namely in insect societies. We selected ants for these investigations, because they live in high number in a limited space and because their reduced flying capacity allows experiments under controlled laboratory conditions.

We wish to point out here, that human society and an ant colony are completely different in their structural organization. But as Edward O. Wilson (1971) mentioned in his prospect for an "unified sociobiology", there are some functional similarities between them. Both societies are characterized by an extensive division of labour, which needs a good coordination of activities. The members of cooperative groups have to communicate among one another, using signals perceptible by the existing sensory channels: Man has his unique language with its enormous potential and plasticity, transmitted through generations by learning. In contrast, the communication system of ants is genetically fixed and highly stereotyped.

The social behaviour of ants is well known and was summarized e.g. by Sudd (1967), Wilson (1971) and Dumpert (1978). However, pharmacological experiments on the behaviour of ants are scarce. Kostowski and his group (1965-1975) investigated the actions of many substances on the aggressive behaviour of ants against beetles.

We examined:
(1) the behavioural and toxic effects of lysergic acid diethylamide-tartrate (LSD) and delta-9-tetrahydrocannabinol (THC) in individual ants (Frischknecht and Waser, 1978b).
(2) We investigated the uptake of LSD and THC into the ants' brain (Frischknecht and Waser, 1978a).
(3) We established their effects on the social behaviour of ants, which will be mainly discussed today. (Frischknecht and Waser, 1980)

Ants

All our investigations were done with ants of the species Formica pratensis Retz. They were taken out of a colony 13 km south-east from Zurich and kept in a formicar equipped with their genuine nesting material. During this period in our laboratory, we made sugar-water (250 mg/ml sucrose) available to them.

Drugs: Application and Dosage

In all our experiments reported here, the drugs were administered to the ants orally in the food.

First, we evaluated in individual ants the suitable drug concentrations to be used in the food, according to the toxicities. LSD was mixed with sugar-water containing 250 mg/ml sucrose. A drug concentration of 100 µg/ml sugar-water seemed to be reasonable. A single uptake of this food caused typical LSD-effects in a good many ants. These ants often displayed fumbling movements with their forelegs in the air at each step. The head was kept elevated and the antennae were stretched out and agitated. In addition these ants often avoided the approach or touch with a probe instead of showing threat-posture and seizing. These drug effects eventually disappeared in almost all ants. Even in ants having free access to this diet for 10 days, the mortality rate was increased only slightly. Because of the low water solubility of THC, we suspended this drug in sugar-water containing 10 vol.% of

Tween-80 as an emulsifier.The uptake of a solution containing 1 mg/ml THC was very good. But subsequently, no distinct changes in the individual behaviour of ants could be observed, not even after repeated drug administrations. Because of increasing difficulties to dissolve the drug in higher concentrations, we used 1 mg/ml THC for further experiments.

Drug Levels in Ant Brains

A fundamental question for a centrally acting drug is its rate of uptake into the brain and the accumulation in the central nervous system. These mechanisms have not been examined before in ants. Similar investigations in other species, e.g. small laboratory mammals are not compatible, because of the morphological differences in the anatomy of the alimentary tract and of the blood-brain barrier.

At increasing time intervals from a single feeding with (^{3}H)LSD, or (^{3}H)THC, ants were taken for analysis of incorporated radioactivity. The brains were dissected and their radioactive content evaluated by means of a liquid scintillation counter. To obtain comparable results, all values were normalized to one brain and a drug concentration of 100 µg/ml food. The results of these experiments are shown in Fig. 1.

The (^{3}H)LSD content in the brain increased gradually during 12 hr after feeding to about 150 pg/brain. After 42 hr however, the drug content decreased to 1/3 of its maximum value. After feeding with (^{3}H)THC, the drug content in the brain increased steeply during 24 hr. The maxima, corresponding to about 800 pg THC per brain were not reached before 24 - 48 hr. Subsequently, the drug content decreased to about 1/3 only after 6 days.

Fig. 1

DRUG CONTENTS IN ANT BRAINS AT DIFFERENT TIMES AFTER A SINGLE INTAKE OF FOOD CONTAINING 100 µg/ml OF EITHER ^{3}H·LSD (Δ x̄, n = 10), OR ^{3}H·THC (♦ x̄ ± s.e.m., n = 6, or 8).

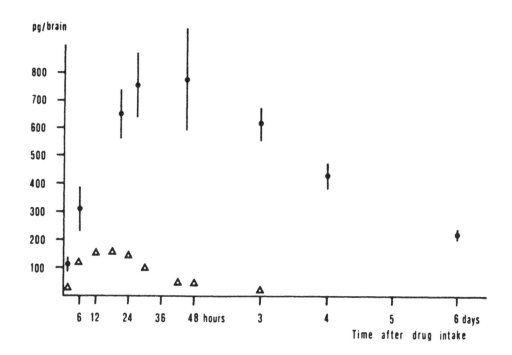

THC is taken up by the nervous system of ants to a larger extent than LSD and remains there for a longer time. This finding corresponds with results in mammals. The THC-molecules are lipophilic and cross the blood-brain barrier much more easily than LSD-molecules with their polar carboxy- and amino-groups. But it should be noted, that a strikingly long time is required for the peak levels of drug concentration in the ant's brain to be reached. This might be connected to some pecularities of the ants intestinal tract. .

Acute Drug Effects on Interactions of Ants

Taking account of these findings, we started our investigations on the social behaviour of ants under the influence of LSD and THC.In a first series of experiments,we always introduced <u>one drug-or control-fed ant to ten hungry nestmates</u> and noted its interactions during 10 min. The following patterns of social behaviour were observed: aggressive behaviour (threat, seizing and biting), social grooming, food-sharing behaviour (frequency and duration) and the number of social interactions per encounter. At the beginning, we used sugar-water alone and sugar-water containing 100 μg/ml LSD. At intervals of 2, 9, 18, 24 and 48 hr from feeding, always 10 drug and 10 control experiments were done.

After introducing a control-fed ant to ten hungry nestmates, aggressive elements were scarce and limited to the first encounters. Soon the test ants calmed and began to share food among the hungry nestmates. In experiments up to 24 hr after feeding, each food-sharing behaviour took generally more than 10 sec. Subsequently, the acceptors often groomed the donors. Later on, many acceptors proceeded in secondary food-sharing behaviour as donors.

Tab. 1

DIFFERENCES IN SOCIAL INTERACTIONS BETWEEN AN ANT FED WITH SUGAR-WATER CONTAINING 100 μg/ml LSD AND 10 HUNGRY NESTMATES. COMPARED TO SUGAR-WATER CONTROLS (n = 10)

Behavioural Elements	Time after food intake				
	2 hours	9 hours	18 hours	24 hours	48 hours
Aggressive Behaviour a) threat	n.s.	n.s.	n.s.	n.s.	n.s.
b) biting	n.s.	n.s.	n.s.	n.s.	n.s.
Social Grooming	n.s.	n.s.	n.s.	n.s.	n.s.
Food-Sharing a) exchange ≤ 10 sec.	↑↑	n.s.	n.s.	↑	n.s
b) exchange > 10 sec.	n.s.	n.s.	↓↓	↓↓↓	n.s.
Duration of b)	n.s.	↓↓↓	↓↓	↓↓↓	n.s.
Interactions per Meeting	n.s.	↓↓↓	↓	↓↓	↓

Significant Differences (Wilcoxon–Test) at P≤ 0.1. 0.05. 0.01: ↑ ↓. ↑↑ ↓↓. ↑↑↑ ↓↓↓.

Table 1 shows the effects of feeding the introduced test ant with sugar-water containing 100 µg/ml LSD, compared to sugar-water alone, on their social interactions with hungry nestmates: 2 hr after drug intake, food-sharing behaviour of short duration was increased. At 9,18 and 24 hr after drug feeding, the ants often moved irritatedly and showed the typical LSD-effects in locomotion mentioned before. In addition, these ants often ceased, or even avoided social touchings. This resulted at 18 and 24 hr after drug feeding in a decrease of food-sharing behaviour lasting more than 10 sec.Likewise the duration of food-sharing behaviour and the number of interactions per meeting were diminished at 9,18 and 24 hr after drug intake. 48 hr after LSD-feeding, the drug effects faded, showing just a slightly decreased number of interactions per meeting.

The period of most intense drug effects (9-24 hr after LSD-feeding) coincided with the period of highest LSD-content in the brain. The predominant alterations in social behaviour of worker ants included a diminished food-flow among nestmates and a reduced number of interactions per meeting. Aggressive behaviour towards nestmates was not changed. In an other series of experiments, LSD-fed ants likewise adopted queens from their own nest, from which they were isolated 2-3 months ago. At 9, 24 and 48 hr after intake of food containing 100 µg/ml LSD, the adoptions were even less dramatic compared to the corresponding controls. This point contrasts with findings of Floru, Ishay and Gitter (1969) in hornets, where LSD in a concentration of 10 µg/ml glucose solution not only impaired locomotion and decreased social activities, but also increased aggression towards members of the same colony and towards foreign wasps. Kostowski, Wysokowski and Tarchalska (1972), however, after injecting LSD (30 µg/g) into the abdominal cavity of ants, couldn't find any significant changes in intraspecific aggression, whereas attacks towards beetles were decreased at 1 and 3 hr following drug administration.

In an analogous series of experiments, where one fed worker ant was added to 10 hungry nestmates, we used THC in a concentration of 1 mg/ml, and the solvent, sugar-water with 10 vol.% of Tween-80 as a control. The occurrence of aggressive elements, social grooming, food-sharing behaviour, and the number of interactions per meeting were not altered at 2, 9, 24 and 48 hr after a single feeding with THC, compared to their corresponding controls. Likewise, at 2, 9, 24 and 48 hr after intake of food containing 1 mg/ml THC*all adoptions of queens in groups of worker ants were successful.

Acute Drug Effects on Group Formation

In a next series of experiments, we investigated the distribution and group formation of ants in bipartite formicars, holding 50 worker ants in either half. The ants in one compartment had free access to control food (i.e. sugar-water, or sugar-water with 10 vol.% Tween) for 1 hr, whereas those in the other half were fed with drugged food containing 100 µg/ml LSD, or 1 mg/ml THC. At 2, 9, 24 and 48 hr after food intake, photographs were taken. To evaluate the localization of ants, either compartment was subdivided into 16 squares.

With all 4 food qualities the ants preferred the 4 corner-squares compared to the 4 centre-squares. Significant differences between control- and drug-compartments were rare: 2 hr after THC intake the gathering of ants in the corner-squares was increased compared to the controls, and 9 hr after LSD-feeding the accumulation of ants in the centre-squares was lower than with sugar-water. The mean number of neighbours in the same square decreased from 9 - 24 hr after LSD-intake and at 48 hr it was reduced compared to the sugar-water controls. Altogether LSD diminished the tendency of group formation 1 and 2 days following drug intake, whereas THC had little influence on the distribution of worker ants in a formicar.

LSD is 4 to 500 times more potent a psychotomimetic than THC. The dose of the later drug administered in these experiments was only 10 times larger than that of LSD.

Effects of multiple Drug Application to large Groups of Ants

Our next experiments concerned multiple drug application to large
groups of ants, consisting of several hundreds of worker ants and
a few queens. These investigations were done in a special arrangement
(Fig. 2), where the nest was spatially separated from the feeding
place in the excursion area.Between these compartments two connecting
tubes existed. Because of their construction (wide entrances late-
rally and narrow exits at the cover) a well functioning one way
trafficflow took place. Photocells were fitted immediately before
the exits to count ants interrupting the light beam. Repeated
readings from the counters yielded activity diagrams between nest
and excursion area and vice versa.

Fig. 2

EXPERIMENTAL ARRANGEMENT FOR ACTIVITY-RECORDINGS BETWEEN NEST AND FEEDING PLACE

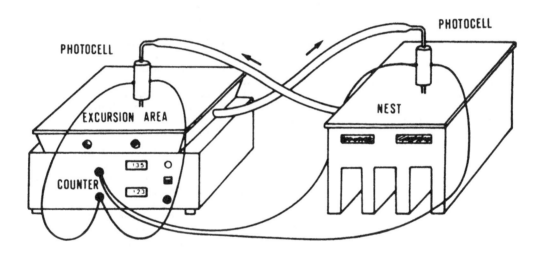

In a first experiment (Fig. 3), during period I of 4 weeks, the ants
had free access to sugar-water with 10 vol.% of Tween-80. Then, during period II of 3 weeks, the food contained 1 mg/ml THC in addition,
and during period III of 4 weeks,the ants were fed again with control
solution. Fresh food was supplied every morning at 9 o'clock, except
for weekends. Observations of behaviour and activity measurements
were obtained from 10 and 15 days of each test-period respectively.

The daily activity back to the nest from 8 a.m. - 5 p.m. was significantly increased during drug feeding, compared to the preceding and
succeeding control periods (Fig. 3). Similar differences were found
from readings on the photocell between nest and feeding place. The
diagram above illustrates the diurnal activity of ants obtained by
hourly readings from the counter of the photocell between the excursion area and the nest. To compare activities during control-
period I and III and during drug feeding (II), the total activity

<u>Fig 3</u>

FEEDING PROCEDURE	DAILY ACTIVITY (8 am to 5 pm) BETWEEN EXCURSION AREA AND NEST
I Sugar–water + 10 Vol.% Tween-80 (4 weeks, n = 10)	750 + 69 ⎫ ⎫
II do + 1 mg/ml THC (3 weeks, n = 15)	1055 + 66 ⎬*** ⎬ n.s.
III Sugar–water + 10 Vol.% Tween-80 (4 weeks, n = 15)	603 + 52 ⎭*** ⎭

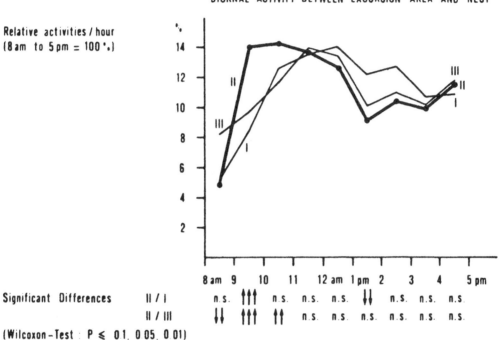

DIURNAL ACTIVITY BETWEEN EXCURSION AREA AND NEST

Relative activities / hour
(8 am to 5 pm = 100 %)

Significant Differences   II / I   n.s. ↑↑↑ n.s. n.s. n.s. ↓↓ n.s. n.s. n.s.

II / III   ↓↓ ↑↑↑ ↑↑ n.s. n.s. n.s. n.s. n.s. n.s.

(Wilcoxon–Test : P ≤ 0.1, 0.05, 0.01)

from 8 a.m. - 5 p.m. was taken as 100% and the relative activities per hour are shown. During either control period I and III, the relative activities/hour increased gradually from 8 o'clock until noon. With THC (II) a large increase of activity above control-levels was found for 1 and 2 hours respectively following the supply of fresh food. In the afternoon, the relative activities/hr were slightly decreased during drug feedings. From observations of the ants' movements it became clear, that foraging workers returned very promptly to the nest after uptake of THC-solution. However, these ants lacked distinct behavioural impairments, and the domestic workers nursed eggs, larvae and pupae even after a drug period of 4 weeks.

In a similar experiment (Fig. 4), during period I of 4 weeks, the
ants had free access to sugar-water. Afterwards, during period II of
2 weeks, the food contained 100 µg/ml LSD in addition. Behaviour and
activities were observed for 16 and 8 days respectively.

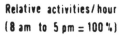

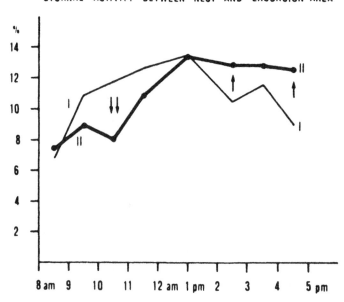

The daily total activity from 8 a.m. - 5 p.m. was not altered by
changing from sugar-water to drugged food. The distribution of the
relative activities/hr in the day-time however, showed a distinct
shift: During control feeding (I), the activity increased by steps
from 8 a.m. until noon, and in the afternoon the activity decreased.
In contrast, during LSD-feeding (II) the activity did not increase
before 11 o'clock, and no distinct decrease from maximum occurred in
the afternoon. Behavioural observations showed, that domestic workers
proceeded to care for the queens and the brood during LSD-feeding
period. Foraging workers however, in the excursion area, often
showed distinct drug effects. The locomotion of these ants exhibited
sporadic periods of unsteadiness and sometimes they displayed fumbling movements with their forelegs in the air at each step. Foraging workers with extremely filled gasters in part isolated themselves socially and didn't return to the nest. Altogether, a distinct
gradient of LSD-effects was found from foraging to domestic worker
ants. The following reasons might account for this observation:
Domestic workers don't procure their food themselves, but receive it
by food-sharing behaviour from foraging ants. Our experiments
suggest, that LSD influenced some foraging workers to return *no more*

to the nest, and LSD decreased food-sharing behaviour among worker ants. Either effect prevented an equal distribution of the drug among the whole colony. Nevertheless, feeding with sugar-water containing 100 µg/ml LSD has detrimental effects for the entire colony. The amount of food available to the queens, larvae and domestic workers decreased. This deficiency has to be made good by recruiting additional foraging workers, which is achieved only by depleting the number of nursing ants. This shift in group population stunts an ant colony, even if their existence is not immediately endangered.

References

Dumpert K. Das Sozialleben der Ameisen. Paul Parey, Berlin (1978).

Floru L., Ishay J. & Gitter S. The influence of psychotropic substances on hornet behaviour in colonies of Vespa orientalis (Hymenoptera). Psychopharmacology, 14, 323-341 (1969).

Frischknecht H.R. & Waser P.G. Actions of hallucinogens on ants (Formica pratensis) I. Brain levels of LSD and THC following oral administration. Gen.Pharmac., 9, 369-373 (1978a).

Frischknecht H.R. & Waser P.G. Actions of hallucinogens on ants (Formica pratensis) II. Effects of amphetamine, LSD and delta-9-tetrahydrocannabinol. Gen.Pharmac., 9, 375-380 (1978b).

Frischknecht H.R. and Waser P.G. Actions of hallucinogens on ants (Formica pratensis).III. Social behavior under the influence of LSD and Tetrahydrocannabinol. Gen.Pharmac., 11, 97-106 (1980).

Kostowski W. A note on the effects of some psychotropic drugs on the aggressive behaviour in the ant, Formica rufa. J.Pharm.Pharmac. 18, 747-749 (1966).

Kostowski W. A note on the effect of some cholinergic and anticholinergic drugs on the aggressive behaviour and spontaneous electrical activity of the central nervous system in the ant, Formica rufa. J.Pharm.Pharmac. 20, 381-384 (1968).

Kostowski W., Beck J. & Meszaros J. Drugs affecting the behaviour and spontaneous bioelectrical activity of the central nervous system in the ant, Formica rufa. J.Pharm.Pharmac. 17, 253-254 (1965).

Kostowski, W. Tarchalska B. & Wanchowicz B. Brain catecholamines, spontaneous bioelectrical activity and aggressive behaviour in ants (Formica rufa). Pharmac.Biochem.Behav. 3, 337-342 (1975).

Kostowski W.,Wysokowski J. & Tarchalska B. The effect of some drugs modifying brain 5-hydroxytryptamine on the aggressiveness and spontaneous bioelectrical activity of the central nervous system of the ant Formica rufa. Dissert.Pharm.Pharmac. 24, 233-240 (1972).

Sudd J.H. An Introduction to the Behaviour of Ants. Arnolds, London, (1967).

Wilson E.O. The Insect Societies. Belknap Press, Cambridge, Massachusetts (1971).

Drug use and the lognormal distribution

W.D.M.Paton

Department of Pharmacology, University of Oxford,
Oxford OX1 3QT, UK

ABSTRACT

The use of the lognormal distribution in the study of drugs of dependence has been debated, with some force, particularly in connection with alcohol and Ledermann's theory relating the incidence of alcoholism to the mean consumption of alcohol in a population (e.g.Davies,1977). It has also been discussed in relation to psychodynamic theories of the aetiology of alcoholism. The association of a particular mathematical pattern of drug use with theories which can have far-reaching implications for methods of social control presents a problem; for, depending on the attitude to such control, it can lead either to an over- or an under-estimate of the generality of the pattern itself, and of its mathematical meaning.This paper is intended to provide a general context, in which the pattern of drug use may be allowed to speak for itself.

KEYWORDS

Lognormal distribution, dispersion, alcohol, cannabis, illicit drugs

Lognormality is an old concept in pharmacology and bioassay. It underlies the procedure of probit analysis and bioassay that, following Trevan's work, was introduced by Gaddum and Bliss. It was found that the relation between the dose of a drug and the proportion of a population responding to it that reached some defined endpoint was well described on the basis that it was not sensitivity but the logarithm of the sensitivity that was normally distributed. (A similar fit could often be obtained, using the response of a tissue to a drug; and this contributed to the widespread use of log-dose response curves). This "lognormal" relationship meant that, with a suitable transformation of the response measure (the probit transformation) the presence of a lognormal distribution could be readily recognised by the linearity of a log-probit plot. Figure 1, from Wilson and Schild's textbook of 1961, illustrates the construction of such a plot in the context of clinical pharmacology. To apply it to patterns of drug use, it is only necessary that data are available over the whole range of rates of drug use for the whole of the population studied, even if the subdivisions are few. In the first step, the construction of a cumulative distribution, there is a choice whether to "cumulate" to the left or to the right. In the present case, there is interest in the proportion of the population with a given rate of drug use <u>or higher</u>, so that we wish to examine the right-hand tail of the distribution and cumulate to the left.Figure 2 (from Paton,1975) shows a number of such plots, for alcohol, cannabis, illegal drugs as a group, tobacco, tranquillizers, opiates, and "narcotics", from a range of countries, (France, Canada, U.S.A., and England),and a range of age-groups (schoolchildren, undergraduates, service personnel, and general public).

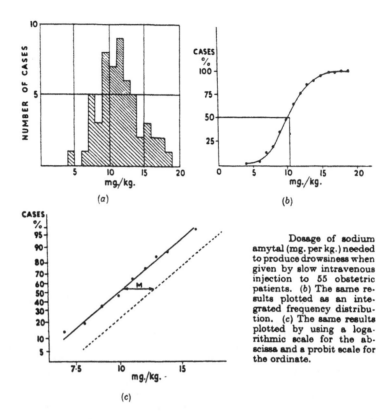

Dosage of sodium amytal (mg. per kg.) needed to produce drowsiness when given by slow intravenous injection to 55 obstetric patients. (*b*) The same results plotted as an integrated frequency distribution. (*c*) The same results plotted by using a logarithmic scale for the abscissa and a probit scale for the ordinate.

Figure 1. Construction of a log-probit plot, illustrated for amytal. (From Wilson & Schild, 1961).

Here the data are "cumulated" to the right. For analysis of drug use, cumulation to the left shows the proportion of high users.

Ledermann, probably the first to use this approach in considering the use of a drug, used a more complicated procedure. On the basis that there is a dose of alcohol that will kill virtually any subject, he thought it necessary to introduce a particular constraint on the plot, namely that the line should pass through a particular point at the right-hand extreme of the plot. This in fact resulted in some rather poor fits, and laid him open to the charge of arbitrariness in allocating a particular lethality to a particular dose. The simple plot outlined above, or a simple probit analysis, avoids this difficulty, but it is also vulnerable; for it suggests that there will be a measurable incidence of drug consumption, however high a level of consumption is proposed. The problem, however, is essentially that of extrapolation. It is not necessary to extrapolate; alternatively procedures may be followed appropriate for a "truncated" distribution (v. Aitchison and Brown, 1969).

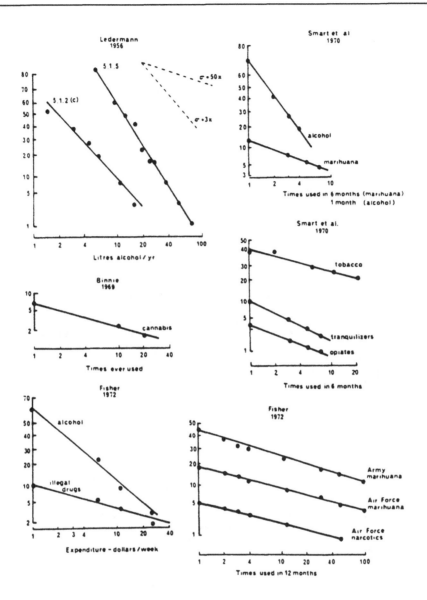

Figure 2. The lognormal distribution of drug use. (From Paton,1975).
    For alcohol,examples are from France (Ledermann, Tables 5.1.2c and
5.1.5),Canada(Smart et al.) and U.S.A. (Fisher).
    For other drugs, from U.S.Armed Services (Fisher), a British
University (Binnie) and Canada (Smart et. al.)
    The survey by Smart et al. was among schoolchildren.

Inspection of Figure 2,with other such data, suggests that there are two common patterns of drug use: that shown by alcohol, and that shown by other drugs. This is best characterised in terms of the dispersion of the log-normal distribution, i.e. its standard deviation. This will, of course, be in logarithmic units. That implies, for instance, that if one wishes to express the standard deviation directly in terms of drug use (not the logarithm of it) then its value gives a dose-ratio (that which, applied either side of the median rate of drug use, includes about two-thirds of the population). In figure 2, therefore, are also shown the slopes corresponding to standard deviations of 3x and 50x. More data of this type are needed, but from the surveys available (including those cited in Paton, 1975 and Evans et.al,1974) the following have been calculated:

(a)for alcohol, the values of the s.d. lay between 0.558 and 1.520: mean =1.186. The standard deviation of the s.d. was 0.298,and s.e.m.(n=12)=0.086.

(b)for cannabis consumption(including hashish), values of the s.d. lay between 2.42 and 6.12: mean=4.293;the standard deviation of the s.d. was 1.302;s.e.m. (n=13)=0.361.

(c)for other drugs generally, there were insufficient data for separate analysis; but for "glue", barbiturates, opiates, stimulants, tranquillizers, LSD, and tobacco, the values of the s.d. fell between 2.56 and 5.28.

It must be remembered that logarithmic transformations generally tend to minimise non-linearities; yet the consistency of the two patterns is notable, and clearly they are significantly different. Some care is needed, however, in interpretation. Question may also be raised about the reliability of survey data; these usually provide users' self-reported drug consumption, which is known to be unreliable. It must be noted, however, that some of Ledermann's data showing the typical pattern were, not reported use, but alcohol blood-levels.

The question of special proneness to drug use. Experience of the fully developed alcoholic or drug addict makes it easy to postulate a special group of addiction-prone personalities. But it is equally the case that there has been a consistent failure to find any physical or psychological characteristic usefully predictive of addiction before it has developed. Once developed, one may point to a special group of apparently vulnerable individuals,but that vulnerability looks to be a result of drug use, not a cause. The pattern of lognormality of drug use is consonant with this; for if there were a separate substantial group, more liable than the norm to take some drug, then there would be a discontinuity in the plot. Its linearity suggests,on the contrary, that there is a continuous spectrum of drug use; and if there is such a thing as predisposition to it, then that, too, is smoothly distributed.

It was mentioned above that logarithmic transformations tend to minimise non-linearities; so it should be noticed that if there were, for instance, four or five distinguishable groups in the general population, each with its own pattern of use, then an approximation to a simple linearity could well be observed; for the distribution of the sum of a number of normal distributions would not be easy to distinguish by these methods from normality. The practical issue, however, is probably that of whether there is a single substantial identifiable group, needing help. It seems that none can at present be recognized; and a prudent conclusion might be, "There, but for the grace of God, go we all"!

The possibility of predicting rates of use from mean use. Ledermann's hope was that accurate estimates of alcoholism could be deduced from general figures of consumption, facilitating social control. But it is sufficiently clear from Ledermann's own data for alcohol, and from the data for other drug use, that even if the pattern is consistent, there is sufficient variation in the dispersion parameter to make the limits of error around any estimate of alcoholism rates for a given mean or median rate of alcohol consumption far too wide to be usable. The special difficulty is that fairly small (say 20%) changes in dispersion parameter can lead to substantial (c.100%) changes in size of the right-hand tail.

At the same time, the consistency of the patterns does support the general view, that the higher the average rate of use, the higher will be the incidence of over-use; and that to avoid an unacceptable burden of alcoholism or drug abuse requires society as a whole to be temperate. The pattern is too common to allow one to set aside any link between general and individual use, even if that link cannot be quantified for legislation.

Lognormality as a pointer to mechanisms of dependence. It is not easy to resist building models of the addicting process, when the linear log-normal plot is encountered; and theories of infection, spread, or "contagion" can readily be formulated. But the normal distribution has a rather general origin; any quantity that is formed as the resultant of a number of small, independent, random forces, will tend to fluctuate according to the normal distribution. The pattern of distribution is evidence, not of mechanism, but that the determinants of drug use are numerous; i.e. that drug consumption is "multifactorial". That does not falsify any aetiological theory that also predicts lognormality; it merely makes it redundant.

The fact that the distribution is logarithmic implies that the variations that are occurring are not "arithmetical" but "proportionate". This seems reasonable to experience; for to a subject used to taking, say, one glass of sherry a week, a second glass would represent a significant change, yet such an increment would be negligible to a "bottle a day" man. It appears that drug use follows the patterns of a good deal of psychophysical data, as well as those of pharmacology generally.

The origin and significance of the differences in dispersions between drugs. The variability in rate of use of a drug could be envisaged as arising from many causes:- variations in income, in social milieu, in sensitivity to the drug, and from the dose-response relation of the drug itself. Further if a high degree of tolerance can develop, that could be a major factor; equally, if there were a wide span between the dose producing the first detectable effect and that producing the highest (but still tolerable) intensity of action, dispersion would be increased. Variations in metabolism would be another factor. It is not easy to see any very cogent reason for the difference between alcohol and other drugs in any of these directions, although it may be that the degree of tolerance achieved in ordinary use is rather higher with cannabis, opiates, and tobacco than is obtained with alcohol.

The principal implication of the more gradual slope with cannabis and other drugs, is that, for a given mean rate of use, there is a higher proportion of users at high rates. It could be argued that this brings in a greater danger of adverse secondary effects, as distinct from any consequences of the primary action of the drug .

Strategies for the control of over-use. A solution that might appeal to all
parties would be achieved if it were possible deliberately to narrow the
dispersion of drug-use, i.e. make the plot steeper; for that would reduce
the incidence of heavy users while possibly allowing the same mean level of
consumption. For the case of alcohol, that might appeal to tax-hungry
governments. But,in general, success on such lines looks difficult to
achieve. In Figure 2, there are a variety of rates of use, varying by a
factor of more than 100 for a given proportion of the population. Yet the
slope remains similar, with no consistent change. It is interesting to
examine examples from another lognormal distribution-that of incomes, in
Aitchison and Brown's monograph (1969). Figure 3 shows the consistency of
the slope (considerable but not perfect) over a range of occupations; and

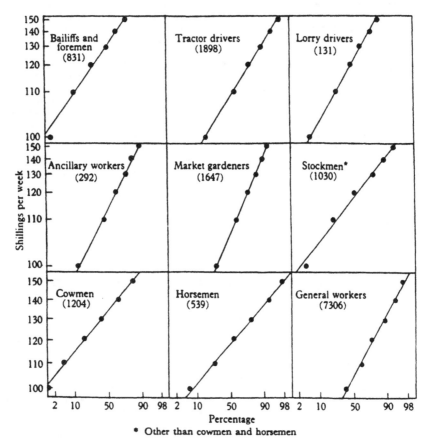

Figure 3. An example of a lognormal distribution is that of income. The
pattern is consistent for various income levels.(From Aitchison &
Brown,1969).

Figure 4 shows the consistency at the national level in the U.S.A. over
time.

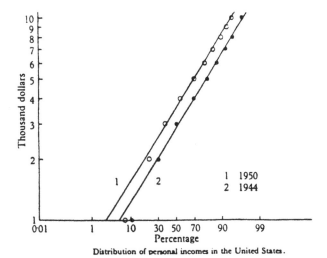

Distribution of personal incomes in the United States.

Figure 4. The lognormal pattern of distribution of income is consistent over time. (From Aitchison & Brown,1969)

On the other hand, as noted above, quite small changes in dispersion can have a considerable effect on the proportion of high users; so perhaps there is some scope for modulating use in this direction. The implications of the two possible approaches, changing the mean consumption or changing the dispersion, have not been studied fully, particularly as regards what it would imply for other regions of the distribution; and we are almost entirely lacking in specific knowledge of how selective change could be brought about.

CONCLUSION

The lognormal distribution appears to characterise the pattern of use of a wide range of drugs in a remarkably compact way. Lognormality in itself says little about the causes of drug use and abuse; but the difference in the pattern for alcohol from that for other drugs is suggestive, and analysis on these lines defines two main approaches. Further analysis and surveys to clarify how far the pattern of use can be modulated would be of value.

REFERENCES

Aitchison,J. & Brown,J.A.C. (1969) "The Lognormal Distribution". Cambridge University Press.
Davies,D.L. (1977) Ed. "The Ledermann Curve". Alcohol Education Centre, Denmark Hill, London.
Evans,M., Stevens,S., & Samuel,P.(1974) Brit.J.Addiction 69 231-236
Paton,W.D.M. (1975) In "Cannabis and Man",ed. P.H.Connell & N.Dorn, Churchill Livingstone
Wilson, A. & Schild, H.O. (1961) "Applied Pharmacology". Churchill, London.

THE DISTRIBUTION OF CONSUMPTION OF
ALCOHOL AND MARIHUANA

Gabriel Nahas[1,2], Renaud Trouvé[2] and Colette Latour[2]

[1]Columbia University, College of Physicians and Surgeons, New York and
[2]Laboratoire de Toxicologie Cellulaire, INSERM, Hôpital F. Widal, Paris.

ABSTRACT

Lederman reported in 1956 that the distribution of consumption of alcohol in different populations could be related as a first approximation to a normal logarithmic mode. In the present study, the distribution of alcohol consumption (in dl per week) in a group of 180 oil rig workers was computed using probability logarithmic scales. A cumulative distribution of rates was established by calculating the proportion in the population which used alcohol in increasing amounts. The weekly consumption of alcohol in this population (1983) follow a log normal distribution similar to that reported by Ledermann on a group of 93 consumers (the slope of the two curves are -0.75 and -0.73 respectively). The same method was used to analyze the cumulative frequency of marihuana use among American high school seniors during the period 1975 to 1981. When plotted on probability logarithmic scale, the fit of a straight line is excellent for frequency of marihuana use among those who had ever used marihuana. The percentage of daily marihuana smokers represent 12 to 17 % of the population of consumers. Other studies of life time prevalence of cannabis use in French or Canadian adolescents also display a log normal distribution. The weekly consumption of cannabis among the male population over age 15 in 3 Jamaican villages follow a log normal distribution with 64 % of the sample of 418 smokers consuming daily 4 to 10 grams a day of the drug. A survey of coca leaf chewing among Peruvian miners reports that over 90 % of them chew the leaf and 80 % are daily chewers consuming 40 to 60 grams of leaves (400 to 600 mg cocaine base). These data suggest the consumption of psychoactive euphoriant drugs more potent than alcohol (cannabis, cocaine) may be associated with a significantly higher rate of intake (daily intoxicating dose) when they are socially acceptable and freely available.

KEY WORDS

Log normal distribution, Alcohol, Marihuana, Cocaine, Ledermann.

A knowledge of the actual rates of consumption of dependance producing drugs in a population is most useful, in order to predict the medical or social risk associated with their use. Such information is fragmentary, except for alcohol.

The French mathematician Sully Ledermann was the first epidemiologist to study the frequency distribution of individual consumption of alcohol in France (1). His fundamental observation was that diverse frequencies did not occur randomly, but that their distribution could be related as a first approximation to the statistical law known as the normal logarithmic law : the distribution curve is sharply skewed : the average consumption does not divide the population into two equal groups, one consuming less than the average, the other more; in France, the distribution of consumption is roughly two-thirds below and one-third above (Fig. 1).

Ledermann also noted in studying different groups of consumers, that if the average consumption is 10 liters a year, the proportion of daily consumers of 200 ml or more ("excessive consumers") will be 1.5% of the population of users. If average consumption is 30 liters of pure alcohol per year the percentage of "excessive consumers" will be 7%. Within the range of average consumption considered, the proportion of excessive consumers "alcoholics" tends to increase geometrically rather than arithmetically. Ledermann concluded that there was co-variance between average "reasonable" consumption and heavy "unreasonable" consumption, associated with pathology. Therefore, average and heavy consumption are not independent from each other, and an increase or a decrease in average consumption should be associated with parallel changes in heavy consumption. Roughly, the percentage of heavy drinkers in a population of alcohol consumers remains constant and is related to the total number of consumers.

Since Ledermann's pioneering work, additional data on the distribution of alcohol consumption were analyzed for a number of populations in North America and Europe (2,3). These distributions are for the most part reasonably approximated by a log normal curve. In all cases, they exhibit a single mode and are markedly skewed to the right, with the top 10% of drinkers consuming 40 to 50 % of the total consumption. Other authors (4,5) expressed their conclusions in terms of tendencies rather than precise mathematical relationships, but still concluded that a substantial increase in mean consumption of alcohol is likely to be accompanied by an increased prevalence of heavy users. All of the conclusions strongly suggest "that a close link exists between the mean consumer's drinking level and the consumption level of the consumer at the 90 through 95th percentile of the distribute ."(6).

Other studies extended the Ledermann analysis to the consumption of dependence producing drug, such as marihuana, among Canadian and British university students, (7,8) and in a randomly selected sample of subjects from Kansas (9). In these later studies, based on questionnaires instead of amount of drug used, prevalence or frequency of intake within a year was computed. The authors reported that the frequency of drug intake "could be plotted on a normal logarithmic curve".

In the present study we have analyzed the distribution of alcohol

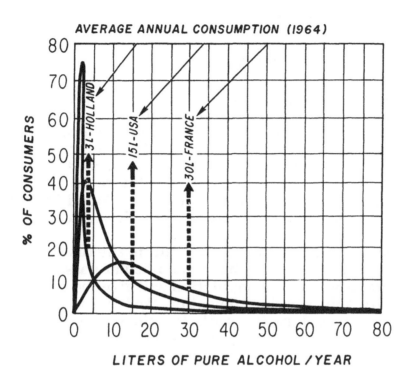

Figure 1. The log normal distribution of the consumption of alcohol in different populations (after Ledermann, 1956).

consumption in a sample of alcohol consumers recently reported (10) and compared it to the distribution computed earlier by Ledermann in 1956. We have also used the same method to analyze the frequency of use of marihuana reported by Jonhston (11) among American high school students and in other populations of consumers (12, 13).

METHODS

We have used the method suggested by Paton (14) and plotted the data on logarithmic-probability scales. Accumulative distribution of rates was established by calculating the proportion in a population which used alcohol or marihuana in increasing amounts or frequency within a certain time frame. These proportions of the population were then plotted against the rates of use. A probability scale (probit) was used on the ordinate for the cumulated percentage of the different groups of consumers and a logarithmic scale was used on the abcissa for the quantities consumed at a given rate or higher.

If a straight line runs across the plotted points the distribution is log normal. The intercept of the curve with a perpendicular passing through the point corresponding to 50 % of the population will give the median and the slope of the curve gives the standard deviation. We have selected dl of pure alcohol consumed per week as the units for alcohol consumption. For marijuana we have used the frequency of consumption, per week, month or year, or the life time prevalence.

RESULTS

The weekly consumption of alcohol in a group of 180 oil rig workers (table I) in the North Sea (10) (out of a total sample of 213) follows a log normal distribution (Fig. 2). This distribution is very similar to that reported by Ledermann in 1956 in a group of 93 consumers. In this case, the consumption of alcohol originally reported in liters per year was converted in dl per week. The slopes of these two curves are - 0.75 and - 0.73 respectively.

The cumulative frequency of marihuana use among American high school seniors during the period 1975 to 1981 was computed from the data reported by Jonhston et al. (Table II). When plotted on probability logarithmic scale, the fit of a straight line to the data is excellent for frequency of marihuana consumption among those who had ever used marihuana. Three of these plots are represented on Fig. 3. The percentage of daily users of marihuana represents 12 to 17% of the population of consumers.

Life time prevalence of hashish use among 109 French adolescents 14 to 18 years old reported by Kandel (12) and yearly prevalence among 3,723 Canadian students studied by Smart (13) also display a log normal distribution. In these two studies the highest frequency of use was reported in 20 % of the consumers (Fig.4). Finally, monthly consumption of marihuana reported by Gallegos (15) in a group of 82 urban youth from Lima (average age 20) is tabulated in Table III. In this group the frequency of daily use is 29 %.

TABLE I

DISTRIBUTION OF CONSUMPTION OF ALCOHOL IN A
POPULATION OF OIL RIG WORKERS.
(from Aiken and Lance, 1983)

Consumers (C) = 180
Non-consumers (nC) = 33
Total Sample (T) = 213

ml/week	No	% T	% C	Cumulated* No.	Cumulated* %
0 (nC)	33	15	-	-	-
7-143	56	27	31	180	100
150-285	43	20	24	124	69
293-427	26	12	14	81	45
434-570	24	11	13	55	31
577-856	17	8	9	31	17
836-1140	8	4	4	14	8
1147-1661	6	3	3	6	3

* Cumulated consumption represents the number and percentage of subjects consuming a given amount of alcohol or more (ml/week).

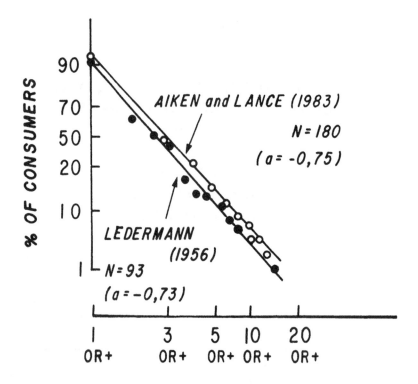

Figure 2. Probit-logarithmic plot of the distribution of consumption of alcohol in two populations : oil rig workers from Scotland (Aiken and Lance, 1983) and Paris residents (Ledermann, 1956). The linear plot indicates a log normal distribution.

TABLE II

PREVALENCE OF MARIHUANA USE AMONG
U.S. HIGH SCHOOL SENIORS *

Class	1975 (9,400)	1976 (15,400)	1977 (17,100)	1978 (17,800)	1979 (15,500)	1980 (15,900)	1981 (17,800)
% ever used	47.3	52.8	56.4	59.2	60.4	60.3	59.5
% at least once in last 12 months	40.0	44.5	47.5	50.2	50.8	48.8	46.1
% at least once in last 30 days	27.1	32.2	35.4	37.1	36.5	33.7	31.6
% daily in last 30 days	6.0	8.2	9.1	10.7	10.3	9.1	7.0

* After Johnston at al, 1981

CUMULATIVE FREQUENCY OF MARIHUANA USE IN THE
POPULATION OF USERS**

	1975	1976	1977	1978	1979	1980	1981
1 or more days in year	84.57	84.28	84.40	84.80	84.11	80.93	77.47
12 or more days in year	57.29	60.98	62.77	62.67	60.43	55.89	53.10
360 days in year	12.68	15.53	16.13	13.07	17.05	15.09	11.75

** Each percentage is the ration of the percentage of students who reported using marihuana at a given yearly frequency to the percentage who reported ever using marihuana. For example, 84.57 % for the class of 1975 is equal to 40.0 / 47.3.

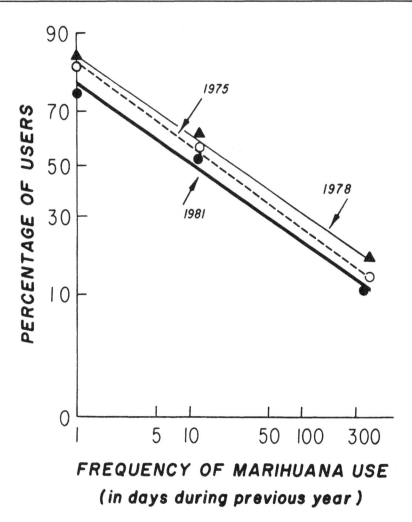

Figure 3. Probit-logarithmic plot of the distribution of frequency of daily
marihuana use among U.S. high school students (data from
Table I).

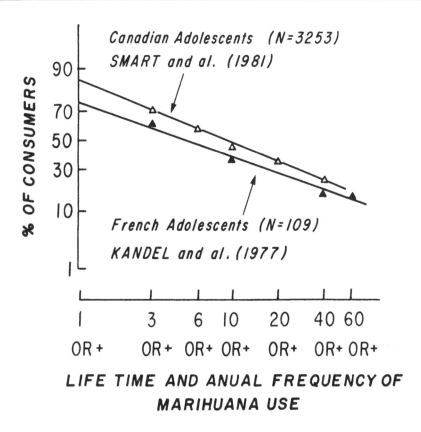

Figure 4. The probit-logarithmic plot of the distribution of the lifetime
(Kandel) and annual (Smart) frequency of marihuana use in
Canadian and French adolescents.

TABLE III

FREQUENCY OF UTILISATION OF MARIHUANA
AMONG 82 URBAN YOUTH IN LIMA, PERU
(From Gallegos, 1983)

	N	%	Cumulated Frequency	
Once a month or more	32	39	82	100 %
5 times a month or more	26	32	50	61 %
30 times a month or more	24	24	24	29 %

DISCUSSION

The two samples of consumers of alcohol analyzed in this study were reported by different authors in different populations, Scotland and France, and at 27 years interval. The fitted straight lines, based on a coarse grouping of weekly alcohol consumption in the two groups have similar slopes and run very close to each other. In both cases consumption of 7 dl of pure alcohol or more a week (a consumption which has been associated with alcohol intoxication) is observed among 7 to 8 % of the consumers. This percentage of "heavy" users of alcohol is comparable to that observed in the U.S.

Another survey (16) concludes : "Out of 394 American physicians who took the self administered alcoholism screening test, 12% were abstinent. Among the 350 who consumed alcohol, 8.6% were classified as "possible alcoholics"... This proportion was similar to that prevailing in a non-physician general medical patient population."

The relative constancy of the percentage of heavy consumers of alcohol in different populations might indicate that some proneness to alcohol abuse is present in certain individuals. However, the unimodal distribution of consumption does not permit one to define statistically two populations of "moderate" and heavy drinkers. If there were two distinct groups, a discontinuity in the distribution would appear. The log normal analysis does not eliminate the proneness to alcohol use, it suggests that such a feature is not limited to a sharply defined group but randomly distributed to the whole population. Such an interpretation would account for the fact that an increase in average consumption of alcohol is accompanied by an increase in heavy use. Lederman (1) did suggest that the log normal distribution of alcohol consumption did result from a snow-ball, recruiting effect which was a function of two factors, one biological related to the action of the drug on the individual and the other environmental related to the socio-cultural milieu. The log normal distribution of alcohol consumption would be another example of a nature-nurture interaction.

One would therefore expect that a similar log normal distribution might be observed with other dependence-producing drugs as was suggested by Paton (14). Consumption of these drugs are related to socio-cultural factors, and they also have a reinforcing effect on brain and behavior.

The frequency of marihuana use in the populations studied also follows a log normal distribution with a rather high incidence of daily or frequent intoxication among high school seniors and adolescents (12 to 29%) depending on its availability and social acceptance. The present data does not permit one to relate frequency of consumption to amount consumed and to perform a quantitative analysis of the distribution of consumption of marihuana similar to that of alcohol. However, frequency of consumption of marihuana is related to a similar frequency of the psychoactive, intoxicating effect after each exposure to the drug. There is also some data concerning the amounts of cannabis consumed in parts of Jamaica where the drug is freely available ad has wide social acceptability. The weekly consumption of "ganja" (cannabis) among the male population over the age of 15 in 3 Jamaican villages (17) indicate that 64 % of the sample of 418 smokers of ganja were heavy daily users

(Table IV). They consumed 4 or more "spliffs" a day which equals 4 to 10 gr of cannabis (ganja) equivalent to 60 to 120 mg delta-9-tetrahydrocannabinol (THC).

On the basis of the data presently available, it would appear that easy access and social acceptance of marihuana as prevailing in Jamaican rural villages are associated with a high proportion of heavy use (over 50% in the population of consumers). This proportion is significantly greater than that observed among consumers of alcohol. THC, the psychoactive substance in marihuana, has been reported to induce a pleasant euphoric experiences (18), and to interact with limbic structures (19) associated with pleasure reward (20). Such reinforcing psychopharmacological properties might account for its heavy use in a climate of social acceptability.

The consumption of other dependence producing drugs has not been systematically reported in a fashion which could be analysed according to the log normal distribution. However, may clinical observations indicate that the proportion of heavy users of opium and cocaine is very high among the consumers of these drugs. And a survey of coca leaf chewing among miners and farmers in Peru (21) reports that 96 and 92 % of them chewed the coca leaf and 88 and 82 % were daily chewers ("acullicadores"), consuming 40 to 60 grams of coca leaves (corresponding to 400 to 600 mg of cocaine base) (Table V). Blood levels exceeding 500 ng/ml of cocaine have been measured on these subjects, concentrations which produce psychoactive effects. This data would indicate that the more potent dependence producing psychoactive drugs are associated with a higher frequency of consumption. Such tentative conclusions should be confirmed by systematic epidemiologic studies of the distribution of consumption of dependence producing drugs in different populations. Selected well-studied, relatively small samples of a few hundred subjects reporting consumption over a period of one month in different populations should yield meaningful information.

These studies will provide a clearer definition of the magnitude of vulnerable groups : those with a consumption greater than some limit. The log normal distribution of consumption gives a concise description of an entire distribution because it may be defined by only two parameters, the mean and standard deviation. Given the value of those parameters, one may compare the results of different studies. Furthermore, the general effects of intervention methods may be quantitatively expressed. The log normal distribution also defines more clearly the choices confronting those who wish to control the use of dependence producing drugs and limit their spread: namely, either to steepen the slope of the k-line by reducing consumption at high rates, or to reduce the overall rate of use at all levels. So far the latter solution, though fraught with controversy, has been the only one which has been proven to be effective.

The authors acknowledge with thanks the invaluable assistance of Dr. Joseph Fleiss, Professor of Biostatistics, at the College of Physicians and Surgeons of Columbia University.

TABLE IV

FREQUENCY OF GANJA SMOKING IN 3 JAMAICAN VILLAGES IN MALE POPULATION OVER 15 (1971-1973)

(After Dreher M.C., 1983)

FREQUENCY	LEYBURN (S =87, NS =96, T =183)			CUMULATED*		BUCKLAND (S =149, NS =91, T =240)			CUMULATED*		DERFIELD (S =182, NS =100, T =282)			CUMULATED*		TOTALS (S =418, NS =287, T =705)			CUMULATED*	
	NO.	%T	%S	NO.	%	NO.	%T	%S	NO.	%	NO.	%T	%S	NO.	%	NO.	%T	%S	NO.	%
INFREQUENT	20	11	23	87	100	24	10	16	147	100	12	4	7	182	100	56	8	13	418	100
OCCASIONAL	22	12	25	67	77	44	18	30	125	83	30	11	16	170	93	96	14	23	362	86
HEAVY	45	25	52	45	52	84	34	54	81	54	140	50	77	140	77	266	38	64	266	64

No. = number of ganja smokers in each category

S = Ganja Smokers   NS = Non-Smokers   T = Total population sample

Infrequent = One or twice a month   Occasional = Once or twice a week   Heavy = 4 times a day or more

* "Cumulated frequency" represents the frequency of ganja smoking at or above a given rate (once or twice a month or more, once or twice a week or more, 4 times a day or more. Amount smoked each time = 3 to 5 grams of ganja equivalent to 60 to 100 mg delta-9-hydrocannabinol).

TABLE V

FREQUENCY OF COCA LEAF CHEWING
AMONG MALE BOLIVIAN FARMERS AND MINERS.

	Miners (N = 227)		Farmers (N = 2712)	
	% sample	% consumers	% sample	% consumers
Overall	96	100	92	100
Occasional	92	94	88	96
Daily ("Acullicadores")	88	92	82	89

REFERENCES

1. Ledermann S., Alcool, Alcoolisme et Alcoolisation. Presses Universitaires de France, Vol. 1 and Vol 2, 1956 and 1964.

2. De Lint J.E. and Schmidt W., Estimating the prevalence of alcoholism from consumption and mortality data. Quart.J.Stud.Alc., 1970, 31, 957-964.

3. Bruun K., Griffith E., Lumio M., Makela K., Pan L., Popham R.E., Room R., Schmidt W., Skog O.J., Sulkinen P. and Osterberg E., Alcohol Control Policies in Public Health Perspective, Rutgers University Center of Alcohol Studies, New Brunswick, N.J., 1975.

4. Skog O.G., The collectivity of drinking cultures: A c-theory of the distribution of alcohol consumption. National Institute for Alcohol Research., 1983, Oslo.

5. Popham R.E. and Schmidt W., Words and deeds: the validity of self-report data on alcohol consumption. Journal of Studies on Alcohol, 1979, 42(3):355-358. In the same volume: comments by J.de Lint, Merton M. Hyman, H.A. Mulford, J.L. Fitzgerald, and Henri Weschler.

6. Alcohol and Public Policy: Beyond the Shadow of Prohibition. Moore M.H. and Gerstein D.R., Editors, p.67. National Academy Press, Washington, D.C., 1981.

7. Smart R.G. and Whitehead P.C., The prevention of drug abuse by lowering per capita consumption: distribution of drug use in samples of Canadian adults and British university students. Bulletin on Narcotics, 1973, XXV, 49-55.

8. Smart R.G., The distribution of illicit drug use correlations between extent of use, heavy use and problems. Bulletin on Narcotics, 1978, XXX, 34-41.

9. McDermott D. and Scheurich. The log normal distribution in relation to the epidemiology of drug abuse. Bulletin on Narcotics, 1977, XXIX, 14-19.

10. Aiken G.J.M. and Lance C. Alcohol consumption by offshore oil-rig workers. U.S.N. Science News, 1983, 37:195-196.

11. Johnston L.D., Bachman J.G. and Malley P.M., Drugs and the Nation's High School Students. N.I.D.A., Rockeville, Md., 1982.

12. Kandel D., Adler I. and Sudit M., The epidemiology of adolescent drug use in France and Israel. Am.J.Public Health, 1981, 71:256-265.

13. Smart R.G., Goodstadt M.S. and Sheppard M.A., Preliminary Report of Alcohol and Other Drug Use Among Ontario Students in 1981. Alcoholism and Drug Addiction Research Foundation, Toronto, 1981.

14. Paton W.D.M, The uses and implications of the log-normal distribution of drug use. In Cannabis and Man: Psychological and Clinical Aspects and Patterns of Use. Churchill Livingstone, Connell and Dorn, Editors, 1975.

15. Gallegos M.F., The Problem of Drugs in Peru. Historical and Epidemiological Consultations Symposium on Alcoholism and Drug Dependence, Sao Paolo, 1982.

16. Niven R.G., Hurt R.D., Morse R.M., Swenson W.M., Alcoholism in physicians, Mayo Clin.Proc., 1984, 59:12-16.

17. Dreher M., Working Men and Ganja, p.34, I.S.H.I., Philadelphia, 1982.

18. Moreau L.(1844), Hashish and Mental Illness. Translated (p.27), Raven Press, New York, 1973.

19. Olds J. , Drives and Reinforcements. Raven Press, New York, 1977.

20. Heath R.G., Marihuana: Effects on deep and surface electro-encephalograms of man. Arch. Gen. Psychiat., 1972, 26:577-584.

21. Carter W.E., Parkerson P. and Mamani M., Traditional and changing patterns of coca use in Bolivia. In Cocaine, Proceedings of the Intra-American Seminar on Coca and Cocaine, Jeri F.R. Ed., Lima, Peru, 1980, 159-164.

FROM THE LONDON TIMES OF FRIDAY, JULY 24, 1992.

"25 years ago on the 24th of July 1967 a full page advertisement appeared in The London Times calling for a reform of the law against marijuana".

On the 25 of July 1992, a similar full page advertisement was published by the London Times.

The editors of this monograph thought that the reproduction of this new appeal might be informative.

THE TIMES, FRIDAY JULY 24, 1992

JUST SAY NOW

The signatories to this petition call upon the Home Secretary to recognise that the overwhelming weight of evidence demonstrates that the prohibition of cannabis has promoted criminality, conflict and more harm to the individual and society than its use ever has. On behalf of the citizen and the community we therefore call on him to introduce a programme of reform of the law which will include:

1. The abolition of the possession of cannabis as a criminal offence.

2. A thorough review to examine appropriate measures for the establishment of legal and properly regulated sources for the supply of cannabis.

.../...

THE TIMES, FRIDAY JULY 24, 1992 (.../...)

JUST SAY NOW

Douglas Adams
Richard Adams
Don Aitken
*Tariq Ali
William F Annesley
Lord Avebury
*David Bailey
Desmond Banks
Tony Banks MP
Edwin Belchamber
Tony Bennett
Grace Berger
Dr Joseph H Berke
*Humphry Berkeley
Sally Berriff
James Birch
Celia Birtwell
J H Blackham
Chris Blackwell
Revd Eric Blakebrough MBE
Alan Bleasdale
*Anthony Blond
Sir Hermann Bondi
*Derek Boshier
Joe Boyd
Philip Boye-Anawomah
Billy Bragg
Tony Brainsby
Trudi Braun
Dr Colin Brewer
Anthony Burton
Tony Burton
Bernard Carnell
Michael Cashman
Julie Christie
Margi Clarke
Anne L Clarke
Dr George Cohen
Nigel Coke
Dr Vernon Coleman
Gail Colson
Colin Consterdine
Caroline Coon
Frank Crichlow
Janet Daley
Bob Davenport
Rick Davey
Jeff Dexter
Alison Downie
Andrzej Dudzinski
Kirsty Dunlop
Brian Eno
Pamela Esterson

Exeter Drugs Project
H J Eysenck
Duggie Fields
Harry Fletcher
*Tony Garnett
Anthony George
Sara George
Terry Gilliam
Dave Gilmour
Bill Godber
Ray Giltrow LLB
Jane Goodsir
Jonathon Green
Timothy Greene
Sue Hall
*Richard Hamilton
John Hanson
Tim Harding
Derek Harper
Brigitte T Harris
Kevin Harris
Lee Harris
Jillian Harris
*Michael Hastings
Neil Henfrey
Anthony ('Antonil') Henman
Peter Herbert
*David Hockney
Howard Hodgkin
John 'Hoppy' Hopkins
*Jeremy Hornsby
*Michael Horovitz
Tony Howard
Roger Hutchinson
*Dr Sam Hutt
(Hank Wangford)
*Brian Inglis
Peter Jenner
Matthew Johnson
MBchB MRCGP
Jennifer Kane
James Kay
Ludovic Kennedy
Andrew Keogh
Andy Kershaw
*George Kiloh
Debbie Knight
Philip Knightley
Marek Kohn
Hanif Kureishi
Irma Kurtz
Nick Laird-Clowes
Fran Landesman
Paul Lashmar

Helen Laval
Roger Law
Ann Marie Legge
Rowley Leigh
David Leitch
Don Letts
Robert Lomas
David Longmoor
Neil Lyndon
John MacDougall
Caroline MacKechnie
Tim Malyon
Philip Manley-Reaves
Steve Mann
Michael Mansfield QC
Alan Marcuson
Howard Marks
Dr John Marks
Rita Marley
*Tom Maschler
Gaz Mayall
Scarlett MccGwire
William McIlroy
Fiona Mactaggart
Darin Marsh
Jonathon Meades
*George Melly
Paul Merton
Sue Miles
*Adrian Mitchell
P W R Monahan FRCS
Cllr Robin Moss
Edward Muhammed
Richard Neville
Dr Russell Newcombe
Suzette Newman
*Professor Nowell-Smith
David Offenbach
Steve O'Rourke
John Pearse
Professor Geoffrey Pearson
Gareth Peirce
Rupert Pennant-Rea
John Phillips
Martin Polden
Claire Powell
*Patrick Procktor
Andrew Puddephatt
Barbara Pukwana
Ramus
William Rankin
Mike Reed LLB
Ron Reid

David Reynolds
Danny Roche
Tom Robinson
Julian Rothenstein
Chris Salewicz
Craig Sams
Gregory Sams
Bill Sanderson
Carol Sarler
Jon Savage
Professor Wendy Savage
FRCOG
Eugene Schoenfeld MD
*Michael Schofield
W G & Jo Sno Serpliss
David L Shaw
Willy Slavin
Larry Smart
Pennie Smith
Arthur Smith
Barbara Smoker
*Tony Smythe
Cathy Snipper
Joyce Stanbridge
Lindi St-Clair
Sue Stapely
*Dr Anthony Storr
A J Taylor
Joyce Taylor
Michael Thomas
James Tighe RMN
Peter Till
Jim Tomlinson
Edward Totah
Geoff Travis
Wainwright & Cummins
Dr Tom Waller
†Nicolas Walter
Sandar Warshal
Peter Webb
Cass Wedd
George Weiss
Edward Welsh
Dr David Widgery
John Wilcock
Richard Williams
Mark Williams
Anthony Wilson
Christian Wolmar
Rowdy Yates
Professor Jock Young
Caroline Younger

*Original signatory   †Son of original signatory

.../...

THE TIMES, FRIDAY JULY 24, 1992 (.../...)

JUST SAY NOW

Cannabis smoking is a common feature of British life. The number of people estimated to have used cannabis in this country is now generally recognised to be in excess of 5 million. More people smoke cannabis than go to football matches, visit art galleries or go to church on Sunday. The people who use it are from all walks of life, all age groups, all social classes and all sections of the community. They do not fit any conventional stereotype — in fact the only thing they have in common is that they are all breaking the law. By participating in a recreational activity for which there is still no conclusive evidence to demonstrate that it causes any significant harm, they are branded criminal. The only time people who use cannabis should be subject to criminal law is when their drug use causes harm to others.

The argument that cannabis leads people inevitably to addiction to harder drugs has proved worthless. The only link of any kind lies with the law's insistence on bracketing cannabis with other drugs thereby seducing some who will use and enjoy it into the mistaken belief that other drugs are equally harmless.

Not only is the prohibition of cannabis a law which lacks justification and credibility it is also an affront to individuals' liberty and a constant threat to the welfare of significant sections of society. It is a law which has proved immoral in principle and unworkable in practice.

Reform of the law on the use of cannabis would involve an effective legal framework of controls and licensing. Reform of the law would ensure proper information and regulation around cannabis use in a way that already exists for many other substances and commodities. Reform of the law would help restore confidence and credibility in response to drug use. Reform of the law would show a responsibility and maturity which we have yet been unwilling to display.

In the light of overwhelming evidence that our present law is iniquitous and unjust we call upon the Government and citizens of the UK to abolish the prohibition of cannabis and legalise its use now.

.../...

THE TIMES, FRIDAY JULY 24, 1992 (.../...)

The Experts Say

"Having reviewed all the material available to us we find ourselves in agreement with the conclusions reached by the Indian Hemp Drugs Commission appointed by the Government of India (1893-1894) and the New York mayor's committee on marijuana (1944) that the long-term consumption of cannabis in moderation has no harmful effects."
- THE ADVISORY COMMITTEE ON DRUG DEPENDENCY (THE WOOTON COMMITTEE), 1988

"There is insufficient evidence to enable us to reach any incontestable conclusions as to the effects on the human body of the use of cannabis, but that much of the research undertaken so far has failed to demonstrate positive and significant harmful effects in man attributable solely to the use of cannabis." THE ADVISORY COUNCIL ON THE MISUSE OF DRUGS, REPORT OF THE EXPERT GROUP ON ON THE EFFECTS OF CANNABIS USE, HOME OFFICE, 1981.

"On any objective reckoning cannabis must at present get a cleaner bill of health than our legalised 'recreational drugs'."
- A REPORT ON DRUGS AND DRUG DEPENDENCE BY THE ROYAL COLLEGE OF PSYCHIATRISTS, 1987.

"Medicines often produce side effects. Sometimes they are physically unpleasant. Many doctors consider marijuana effective in relieving the nausea of chemotherapy, treating glaucoma and helping Aids patients to gain weight. It too has discomforting side effects, but these are not physical they are political."
-THE *ECONOMIST* MARCH 28th, 1992, MEDICAL MARIJUANA THE LAST SMOKE.

"These surveys would suggest that cannabis smoking seems to be a well established leisure activity of up to 10% of young adults. It is certainly no longer true to say, if it ever was, that cannabis smoking is a sign of affiliation to an 'alternative' life style. Clearly, in the light of its popularity and to a degree its apparent social acceptability, questions are raised about the legalisation of cannabis." - THE MISUSE OF DRUGS, OFFICE OF HEALTH ECONOMICS, 1992.

"The only solution to the drug problem is the legalisation of all drugs. If this is too radical and too much opposed to received wisdom and decades of conditioning then perhaps we could begin by repealing the laws against the drug which has been shown by several impartial investigations to be at least no more harmful than alcohol, and whose use causes the most conflict between users and the law: cannabis." - SERGEANT GORDON PAYNE THE POLICE REVIEW, 28/2/92.

"Penalties against possession of a drug should not be more damaging to an individual than the use of the drug itself: and where they are, they should be changed. Nowhere is this more clear than the laws against possession of marijuana in private for personal use."
- U.S. PRESIDENT JIMMY CARTER, QUOTING FROM THE N.C.M.D.A., 1977.

"It seems likely that if there were any hazards associated with the use of cannabis, they would be fairly well known by now, but all the available evidence suggests that cannabis is no more damaging a drug than tea or coffee. Indeed it is probably less dangerous than drinks containing caffeine." - - DR VERNON COLEMAN MB CHB, 1992.

"Drug addiction and drug misuse should primarily be treated as a subject of health and welfare and not as one of police and justice. Possession of illicit drugs in small quantities for personal use should not be considered as a criminal offence."
- EUROPEAN COMMUNITY COMMITTEE OF ENQUIRY ON DRUG TRAFFICKING, 1991.

THE SWEDISH ADDICTION EPIDEMIC IN GLOBAL PERSPECTIVE

Professor Nils Bejerot[1]

The Swedish Carnegie Institute, Stockholm.

ABSTRACT

The Swedish epidemic of intravenous amphetamine injection which started in 1945, was surveyed annually in Stockholm from 1965 to 1987. During that period, approximately 250.000 arrestees were examined for needle marks from intravenous drug injections that they presented in their cubital regions.The progression or regression of the epidemic was gauged by calculating the percentage of addicts (marked with needle scars) among the population arrrested for any kind of criminal or civil offense. This epidemiological study using an objective marker demonstrated that a permissive drug policy leads to a rapid spread of drug use. A restrictive policy not only checks the spread of addiction but brings about a considerable reduction in the rate of current consumption. The restrictive policy is based on a general consensus of social refusal of illicit drug use, and strict law enforcement. All countries which have adopted this model such as China, Japan, Korea, Singapore and Taiwan have succeeded in controlling epidemics of amphetamine or heroin addiction. By contrast, Western industrialized nations which have accepted permissive policies have seen their epidemics of drug addiction grow steadily since World II War and erode their democratic institutions. The author concludes that such a trend may only be reversed by adopting a restrictive model validated by epidemiological and historical facts.

KEYWORDS

Amphetamine epidemic, Sweden, evolution, restrictive and permissive policy, criminality, objective marker (needle scar).

---

[1] *This paper is the last one written and presented by the Swedish epidemiologist at an International Colloquium held in Paris at the French Senate in March 1988 ("Drogue et Société, Masson, Paris, 1990). N.Bejerot died a few months later.*
We are undebted to Mrs Carol Bejerot and to the Swedish Carnegie Institute for the permission to reprint this incisive and timely adress.

Sweden was the first country in Europe to be afflicted by drug abuse of epidemic type immediately after the end of the Second World War. The Swedish epidemic has been extremely extensive, it has spread to neighbouring countries and to the continent, and it has presented dramatic phases during its development. In addition it is probably the drug epidemic which has been most closely studied and documented. Therefore Swedish experience is of considerable international interest.

There are several different types of drug abuse, regardless of the nature of the drug. It is important to differentiate between these types or patterns of abuse, since they differ fundamentally in regard to prevention and control.

THERAPEUTIC TYPE

First we have the classical medical use of dependence producing drugs which may give rise to abuse and addiction of therapeutic type. Those affected are usually middle aged, socially stable people who developed a drug abuse as a result of an error in medical treatment. These people are ashamed of their drug abuse, they try to keep it hidden, even from their physicians and relatives, and they rarely draw others into their abuse.

CULTURAL TYPE

The other main type of use and abuse is coupled to the culturally accepted consumption of certain inebriates - a consumption that often stretches back to prehistoric times. It is no breach of norms within a culture to enjoy these drugs, but severe cases of dependence may arise, even though the use is ritual in accordance with ancient rules and traditions. This cultural form of abuse may be exemplified by the coca chewing of South American Indians, cannabis smoking in certain Muslim countries, opium smoking in the Far East and alcoholism in the Christian world.

EPIDEMIC TYPE

The third main type of drug dependence is epidemic abuse. Characteristically it arises in bohemian circles where small groups of romantic dreamers or risk-taking norm breakers experiment with exotic or new intoxicating drugs in the pursuit of novel experiences. After years or decades of use of the drug in isolated groups, the first phase of the epidemic, there is a spread in the second phase to new categories, often to other groups of norm-breakers, and then particularly to criminal circles. In the third phase, drug consumption spreads to broad groups of the normal population, and then first to those which have the weakest impulse control and the least stable system of values, that is the youth. In the fourth phase the epidemic abuse tends to spread upwards through the age groups, and may begin to resemble drug use of cultural type: That is, it is no longer considered to be a breach of norms. A new, permanent drug problem has now been added to those already existing in the culture. Regardless of the country and the drug, these epidemics present a number of characteristics in common.

SPREAD

Spread of drug abuse occurs almost without exception through personal, psychosocial contact between an established abuser and a novice in very close friendship relation, often between sexual partners. Initiation usually occurs in an early phase of the initiator's abuse, during the period which is commonly called the honeymoon of addiction, before the negative physical, psychological, social, economic and legal complications have commenced. The honeymoon is short in the case of heroin, usually about a year, but far longer in cannabis abuse. Initiation via pushers and incidental contacts is rare. Pushers enter the scene at a later stage, when they play a very destructive role in maintaining an established abuse or provide for a relapse.

EXPONENTIAL GROWTH

Epidemics of drug abuse often spread very rapidly. In most countries it has been possible to observe an exponential growth for long periods of time. For instance, intravenous abuse of amphetamine in Sweden doubled every thirtieth month during a period of twenty years, 1946-65. In England the number of heroinists doubled every sixteenth month during a period of ten years, 1959-68.

Other characteristics for drug epidemics are their restriction by historic boundaries, and also, for long periods, within small coteries and by age, ethnic, geographical and national boundaries.

YOUTH

Drug epidemics are for long periods checked by such boundaries, but when these barriers are broken through, the abuse spreads in the new population strata. For instance the Jews lived side by side with cannabis smoking Muslims in the Middle East for a thousand years without, as far as I know, any Jew smoking hashish. It was not until young American-Jewish cannabis smokers came to visit Israel that Jewish youth began to smoke the drug.

FASHION

Drug epidemics are extremely sensitive to fashion regarding the type of drug and method of administration, with sometimes rapid changes in the panorama of abuse. An example of this is cocaine, which, for a long period, was only consumed in the traditional way by chewing. With the production of pure cocaine, sniffing was introduced, later followed by intravenous injections, and finally by smoking the free base and coca paste.

The more drug epidemics spread in a society the more common will be the occurrence of mixed abuse with different drugs and varied mode of administration.

INTERACTION

Exposure and susceptibility interact in a predictable way.

The fact that there was no one in Europe before the Second World War who injected drugs intravenously was due to the same simple reason that we had no syphilis or tobacco smoking before Columbus. Nor was there any tuberculosis or alcoholism among the Eskimos before they were colonized by the Danes. There had been susceptible individuals before, but they had not been exposed to these factors.

MASSIVITY

The pressure of exposure, also called massivity, causes people to react differently: Some are affected immediately, others after a time, some only after the pressure from the drug culture has become very great, while many manage to resist throughout their whole lives, despite prolonged and intensive exposure. Thus, susceptibility varies between different individuals, but also in the same individual with age and a number of other factors.

We can now express the connection between exposure to drug culture (E), the susceptibility of the individual (S) and the risk that the individual will commence to use the drug, that is the psychosocial contagion (C):

$$C = S \times E$$

The susceptibility of the individual (S) is the result of a large number of individual factors such as sex, age, social situation, previous experience, etc. Since exposure at one point also affects future susceptibility (fS) we can in general write the formula:

$$C = fS \times fE$$

Of all norm breaking forms of drug abuse, intravenous administration is the one which is most suitable for scientific study, since the breach of norms here is distinct and important, and in addition injections leave clear, objective and characteristic diagnostic signs, which cannot be confused with medical injections (Bejerot 1975).

THE SWEDISH EPIDEMIC OF INTRAVENOUS DRUG ABUSE

The Swedish epidemic started through a few coincidental events. Intravenous drug abuse had been reported in USA since 1926, but as far as I know, this did not initiate any drug epidemics in Europe until a young, adventurous Swede in 1946 learnt the injection technique in USA, and introduced it into a little bohemian coterie in Stockholm. In this group a few persons had become amphetaminists through medical treatment for alcoholism, and in this limited group an epidemic of intravenous abuse was established.

Up to 1949 there were a dozen cases within this bohemian coterie in Stockholm, but not a single case outside this group. In 1949 the epidemic spread out of this circle via a couple of artists models who were also prostitutes, and the epidemic thus gained a foothold in social problem groups. In the summer 1954 I diagnosed the first medically documented case of this type in Sweden.

In 1956 the epidemic of intravenous abuse spread to Gothenburg, when an addict of this category moved there, and for the rest of his life was a central figure in addict circles in the second largest city in Sweden. The Swedish amphetamine epidemic spread to Finland in 1965, to Denmark in 1966, to Norway in 1967 and to Germany in 1972.

In the study of the Swedish epidemic of intravenous abuse, I assumed that a breach of norms such as introducing a needle into a vein and injecting illicit drugs was so extreme that it would be expected to coexist with other severe breaches of norms such as traditional criminal conduct. I therefore initiated a study in 1965 where nurses inspected the veins of the arms of persons brought to the central arrest premises in Stockholm. The first five year period is presented in a monograph (Bejerot 1975). From the study of representativity it is apparent that practically all active intravenous abusers are brought to the arrest premises sooner or later for one reason or another, and are included in the study.

The investigation is still continuing, and hitherto we have examined about a quarter of a million arrestees, many of them, of course, on several occasions. Here I will only present one or two graphs from the study.

GRAPHS AND TEXT

A good indicator of the extent of the intravenous epidemic over time in the arrested clientele is the proportion of intravenous abusers among those arrested for all types of crime. In fig. 1, this evolution is shown for men and women during the period 1965-1986.

The study was commenced in 1965 because of an ultra-liberal policy introduced in Sweden that year, which permitted a number of physicians to prescribe amphetamine to addicts for self-administration. This resulted in an increase in the percentage of intravenous abusers in the arrested clientele from 20 to 40 percent during a period of three years! During an extra restrictive policy 1969-70, a direct result of the catastrophic consequences of the prescribing policy, the epidemic was checked for the first time. The epidemic culminated in 1972, when some large drug syndicates were broken up. The system of distribution was reorganized quickly, however, since the demand was intact, and when heroin was introduced it gave rise to a new branch of the drug epidemic. It finally culminated in 1976 when 60 percent of all arrestees were intravenous abusers. Sine then, a number of minor increases in the severity of drug legislation has reduced the percentage to about 40 percent in Stockholm, where the level has remained relatively constant.

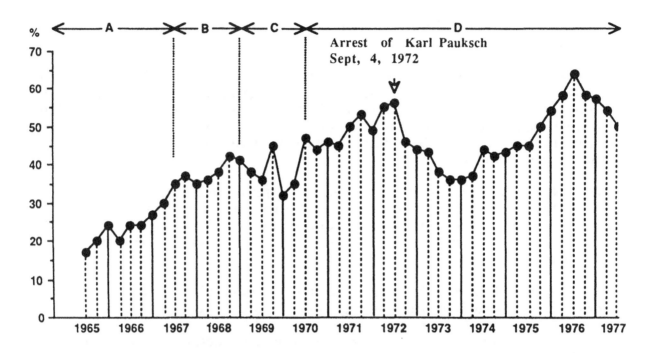

FIGURE 1 : Percentage of intravenous drug abusers among male arrestees (aged 15 - 44) in Stockholm 1965-1976. Legends on top denote drug policy periods :   A = liberal or permissive;

                                                      B = traditionally restrictive;

                                                      C = extra restrictive (police offensive);

                                                      D = traditionally restrictive with somewhat diminishing restrictiveness for possession of smaller amounts of illegal drugs.

(From Bejerot, 1978).

During the seventies the epidemic spread over the whole of Sweden, and cases of intravenous drug abuse now occur even in rural areas.

By means of enquiries as to the year of debut for intravenous abuse, we have reconstructed the incidence for Stockholm (fig. 2). Here the fatal effect of liberalization and prescribing of drugs 1965-1967 appears very clearly.

Prevalence estimations over time in the development of the epidemic in Stockholm for the first 20 year period is shown in fig. 3 & 4. Mortality among drug abusers has proved to exceed by 7 to 15 times that of the normal population, and accounts for a considerable depletion in the population of active drug addicts. In addition, various complications and the increasing difficulty in financing an expensive drug practice cause about a third to discontinue their drug abuse spontaneously after an average of ten years. This means that all estimates of prevalence must be uncertain, unless a central register is drawn up over active intravenous abusers, and in the absence of new reports of abuse, they are removed after five years.

CONTROL

When we have understood how individual and social factors interact to give rise to drug epidemics, we can begin to discuss how we should attack the problem.

Many studies have shown that the individual susceptibility factor is, unfortunately, not easily influenced, composed as it is of everything which has affected the individual up to the present. On the other hand the exposure factor, pressure from the addiction milieu and the drug market, have proved to be highly modifiable by means of certain strategies.

The rate of opiate addiction in USA was reduced by about 90 percent between the years 1923-39 (Harney & Cross 1961), and this without any treatment to speak of, or research. The instrument was a strict drug policy which reduced the exposure factors dramatically.

In the same way an extensive cocaine epidemic in Germany was stopped in the late twenties, and also a widespread amphetamine epidemic in Japan after the Second World War.

JAPAN

The Japanese epidemic deserves special attention. It arose when the Japanese military store of amphetamine went astray after capitulation. Abuse began among people who worked at night: Jazz musicians, artists, bohemians and prostitutes, but it quickly spread to broad strata of the population.

The Japanese authorities introduced a number of countermeasures, but they did this too late and on too small a scale, and with too little energy: It was like operating on a growing cancer which could not be checked since the measures taken were not sufficiently radical.

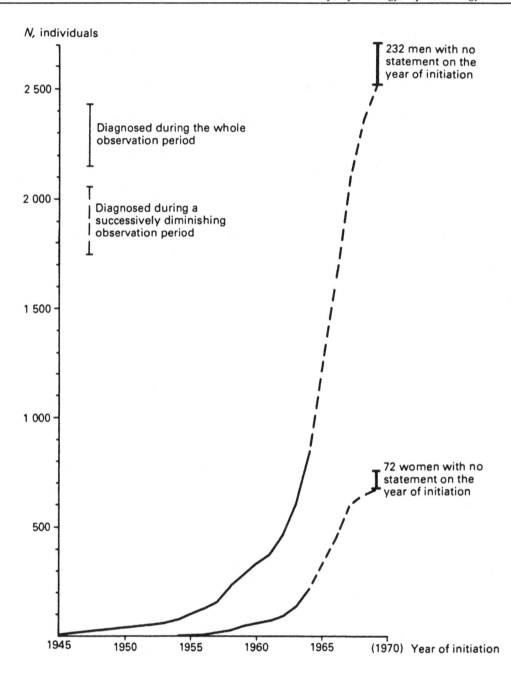

FIGURE 2 : Uncorrected minimum prevalence of abusers of intravenous type reconstructed on the basis of statements on the year of initiation: Swedish abusers diagnosed in the injection mark study 1965-70. *(From Bejerot, 1975).*

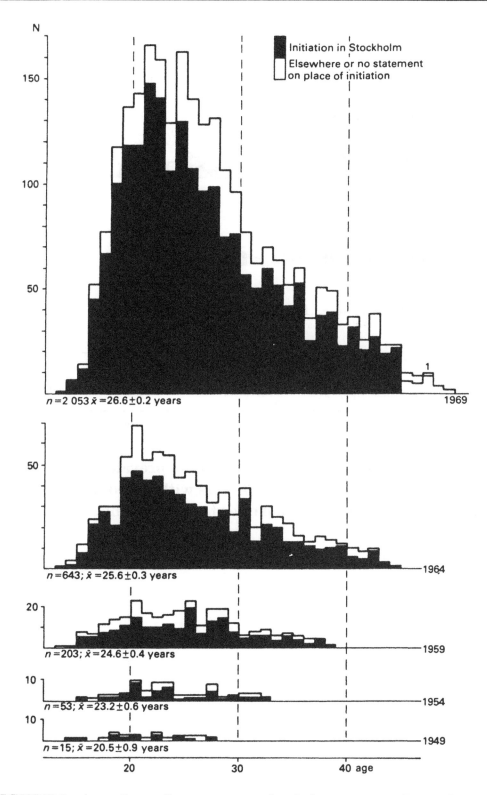

FIGURE 3 : **Age-classed, uncorrected minimum prevalence in 5-years intervals 1949-69 for male Swedish abusers with stated initiation into intravenous abuse in Stockholm and elsewhere or with no statement on place of initiation.   1: Over 45 corresponds to 40-44 in 1969.**
(From Bejerot, 1975).

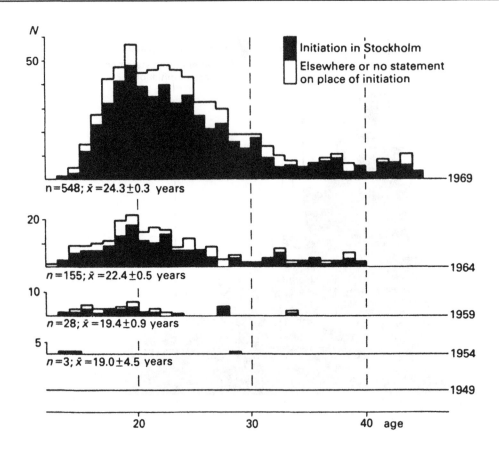

FIGURE 4 : Age-classed, uncorrected minimum prevalence in 5-years intervals 1949-69 for female Swedish abusers with stated initiation into intravenous abuse in Stockholm and elsewhere or with no statement on place of initiation.
(From Bejerot, 1975).

The Japanese epidemic culminated in 1954, when it was estimated that two million of Japan's population of one hundred millions was abusing amphetamine tablets, and over half a million were taking intravenous injections. It was only then that a dramatic increase in the stringency of policy was introduced, with prison sentences of 3-6 months for possession, 1-3 years for drug pushing and five years for illicit manufacture of drugs. There was close surveillance when they were released from prison, and there was an immediate restriction on relapse.

During the first year of the campaign, 1954, 55 600 persons were arrested in Japan for amphetamine offences, but in 1958 the number was only 271, and the whole epidemic was over. Altogether measures had been taken against 15 percent of the estimated number of intravenous abusers. The others stopped from pure fright when the restrictive policy was carried into effect. It should be pointed out that the campaign was drawn up on the basis of broad political consensus, and was carried out with the aid of massive public support.

EXPERIENCE FROM CHINA

The Japanese had learnt from the Chinese the technique of combatting drug epidemics. Between the years 1951-53 China had stopped the 300 year old opium smoking with about 20 million active opium smokers, and this without needing to sentence more than ten percent of the addicts to a year or two in a labour camp, while ninety percent stopped without either medical treatment or psychotherapy.

HISTORY AS A TEACHER

"We learn from history that we never learn from history" said George Bernard Shaw. This applies very much to drug epidemics and control over them.

I have been unable to find any example of widespread drug abuse in any country, which has been overcome without a general restrictive program directed to the drug market and exposure from the addiction milieux, and this regardless of the level of development of the social system.

Nor have I been able to find any example of a voluntary drug-free treatment program which has had more than a marginal effect on the extent of the problem.

BASIC EXPERIENCE

The period 1850-1950 was the age of prevention, when the mechanisms of most of the known infectious diseases were revealed and the great epidemics overcome, not so much by individual treatment as by preventive measures. Even alcohol and drug policy during this period were, in many countries, directed towards prevention. The advances made were often considerable, in Sweden they were epochmaking.

The period after 1950 is the age of therapy. Now preventive strategies and their social necessity have been forgotten, and we have instead an avalanche of different therapeutic schools and programs for the treatment of those already addicted. Most of these programs, unfortunately, have not given better results than no treatment at all.

NOT A DISEASE

Why is this so? Firstly because drug dependence is not a disease, and therefore, by definition, cannot be cured.

Caffeine and nicotine dependence, alcoholism and heroinism are not diseases, even if all these dependence-producing substances may make the individual very ill. The physical dependence, or tolerance, and the very distressing abstinence symptoms following withdrawal of many drugs, are only incidental complications, while true dependence is a learned behaviour where craving for the drug has taken on the character and force of a natural drive. Theoretically, drug dependence is related to such conditions as gambling, pyromania, and kleptomania. The drug acts as a reinforcer.

Drug dependence is, thus, not a a symptom of the factors which originally led to contact with the drug, consumption and dependence. Heavy nicotine dependence at forty is not a late symptom of curiosity in the early teens, but an independent condition which is very difficult to handle.

A common factor in all types of drug dependence is an ambivalence of the addict towards his drug: He is anxious to obtain help for all the complications to his drug consumption, but he is not prepared to sacrifice the drug experience itself.

To combat drug epidemics by means of individual treatment is like attacking malaria by hunting mosquitos. It can occupy an enormous number of people, but the effect is negligible. What is required is drainage of the marshes.

THE LARGE AND THE SMALL DRUG MARKETS

Draining the drug marshes means breaking up drug traffic and reducing general exposure to illicit drugs in society. Enormous efforts have been made by the customs, police and undercover agents all over the world. Despite this, the situation deteriorates very quickly and many countries are on the brink of giving up the fight.

Why were the advances so great in the antidrug campaign in Germany in the twenties, in USA in the thirties, and in China and Japan in the fifties? And why have there been no decisive advances in the Western World during the last two decades? I consider that this is largely because we have forgotten what is of primary and secondary importance on the drug market. The primary factor is not that Nature produces plants such as the opium poppy or coca bush or that international crime syndicates take over the distribution of the drugs. The primary factor is that millions of people are prepared to break norms and laws in order to use these natural inebriates and also hundreds of synthetic preparations.

BREACH OF NORMS

It is thus the personal breach of norms which is the moral basis, and the personal possession of drugs the legal basis of the drug market, and not the international syndicates. These, in fact, are a late consequence of the emergence of a drug market

Naturally the drug syndicates should be combatted just as actively as now, but we must open a new front if we are to win the war. If we were to destroy all the cultivations of narcotic drugs in the world , there would, none the less, still exist substances which are up to 40 000 times as strong as morphine and which can be produced synthetically.

We have to accept the painful fact that we cannot win decisive advances unless drug abuse, the abuser and personal possession are placed in the centre of our strategy.

"The junky merchant does not sell his product to the consumer, he sells the consumer to his product" said William S. Burroughs. I will quote another very astute remark from the foreword to his "Naked Lunch" from 1959;

"If you wish to alter or annihilate a pyramid of numbers in a serial relation, you alter or remove the bottom number. If we wish to annihilate the junk pyramid, we must start with the bottom of the pyramid: The addict in the street, and stop tilting quixotically for the "higher ups" so called, all of whom are immediately replaceable. The addict in the street who must have junk to live is the one irreplaceable factor in the junk equation. When there are no more addicts to buy junk there will be no junk traffic, as long as junk need exists, someone will service it."

This is a brilliant summary of a difficult problem.

STRATEGY AND TACTICS

I consider that democratic, welfare states of western type ultimately stand and fall with the result of the fight against drug epidemics. To win that fight we must have realistic strategies and tactics. We must realize, and dare to affirm, that it is the drug addict who is the motor in the system. But the addict, who is extremely manipulative, and acts as the full time defence lawyer for his dependence, has succeeded in duping so many honest and responsible but naive politicians and journalists, that during the last twenty years he has himself been practically scheduled as a protected monument. This I consider is the most important factor behind our failure.

THEY MUST BE PROSECUTED

This does not mean that I propose a return to the harsh American sentences of the thirties for drug offences. They were unrealistic and undermined their own purpose. We must, however, make it very uncomfortable to abuse illicit drugs if we are to reverse developments. The addict must learn to take the consequences of his behaviour. In regard to Sweden, I have suggested a month clearing the forests for the first offence of possession of illicit drugs, two months for the second, etc.

Society must clearly show that drug abuse is not accepted.We cannot blame the behaviour of our youth on the mountain Indians in Colombia or the peasants in the Golden Triangle. We must, in the first place, put the blame on our own youth, and this may be difficult and painful. In the second place we should put the blame on ourselves for being duped into an inconsequent, permissive, attitude with continual excuses and forgiveness.

POPULAR SUPPORT

No government in a democratic country can manage widespread drug epidemics without strong popular support.This must be achieved through broad political agreement and massive information which leads to something like a popular uprising against drug epidemics.

The near future will be decisive as to whether the Western World will manage to overcome drug epidemics. With a one-sided supply-orientated strategy we will fight a war which we are doomed to lose. Only by opening a new front with a strategy orientated towards demand can development be reversed, and the fight against drugs be won. Otherwise developments will progress towards capitulation and a social chaos which may be the basis for a new period of fascism.

REFERENCES

Bejerot N., Drug abuse and drug policy. Acta Psychiatrica Scandinavica, Copenhaguen, Munksgaard, 1975.

Bejerot N., Missbruk av alkohol, narkotika och frihet (Swedish), Stockholm, Ordfront, 1978.

Burroughs W.S., Naked Lunch, 1959.

Harney M. and Cross J., Narcotics Officer's Notebook, Springfield III; Ch. Thomas, 1961.

INDEX

INDEX

Printed and bound by CPI Group (UK) Ltd, Croydon, CR0 4YY

22/10/2024

01777638-0015